CLIMATE CHANGE ADAPTATION AND DISASTER RISK REDUCTION: AN ASIAN PERSPECTIVE

COMMUNITY, ENVIRONMENT AND DISASTER RISK MANAGEMENT VOLUME 5

CLIMATE CHANGE ADAPTATION AND DISASTER RISK REDUCTION: AN ASIAN PERSPECTIVE

EDITED BY

RAJIB SHAW

Graduate School of Global Environmental Studies, Kyoto University, Kyoto, Japan

JUAN M. PULHIN

Department of Social Forestry and Forest Governance, College of Forestry and Natural Resources, University of the Philippines Los Baños, Philippines

JOY JACQUELINE PEREIRA

Southeast Asia Disaster Prevention Research Institute (SEADPRI-UKM), Institute for Environment and Development (LESTARI), Universiti Kebangsaan Malaysia, Malaysia

Emerald

United Kingdom – North America – Japan
India – Malaysia – China

Emerald Group Publishing Limited
Howard House, Wagon Lane, Bingley BD16 1WA, UK

First edition 2010

British Library Cataloguing in Publication Data
A catalogue record for this book is available from the British Library

ISBN: 978-0-85724-485-7
ISSN: 2040-7262 (Series)

Printed and bound in Great Britain by
CPI Antony Rowe, Chippenham and Eastbourne

Emerald Group Publishing Limited, Howard House, Environmental Management System has been certified by ISOQAR to ISO 14001:2004 standards

Awarded in recognition of Emerald's production department's adherence to quality systems and processes when preparing scholarly journals for print

INVESTOR IN PEOPLE

CONTENTS

LIST OF CONTRIBUTORS

Kou-Chiang Chang	Division Leader of Planning Division, Water Resources Agency, Ministry of Economic Affairs, Taipei, Taiwan, Republic of China
Hoang Duc Cuong	Research Center for Meteorology and Climatology, National Institute of Meteorology, Hydrology and Environment, Hanoi, Vietnam
Rajesh Daniel	Unit for Social and Environmental Research, Chiang Mai University, Chiang Mai, Thailand
Kumiko Fujita	Graduate School of Global Environmental Studies, Kyoto University, Kyoto, Japan
Umma Habiba	Graduate School of Global Environmental Studies, Kyoto University, Kyoto, Japan
Tran Thi Tuyet Hanh	Department of Environmental Health, Hanoi School of Public Health, Hanoi, Vietnam
Chia-Chen Hsu	Chief Leader, Research Center for Soil & Water Resources and Natural Disaster Prevention, National Yunlin University of Science & Technology, Douliou, Yunlin, Taiwan, Republic of China
Shao-Yang Huang	Vice-Executive Secretary, Research Center for Soil & Water Resources and Natural Disaster Prevention, National Yunlin University of Science & Technology, Douliou, Yunlin, Taiwan, Republic of China
Tawhidul Islam	Department of Geography and Environment, Jahangirnagar University, Savar, Dhaka, Bangladesh

Shimpei Iwasaki	Research Department, Research Institute for Humanity and Nature, Kyoto, Japan
Amir Nawaz Khan	Center for Disaster Preparedness and Management, University of Peshawar, NWP, Pakistan
Tran Mai Kien	Climate Change and Adaptation Group, Mekong River Commission Secretariat, Vientiane, Lao PDR
Ibrahim Komoo	Southeast Asia Disaster Prevention Research Institute (SEADPRI-UKM), Universiti Kebangsaan Malaysia, Bangi, Malaysia
Louis Lebel	Unit for Social and Environmental Research, Chiang Mai University, Chiang Mai, Thailand
Phimphakan Lebel	Unit for Social and Environmental Research, Chiang Mai University, Chiang Mai, Thailand
Bing Li	School of Civil and Environmental Engineering, Nanyang Technological University, Singapore
Fuad Mallick	Postgraduate Programs in Disaster Management, BRAC University, Dhaka, Bangladesh
Huy Nguyen	Graduate School of Global Environmental Studies, Kyoto University, Kyoto, Japan
Rekha Nianthi	Department of Geography, University of Peradeniya, Candy, Sri Lanka
Joy Jacqueline Pereira	Southeast Asia Disaster Prevention Research Institute (SEADPRI-UKM), Universiti Kebangsaan Malaysia, Bangi, Malaysia
Rosa T. Perez	The Manila Observatory, Ateneo de Manila University Campus, Loyola Heights, Quezon City, Philippines
Tran Phong	College of Economics, Hue University, Hue, Vietnam

Juan M. Pulhin	Department of Social Forestry and Forest Governance, College of Forestry and Natural Resources, University of the Philippines Los Baños, Laguna, Philippines
M. Aminur Rahman	Postgraduate Programs in Disaster Management, BRAC University, Dhaka, Bangladesh
Vikram Rajola	School of Civil and Environmental Engineering, Nanyang Technological University, Singapore
Rajib Shaw	Graduate School of Global Environmental Studies, Kyoto University, Kyoto, Japan
Prabhakar SVRK	Institute of Global Environmental Strategies, Hayama, Japan
Jishnu Subedi	Department of Civil Engineering, Institute of Engineering, Tribhuban University, Pulchowk, Nepal
Yukiko Takeuchi	Graduate School of Global Environmental Studies, Kyoto University, Kyoto, Japan
Bui Thi Tam	College of Economics, Hue University, Hue, Vietnam
Maricel A. Tapia	Department of Social Forestry and Forest Governance, College of Forestry and Natural Resources, University of the Philippines Los Baños, Laguna, Philippines
Bui Dung The	The College of Economics, Hue University, Hue, Vietnam
Bui Duc Tinh	College of Economics, Hue University, Hue, Vietnam
Tan Ching Tiong	Southeast Asia Disaster Prevention Research Institute (SEADPRI-UKM), Universiti Kebangsaan Malaysia, Bangi, Malaysia

Tran Huu Tuan	College of Economics, Hue University, Hue, Vietnam
Noralene Uy	Graduate School of Global Environmental Studies, Kyoto University, Kyoto, Japan
Jet-Chau Wen	Research Center for Soil & Water Resources and Natural Disaster Prevention, National Yunlin University of Science & Technology, Douliou, Yunlin, Taiwan, Republic of China
Deepthi Wickramasinghe	Department of Zoology, University of Colombo, Colombo, Sri Lanka

LIST OF EDITORS

Rajib Shaw
Graduate School of Global Environmental Studies, Kyoto University, Kyoto, Japan

Juan M. Pulhin
Department of Social Forestry and Forest Governance, College of Forestry and Natural Resources, University of the Philippines Los Baños, Philippines

Joy Jacqueline Pereira
Southeast Asia Disaster Prevention Research Institute (SEADPRI-UKM), Institute for Environment and Development (LESTARI), Universiti Kebangsaan Malaysia, Malaysia

EDITORS' PROFILE

Rajib Shaw is an Associate Professor at the Graduate School of Global Environmental Studies of Kyoto University, Japan. He has worked closely with the local communities, NGOs, governments, and international organizations, including United Nations, especially in the Asian countries. He is currently the Chair of the United Nations Asia Regional Task Force for Urban Risk Reduction. His research interests include community-based disaster risk management, climate change adaptation, urban risk management, and disaster and environmental education.

Juan Pulhin is Professor and Scientist II at the Department of Social Forestry and Forest Governance, College of Forestry and Natural Resources, University of the Philippines Los Baños. He was a Lead Author of the adaptation chapter of the Working Group II Contribution to the Fourth Assessment Report of the Intergovernmental Panel on Climate Change (IPCC) and a Coordinating Lead Author of the human security chapter of the IPCC Fifth Assessment Report. His research interests include integrated assessment of climate change impacts, vulnerability and adaptation using participatory methods, social forestry/community-based forest management, forest and natural resource governance, and forestry and environmental education.

Joy Jacqueline Pereira is a Professor at Universiti Kebangsaan Malaysia. She leads the Malaysian Research Network for Climate, Environment and Development (MyCLIMATE), which is dedicated to making a positive difference to humanity in addressing the challenges of climate change. She also serves the Intergovernmental Panel on Climate Change (IPCC) as Lead Coordinating Author for Asia for the Fifth Assessment Report of the IPCC (AR5), and as Review Editor for the IPCC Special Report on "Managing the Risks of Extreme Events and Disasters to Advance Climate Change Adaptation" (SREX).

BRIEF INTRODUCTION TO THE SERIES

COMMUNITY, ENVIRONMENT AND DISASTER RISK MANAGEMENT

This series connects academic research to field practice, strengthening the links between the environment, disaster, and community. The series will be developed on field evidences and community practices, and thus will provide specific guides to professionals who are grounded in rigorous academic analysis. The series will have specific focus on community-based disaster risk management, urban environmental management, human security, water community, risk communication, climate change adaptation, climate disaster resilience, and community-based practices.

BRIEF INTRODUCTION TO THE VOLUME

Climate Change Adaptation (CCA) and Disaster Risk Reduction (DRR) are closely associated. After the Bali Conference of Parties (COP) 2007, there has been significant awareness at global and national levels to make collective focus on CCA and DRR. There exists some literature on CCA; however, the linkages with DRR are not clear. The Nairobi Work Program (NWP), facilitated by the UNFCCC (UN Framework of Convention on Climate change) and the Parties, has found specific scope of actions to enhance the CCA and DRR agenda in developing countries. Different national and international agencies have started undertaking initiatives on this aspect, but very few concrete measures have been implemented. At this onset, the book provides 19 case studies from 13 countries and regions in Asia. The case studies highlight different aspects of CCA–DRR entry points, like policy interventions, drought risk management, coastal management, agro-forestry, lagoon management, livelihood issues, and risk communication. This book supplements Volume 4: *Climate Change Adaptation and Disaster Risk Reduction: Issues and Challenges*, which focuses on the theory of CCA and DRR and its enabling environment; CCA–DRR linkages framework of governance, education, and technology; and different entry points like urban, coast, forest, river, and housing, and finally provides a regional perspective of developing countries. The primary targets of this book are the students and researchers in the field of disaster risk reduction, climate change, environmental studies, and related risks. The book will help them to have a better idea of the current trend of research in the field, and will provide basic knowledge of this important topic. The other target group is practitioners and policy makers who can apply the collective knowledge into policy and decision making.

PREFACE

Climate change is happening; it is within primary focus and attention. Over years, the importance of climate change adaptation (CCA) has increased, and for many countries, it is becoming one of the prime targets. Climate change impacts are visible in the form of disasters, which have added to this importance. In contrast, the disaster risk reduction (DRR) field is more submissive and has had less attention so far. However, due to recent major disasters in different parts of the world, media and public attention is high. The news becomes more intense when we observe unpredictable rainfall, heat waves, drought, typhoons, or other climate-related disasters. The intensity and frequency of natural disasters are showing an increasing trend.

Asia is arguably among the regions of the world most vulnerable to climate change. The continued melting of glaciers in the Himalayan region is projected to increase flooding and rock avalanches and to adversely affect water resources in the next two to three decades. As a result of the changes, climate change could hamper the achievement of many of the Millennium Development Goals (MDGs), including those on poverty eradication, child mortality, malaria and other diseases, and environmental sustainability. Many of the risk reduction measures particularly those related to hydro-meteorological disasters, such as drought proofing, flood protection, saline embankment and bio-shields, alternative livelihood development, etc., have similarities with CCA programs. Therefore, synergies between DRR and CCA would be necessary not only to avoid duplicities and derive optimal benefits from scarce resources but also to add value to the projects through lessons learnt from the respective perspectives.

Asia has different mechanisms for CCA and DRR activities. At the regional level, intergovernmental initiatives are prominent; at the local level, local governments and local NGOs have their respective network. There also exists a regional network of Asian universities. Thus, there is every reason to be optimistic that Asia will adapt to the climatic change through proactive measures in disaster risk reduction. With this context, the book provides 19 case studies from 13 countries and regions in Asia. The case studies highlight different aspects of CCA–DRR entry points, like policy interventions, drought risk management, coastal management, agro-forestry, lagoon management, livelihood issues, and risk communication.

This book was a joint initiative of different groups, and has had support from different sources. The editors acknowledge the support of Asian University Network of Environment and Disaster Management (AUEDM) and its members. Funding support from Mitsui Foundation and GCOE-ARS is highly acknowledged. The book is dedicated to the practitioners, professionals, and researchers in the environment and disaster field, especially those who are dealing with CCA- and DRR-related issues. We will be very happy if it serves their purpose.

Rajib Shaw
Juan M. Pulhin
Joy Jacqueline Pereira
Editors

CHAPTER 1

CLIMATE CHANGE ADAPTATION AND DISASTER RISK REDUCTION: AN ASIAN PERSPECTIVE

Rajib Shaw, Juan M. Pulhin and Joy Jacqueline Pereira

INTRODUCTION

Climate change is one of the biggest challenges to development. Intergovernmental Committee in response to Climate Change (IPCC, 2007), with majority agreement, has pointed out that climate change is caused by human activities. Intergovernmental Panel on Climate Change (IPCC) also pointed out that these serious impacts by climate change will directly affect the poorer and more vulnerable communities and nations. According to the Human Development Report 2007–2008 by United Nation Development Program (UNDP, 2008), people in rural areas are most vulnerable to climate change, because it directly affects the resources in the ecosystem on which their lives depend. Countries with newly developed economy will be seriously hit, economically and socially, within the next few decades by natural disasters such as flood, draught, and storm, which have been increasing in number and severity. Human health is also adversely affected, which has multiplying effects in different sectors.

Climate change impacts, vulnerability, adaptive capacity, and barriers to adaptation are location-specific and will change over time, but the *processes*

Climate Change Adaptation and Disaster Risk Reduction: An Asian Perspective
Community, Environment and Disaster Risk Management, Volume 5, 1–18

ISSN: 2040-7262/doi:10.1108/S2040-7262(2010)0000005007

needed for adaptation that supports the most vulnerable will be similar (Pettengell, 2010). National adaptive capacity is one part of that puzzle. Community design and implementation of adaptation strategies suited to their location is another. A crucial element of both is the role of local-level government and services. They must be empowered and resourced to act as intermediaries, linking the bottom-up and top-down processes.

Climate change adaptation (CCA) and disaster risk reduction (DRR) are closely linked. More extreme weather events in future are likely to increase the number and scale of disasters, while at the same time, the existing methods and tools of DRR provide powerful capacities for adaptation to climate change (ISDR, 2008). Natural hazards by themselves do not cause disasters. It is the combination of an exposed, vulnerable, and ill-prepared population or community with a hazard event that results in a disaster. Climate change will therefore affect disaster risks in two ways, first through the likely increase in weather and climate hazards, and second through increases in the vulnerability of communities to natural hazards, particularly through ecosystem degradation, reductions in water and food availability, and changes to livelihoods. Climate change will add yet another stress to those of environmental degradation and rapid unplanned urban growth, further reducing communities' abilities to cope with even the existing levels of weather hazards.

Shaw, Pulhin, and Pereira (2010) in their book "Climate Change Adaptation and Disaster Risk Reduction: Issues and Challenges," have discussed different factors related to the CCA and DRR. Through 17 chapters, distributed in four parts, Shaw et al. have focused on the theory of CCA and DRR and its enabling environment; CCA–DRR linkages framework of governance; education and technology; different entry points like urban, coast, forest, river, and housing; and finally provided a regional perspective of developing countries. This book is a complimentary volume of the earlier book, where it provides the examples from Asia, one of the most potential areas of climate change impacts, as well as areas of innovations in adaptation.

ASIAN PERSPECTIVE OF CCA AND DRR

Climate change will affect all countries, but people in the poorest countries and poor people in richer countries are more likely to suffer the most. They tend to live in high-risk areas such as unstable slopes and flood plains and often cannot afford well-built houses. Many of them depend on climate-sensitive sectors, such as agriculture, and have little or no means to cope with climate change, for example, owing to low savings, no property insurance, and poor access to public services. Climate change is expected to reduce already low incomes and increase illness and death rates in many developing countries. Africa, small island states, and the Asian and African mega-deltas are likely to be particularly affected by climate change (ISDR, 2008).

Asia is arguably among the regions of the world most vulnerable to climate change. The continued melting of glaciers in the Himalayan region is projected to increase flooding and rock avalanches and to adversely affect water resources in the next two to three decades. The evidence of climate change in Asia is widespread: overall temperatures have risen from 1 to 3°C over the last 100 years, precipitation patterns have changed, the number of extreme weather events is increasing, and sea levels are rising. One of the main issues will be the availability of adequate fresh water, which by the 2050s will be a concern for possibly more than 1 billion people. Asia's coastal areas, and especially its heavily populated delta regions, will become even more prone to increased flooding because of both rising sea levels and river flooding. Because many of the largest cities in Asia are located on the coast and within major river deltas, they are even more susceptible to the impacts of climate change (WWF, 2009).

Climate change and climatic variability have and will continue to impact all sectors, from national and economic security to human health, food production, infrastructure, water availability, and ecosystems. Asia's sustainable development will be challenged as climate change compounds the pressures that rapid urbanization, industrialization, and economic development have placed on natural resources (IPCC, 2007). As a result of the changes, climate change could hamper the achievement of many of the Millennium Development Goals (MDGs), including those on poverty eradication, child mortality, malaria, and other diseases, and environmental sustainability (World Bank, 2010). Much of this damage would come in the form of severe economic shocks. In addition, the impacts of climate change will exacerbate existing social and environmental problems and lead to migration within and across national borders. In sum, climate change is clearly not just an environmental issue but one with severe socioeconomic implications in different parts of Asia.

Fig. 1 shows the disaster statistics of Asia, compared with the rest of the world. It shows that the world disasters have 44% stake in Asia, which contribute to 76% of death and 92% affected people due to disasters. Increasing trends of natural disasters and their threatening impacts on lives

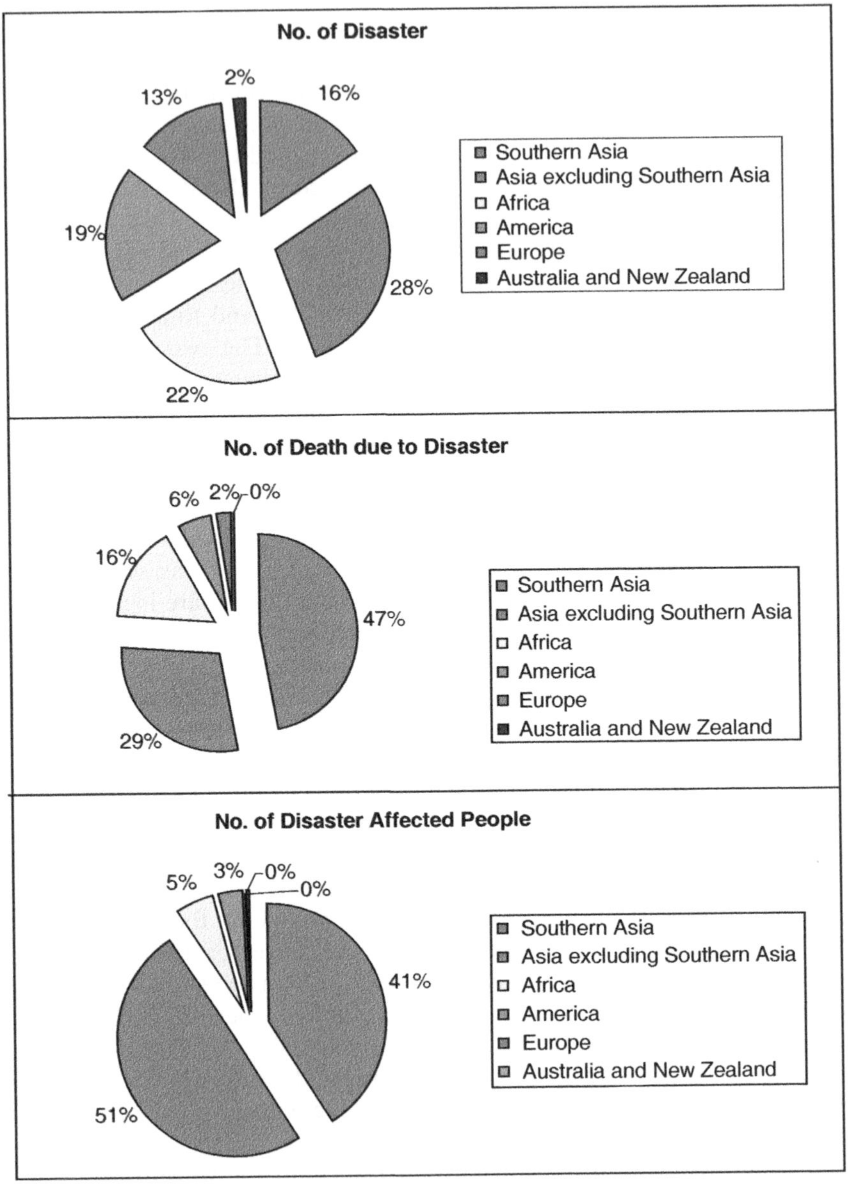

Fig 1. South Asia Compared to Other Regions on Number of Disasters, Deaths, and Affected People Due to Disasters. *Source*: EM-DAT: The OFDA/CRED International Disaster Database – www.emdat.net – Université Catholique de Louvain – Brussels – Belgium.

and livelihood have resulted in a paradigm shift in disaster management in all the countries of Asia – from one postdisaster relief and rehabilitation to holistic management of disasters covering all phase. The focus is clearly on DRR-preparedness, mitigation, and prevention. Many of the risk reduction measures, particularly those related to hydrometeorological disasters, such as drought proofing, flood protection, saline embankment and bioshields, alternative livelihood development, etc. have similarities with CCA programs. Therefore, synergies between DRR and CCA would be necessary not only to avoid duplicities and derive optimal benefits from scarce resources but also to add value to the projects through lessons learnt from the respective perspectives. Factoring climate change issues in disaster risk mitigation projects would enrich the projects and make them more relevant to the emerging concerns just as risk management tools would assess climate change from the perspectives of risks and vulnerabilities over time and the cost-benefit of alternative strategies of adaptation.

REGIONAL INITIATIVES OF ASIA

The financing needs for climate change mitigation and adaptation are uncertain and large. Reflecting the uncertainties associated with potential climate change scenarios and their likely impact, various estimates of financing needs for climate change mitigation and adaptation show wide diversity. According to the United Nations Framework Convention on Climate Change (UNFCCC), mitigation measures needed to return the global greenhouse gas (GHG) emissions to current levels in 2030 require investment and financial flows of $200 billion to $210 billion per annum (World Bank, 2008). Investment needs for adaptation in developing countries in 2030 are estimated at $28 billion to $67 billion. The Adaptation Fund (financed through a 2% levy on revenue generated by the CDM and through voluntary contributions) is a key fund dedicated for adaptation to climate change and is estimated to be $80 million to $1 billion per annum by 2012. Other main adaptation funds include UNFCCC Special Funds (about $270 million) and a portion of the GEF Trust Fund ($50 million until 2010). The Global Facility for Disaster Risk Reduction, with $48 million during 2007–2008, is another source of financing for adaptation.

Establishment of regional financing arrangements for CCA and mitigation in the Asia and Pacific region has started well, but the process is still at a nascent stage. The process needs to be nurtured and supported as part of the larger concern for sustainable development in the region. Other than

ADB, there are no significant financing arrangements on climate change in the Asia and Pacific region (Sharan, 2008), including subregional organizations such as the Greater Mekong Sub-region (GMS), Association of Southeast Asian Nations (ASEAN), Central Asia Regional Economic Cooperation (CAREC), and South Asia Association for Regional Cooperation (SAARC). Asian Development Bank (ADB) provides good examples of regional financing arrangements on climate change. In addition to its regular financing as a regional development bank with increasing emphasis on climate change, ADB has several dedicated funds for financing climate change mitigation and adaptation in the Asia and Pacific region. These perform a variety of functions, including mobilizing concessional resources, catalyzing private capital, and maximizing market mechanisms. All these funds are regional in coverage (with activities limited to the Asia and Pacific region) and based on a mix of regional and global financing.

Kameyama, Sari, Soejachmoen, and Kanie (2008) carried out a review of the existing policy-making institutional interplay of their respective countries on poverty reduction, sustainable development, and climate change. The interplay includes the links between local, project or "field" reality and national policies, and between these national policies and international diplomacy. Among key issues reviewed are the World Bank-led poverty reduction strategy paper (PRSP), the adoption of the MDGs, and the major players and their interactions in these policies. Through analysis, Kameyama et al. (2008) identified the institutional dimensions of climate change and, importantly, the linkages between climate change and sustainable development. Finally, attention is given to the diversity of the Asian region as well as to the many commonalities that exist among the Asian region's countries.

STAKEHOLDERS OF REGIONAL COOPERATION IN ASIA

Intergovernmental Initiatives

In the Asia region, two specific regional initiatives are important to mention, apart from the national initiatives in respective countries. These are the SAARC and ASEAN regional initiatives. SAARC region is highly vulnerable and exposed to natural disasters. The regional efforts to mitigate the risks of disasters and adapt to the climate changes are being taken in ad hoc manners, in two parallel tracks, without much synergy between the two. In the recent months, there have been some serious efforts to integrate these initiatives. One such initiative was the SAARC Action Plan on Climate Change and more recently a Road Map on Regional Cooperation on Climate Change Adaptation and Disaster Risk Reduction in South Asia. In a recent review, Dhar Chakrabarti (2010) examined the contexts, constraints, challenges, and opportunities for these initiatives. Similarly, the ASEAN region has a specific regional framework related to climate change and DRR. Some of them are ASEAN Agreement on Disaster Management and Emergency Response, Regional Framework for Action to Protect Human Health from Effects of Climate Change in South-East Asia and Pacific Region, Regional Climate Change Adaptation Network, ASEAN Climate Change Initiative, and ASEAN Multisectoral Framework on Climate Change and Food Security. Uy and Shaw (2010) made a review of the ASEAN region and argued that high-level government action is required for CCA, supplemented by strengthening adaptive capacity and research and information sharing on CCA.

Local Government Network

It has been argued several times that local governments are the first responders to disaster risks as well as CCA. The key challenge of international negotiations and regional framework is how it is linked to the activities at the local level. CITYNET is a regional network of local authorities for the management of human settlements in the Asia-Pacific region. CITYNET was established in 1987 at the Second Congress for the Development of Human Settlements in Asia and the Pacific organized by Economic and Social Commission for Asia Pacific (ESCAP) and the City of Nagoya, Japan. At that conference, 26 city authorities and 11 NGOs and government agencies adopted the Nagoya Declaration creating the Regional Network of Local Authorities for the Management of Human Settlements in Asia and the Pacific or CITYNET. By bringing together local authorities in the region, the network is envisioned to promote people-friendly cities in the Asia-Pacific that are environmentally sustainable, economically productive, politically participatory, culturally vibrant, socially just, and globally connected (Tjandradewi & Marcotullio, 2009). CITYNET's overarching goal is to facilitate the transfer of best practices between and among its members through City to City (C2C). Membership to the network is divided into two categories, namely full and associate. Full members comprise local authorities in the Asia-Pacific region, while associate members represent national-level organizations (e.g., national municipal associations, development authorities, etc.), NGOs and CBOs, research, training and academic institutions, private companies, and local authorities outside of the region. The uniqueness of CITYNET's approach lies in the composition of its membership. The participation of Associate Members (e.g. NGOs/CBOs, private companies, etc.) greatly complements and augments the capacities of the municipal participants in various areas, from solid waste management to environmental education to transportation planning. Another unique feature is CITYNET's reliance not only on assistance and solutions from developed cities, but also on tested sound practices from those in developing countries. Developed as part of its Technical Cooperation among Developing Countries (TCDC) program, South–South linking is considered to be an effective modality for C2C in Asia, as elsewhere, since the partners are usually similar in one way or another, whether geographically, economically, socially, or culturally. To date, CITYNET has supported approximately 500 municipal exchanges among its members.

Nongovernmental Network

Another key stakeholder is nongovernment organizations, which have direct links with people and communities. The Asian Disaster Reduction and Response Network (ADRRN) is the prominent network in the region, consisting of 34 members from 13 countries (ADRRN website, www.adrrn.net). ADRRN was formed in February 2002 in Kobe, Japan, when "Regional Workshop on Networking and Collaboration among Asian NGOs in Disaster Reduction and Response" was organized by the UN Office for the Coordination of Humanitarian Affairs (UNOCHA) Kobe office and the Asian Disaster Reduction Center (ADRC). The participated NGOs agreed on the need for an Asian network of NGOs for disaster reduction and response in Asia including, in particular, with representation from community-based NGOs (ADRC, 2002). NGO Network can be a platform for capacity development for national and local NGOs. Networks are best known for the capacities that they attempt to build in technical areas such as the creation and dissemination of best practices, improved interventions and approaches, and project/program design. It aims for being a major Asian voice in influencing national and international policies related to disaster reduction response in Asia and the Pacific. The national and local NGOs are most familiar with the hazards, languages, customs, and the environment at the local level and possess a wealth of experiences in disaster reduction and response in the region. The critical role of the national and local NGOs has been acknowledged; however, networking and cooperation among NGOs, in particular, national and local NGOs, were not very common and there was no platform to share the information and experiences for national and local NGOs in the South East Asia when ADRRN was formed (ADRC, 2002). The ADRRN members considered that networking and enhancing the capability of the NGOs were important strategies for increasing the efficiency and effectiveness of disaster reduction and response activities, especially at the local level.

University Network

Asian University Network for Environment and Disaster Management (AUEDM) was formed in 2008 and currently has 21 universities from 17 countries and regions. Education is considered as one of the key mainstreaming tools for any subject. There have been several attempts to mainstream risk reduction in development practices. However, the real mainstreaming starts from formal educational institutions, including schools, colleges, and universities. The mode and method of education should be one of the important issues to make positive impacts. DRR being a multidisciplinary subject needs a combination of theory and practice. Possibly, more emphasis should be given to the field practice and learning from the experiences. Thus, the academic and formal education should go beyond the traditional boundaries of the school and/or university compound, communicate more with the communities, and learn from their experiences. The objectives of AUEDM were to share and work together (bilaterally or multilaterally) in promoting environment and disaster management in higher education (focusing on, but not restricted to, postgraduate education), to seek possibilities of mutual collaboration on field-based action research, and to broaden the scope of education and learning in the environment and disaster management field through collaboration with other stakeholders like NGOs and local governments. The characteristics of AUEDM operation in CCA–DRR research are multidisciplinary approach, field-based action research, and linking academic research to field practice. In a recent study, AUEDM (2010) argues that while there is very popular recent focus on carbon footprints, the concept of ecological footprints has existed for a long time in the academic domain. The fact that the environment has a limited carrying capacity beyond which it cannot support consumption and emission levels has been a well-known fact for very long and has been articulated in scientific terms for many decades. Yet ecological footprints have gone on increasing exponentially with increasing consumption patterns that have accompanied development and economic growth. The study identified the following sectors as the key entry points of CCA–DRR research and higher education: coastal zone management, mountain ecosystem, forest management, urban risk reduction, education and learning approaches, and local governance system.

OVERVIEW OF THE BOOK

This book presents 19 case studies from 13 countries and regions: Bangladesh (2), Cambodia (1), India (1), Japan (2), Malaysia (1), Nepal (1), Pakistan (1), Philippines (2), Singapore (1), Sri Lanka (2), Taiwan (1), Thailand (1), and Vietnam (3). Finally, Chapter 21 presents the summary and way forward of CCA–DRR in Asia (Fig. 2).

Chapter 2 by Islam et al. provides the adaptation processes and framework of Bangladesh. By analyzing the NAPA (National Action Plan

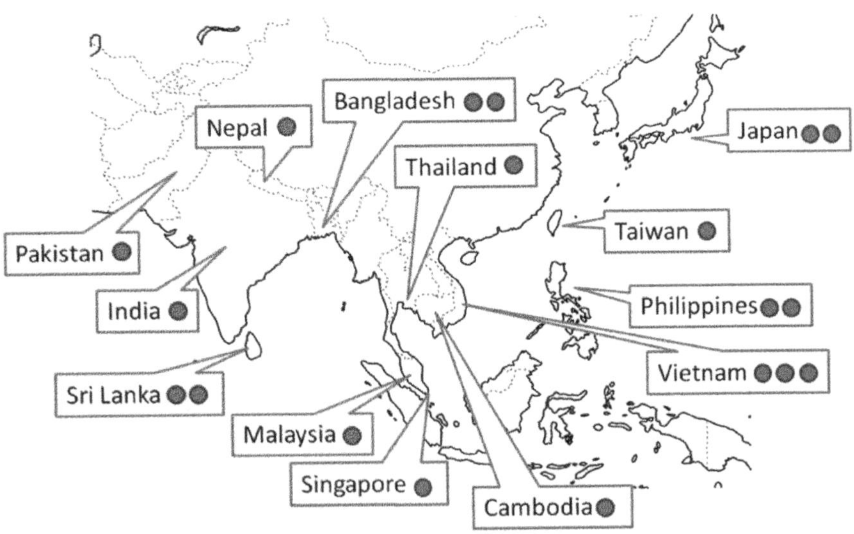

Fig 2. Location and Number of Case Studies in Each Country.

for Adaptation) and BCCSAP (Bangladesh Climate Change Strategy and Action Plan), the chapter shows that future adaptation strategy should focus more on the importance of community resilience, resources (like health and natural resources), knowledge (education and skills), and freedom. Arguing that some of the disasters in Bangladesh are more due to human interventions, the CCA–DRR interventions should also emphasize how to reduce man-made disasters.

Chapter 3 by Habiba et al. identifies a crucial hydrometeorological hazard (drought) of Bangladesh, which is aggravated by climate change impacts. Showing specific examples from the northwestern part of the country, the chapter provides the importance of focusing on drought due to its slow onset and creeping nature. Traditional drought risk reduction approaches have focused on physical issues, by developing better irrigation system. However, the climate change issues have shown that a comprehensive drought resilience mapping needs to be undertaken with focus on socioeconomic, institutional, and physical issues. The chapter proposed a tool to undertake this comprehensive drought risk management in Bangladesh.

Chapter 4 by Nguyen et al. introduces the CCA issues in Cambodia focusing on flood and drought risk management. Through analyzing different scheme of CCA funding mechanisms in the country, the chapter concluded that most of the initiatives focus on external funding. Better governance in all sectors and at all levels, coordination, information sharing, and trained professionals are required to overcome the barriers of sustainable development in the country. Planned adaptation to future climate should be based on current individual, community, and institutional behavior, which, in part, have been developed as a response to current climate. However, common practices used by local people to adapt to existing climate hazards have not received enough attention from government and international agencies.

Chapter 5 by Iwasaki and Shaw provides a case study on livelihood security by citing the example of Chilika lagoon fisheries issues. The chapter argues that the lagoon areas are considered one of the most vulnerable places against climate variability, since the area is fragile to ecological and physical balances. It is important to note that the impacts of climate change cover not just occupational activities but also multifaceted aspects of rural livelihoods that determine the extent of their capacity to adapt to changes in climate. The chapter concludes that fishery livelihoods that encompass numerous patterns of internal relationships among the stakeholders and external effects of fishery governance, market fluctuations, immigration, development of science and technology, etc. are indirectly tied to their resilience against climate hazards.

Chapter 6 by Takeuchi and Shaw provides the adaptation option in Japan. Typhoon and flooding incidents have become an increasing trend in Japan. The coastal and mountain disaster trends are increasing, and the vulnerable communities are the aged population living alone in the coast and the mountain villages. Citing the example from the Hiroshima area, the chapter shows the importance of gender perspective in reducing the damages due to increasing disasters. Risk communication plays a crucial role in bridging the gap between the expert knowledge and local knowledge. Thus, the chapter argues that an appropriate risk communication system is required to link the CCA–DRR in Japan.

Chapter 7 by Fujita and Shaw exemplifies the forest management as the key to CCA in Japan. The chapter addresses the following four questions: (1) What are the impacts of climate change on forest? (2) What are the measures for the impacts on climate change? (3) How does Japan manage to adapt to the climate change in forest sectors? (4) What is the current forest management practice in Japan? Citing an example from the Reihoku region of Kochi prefecture, the chapter argues that due to increasing aged population and lack of forest management, communities in the mountain areas are becoming more vulnerable to increasing disaster risks, which are affected by climate change.

Chapter 8 by Pereira et al. shows the Malaysian approach of CCA and DRR. The chapter argues that one way to achieve mainstreaming at the country level is through policy formulation. Mainstreaming of CCA and DRR through policy formulation can facilitate economic planning and project implementation. The issue of climate change is expected to become an agenda in the mainstream through the implementation of the National Policy on Climate Change. Disaster reduction can be linked and incorporated to CCA goals and indicators when implementing the National Policy, giving both issues more prominence at national, state, and local levels.

Chapter 9 by Subedi shows the dynamic mechanism of strategic entry points, building blocks, driving forces, and achievements (short-term, long-term, and goals) of CCA–DRR in Nepal. The strategic entry point for Nepal in addressing climate change issues can be already existing robust institutions at community level such as community forest users' groups and community owned initiatives such as micro-hydro and community forests. The existing policy, legislatives, and institutional frameworks provide excellent building blocks. In order to give the process enough momentum political commitment and MDGs are thrust forces, whereas research initiatives and forums like UNFCC and regional meetings endow with a dragging force. The process produces short-term achievements in the form of better livelihoods, better prepared communities against adverse impacts of climate change and disasters, improved health and environmental management, and ultimately linked to sustainable development.

Chapter 10 by Khan shows the CCA–DRR issues in Pakistan. By analyzing different climate and disaster related policies and practices in the country, the chapter concludes that local people are traditionally adapted to their environment. Communities know the risks in their surrounding and adopt mitigation measures to reduce their vulnerability. However, due to the climatic variability the number, frequency, density and intensity of extreme weather events are increased, which are beyond the capacity of the communities. As a result, these communities are no more resilient to natural disasters. The worst aspect of the impacts of climate change is that it increases the vulnerability of poorest communities of the poor nations. Some of the ways forward suggested by the chapter include policy formulation, and institutional bases, improvements of irrigation system through related physical measures, preservation of ecosystems in the vulnerable areas, installation of early warning system, and research in the related topics.

Chapter 11 by Pulhin et al. suggests that in the Philippines, while DRR and CCA are closely linked, efforts to address these were done separately, at the national level in particular, with the latter gaining greater political attention. Furthermore, responses are more reactive in nature focusing on rescue, relief, and other postevent measures. Measures in terms of policy, programs/plans, and projects that seemingly integrate the two issues were also found to be more of a product of overlap rather than collaboration. Integration initiatives at local government levels exist, but these are also highly driven by climate change or the expected risks from climate- or disaster-related events. Nevertheless, they demonstrate the key role of political champions in embedding DRR and CCA for minimizing the impacts of natural hazards and increasing the resilience of the local communities.

Chapter 12 by Uy et al. provides one example of community-based adaptation with focus on livelihood issues in the coastal communities of Albay province in the Philippines. The chapter analyzes the coastal livelihoods through the sustainable livelihood framework of natural, social, human, financial, and physical aspects. The fact that coastal communities are at the frontline of climate change necessitates appropriate interventions to reduce the risk to natural and human systems in coastal zones. Effective coastal management entails an integrated approach including CCA and DRR for it to have significant impact. In the case study of the municipality of Bacacay, climate trends and perceptions clearly showed the vulnerability of livelihoods of coastal communities. The assessment of vulnerabilities and strengths affords the local planning process to effectively consider locally relevant resilience-building options. Thus, looking from macro and micro perspectives, the local government can plan and strategize actions in a holistic and integrated manner. In addition, knowledge of individual vulnerabilities and strengths gives each of the barangay (ward office) an understanding of the appropriate approaches and measures the community can undertake on its own.

Chapter 13 by Li and Rajola describes the Singaporean experiences of CCA–DRR. Singapore's climate change strategy is basically focused on assessing and studying the vulnerability to climate change and addressing it through requisite adaptation/mitigation measures, and reducing the greenhouse gas emissions produced in the country through energy efficiency measures. Both the above initiatives are buttressed by the concurrent focus on competency building in Singapore that will enable it to develop and implement technologies and processes that support its efforts in minimizing the adverse impacts of climate change. The chapter concludes that the country is ensuring reliability and resilience of water supply through integrated water resource management and also by tapping on research, development, and technology to develop and test beds new technologies. Singapore hopes to offset the adverse impacts of an increased sea level and changing weather patterns through land reclamation, improved drainage infrastructure, building of reservoir dams, and protection of coastal areas.

Chapter 14 by Nianthi provides an example of CCA–DRR through agro-forestry sector. Citing different examples of agro-forestry practice from the country, the chapter shows that these practices will continue to be a profitable and sustainable way of using land, as it has been almost unnoticed for thousands of years. It shows the potential for great benefits, for soil and water conservation, increased yields, enhanced biological diversity, reduced need for pesticide and fertilizer, and provides greater economic stability of farming communities. Thus, the importance of agro-forestry as a land-use system is receiving wider recognition not only in terms of agricultural sustainability but also as adaptation measures to climate change. This chapter discusses the numerous benefits of agro-forestry systems in Sri Lanka and identifies the usefulness in the absorption of CO_2 as adaptation strategy for climate change issue in Sri Lanka, which needs a specific policy support.

Chapter 15 by Wickramasinghe focuses on the coastal ecosystem of Sri Lanka and its implications to CCA with focus on coastal disasters. It is clear that coastal ecosystems play a key role in maintaining ecological functions while providing economical, social, and cultural services to island nations. However damaging the impacts of anthropogenic activities and climate change would be, there are mitigation measures to be adopted with effective institutional, ecosystem, and community approaches. Institutional issues focus on laws and regulations and specific policy measures at national and local levels. Ecosystem approach focuses on risk assessment of different habitat sectors, designs new natural resources, practices realistic conservational planning, alleviates the effects of other stressors, and promotes ex situ conservation. Community approaches include adapting local agenda, enhancing existing capacities as well as building new capacities, and promoting social coherence.

Chapter 16 by Wen et al. provides the example of adaptation in the water sectors in Taiwan. In order to reduce the level of impacts of climate change on water resources in Taiwan, suitable adaptation strategies are needed in three major areas: water resource conservation, development, use, and management; strengthening flood control and rescue; and drought prevention and emergency response. The chapter suggests strategies to ensure water resource conservation, development, use, and management; strategies to strengthen flood prevention and rescue; and strategies for drought prevention and emergency response. Nevertheless, disaster prevention adaptation strategies devised by the Taiwan government do address the impact from climate change and do focus on sea-level rise and water resources. The conservation of watersheds included in the water resource strategy closely interrelates with debris flow prevention in Taiwan.

Chapter 17 by Lebel et al. shows the water insecurities issues and its relation livelihood and infrastructures due to climate change in Thailand. In Thailand, DRR initiatives and the emerging policies for adaptation to climate change have had parallel histories with little constructive interaction. Reducing the risks of disaster is important to sustainable development and is only likely with long-term adaptive planning. Given large uncertainties about key aspects of future climate and the effectiveness of different policies and management interventions, it is clear that learning processes and public engagement should be an integral part of adaptation governance. Developing effective institutions and capacities requires broad engagement with stakeholders and a paradigm shift from reactive to adaptation and proactive risk management.

Chapter 18 by Nguyen and Shaw analyzes the policy dimensions of CCA and DRR in Vietnam. Climate change brings closer together the three areas of the disaster management community, the environmental community, and the poverty reduction community – mainstreaming is the key to doing so but it is important to stay focused. The threat of climate change is beginning to be acknowledged but information and awareness remain at a low level. Vietnam does not yet have national or local CCA strategies, and national and local capacity building is urgently needed to ensure that policy responses are adequate and effective. Coordination between line ministries also needs to be urgently improved, and cooperation with international agencies and NGOs enhanced so that climate change can be addressed in an integrated way with long-term socioeconomic and poverty-reduction efforts. Crucially, those most at risk from climate change, the rural poor living in provinces most affected by climate change have limited information or financial and technical support to adapt to their changing world, despite some localized successes.

Chapter 19 by Tinh et al. provides example of local vulnerability and adaptation in Quang Nam province of central coastal area of Vietnam. The chapter shows that there is a significant association among risk perception, adaptation measures, and impact reduction, in which risk perception plays an important role in shaping the climate change induced event coping mechanism and adaptation to climate change. Risk perception can influence both the design and operational aspects of disaster risk management. If disasters were perceived as hazard-led events, then the coping and management mainly focused on the physical aspects, structure measures, and external relief. In contrast, if disasters were considered as a product of hazard and vulnerability, then the adaptation measures recognized human behavior as important aspects and the copying and management focused on nonstructural measures. Thus, the chapter argues that understanding how people perceive the disaster risk is important to formulate a disaster risk management policy and plan.

Chapter 20 by Kien et al. provides a case study of health issue (water-borne diseases and influenza) in Vietnam, which is related to the climate variability. Through data and policy analysis, the chapter concludes that the crucial step to prevent and reduce the adverse health effects of extreme events and empower the adaptive capacity is raising public awareness and knowledge about the nature of the risk. However, according to practical observation, especially from the low-income developing countries with low education index, the public awareness, which always increases after any serious natural disasters, is often short lived if the situation returns to previous state and if no "reminding" or educational measures were maintained.

Finally, Chapter 21 by Pulhin et al. provides an analysis of the key findings from all the chapters. The chapter shows that enhancing CCA and DRR integration needs to be done at various levels, through functional, spatial, temporal ways, integrating CCA to DRR cycle, harnessing existing platform for integration, and developing assessment methods and tools. Some of the key findings include removing institutional barriers and promoting adaptive governance, building on sectoral success and promoting social learning, enhancing local resilience and building capital assets, investing on innovative education and research, and championing the cause of CCA and DRR.

ACKNOWLEDGMENTS

The first author acknowledges the support from Mitsui Foundation and GCOE ARS. The authors also acknowledge the support from the colleagues of the Asian University Network of Environment and Disaster Management (AUEDM).

REFERENCES

Asian Disaster Reduction Center (ADRC). (2002). *Proceedings of Regional Workshop on Networking and Collaboration among Non-Governmental Organizations of Asian Countries in Disaster Reduction and Response*, February 20–22, 2002, Kobe, Japan.

AUEDM. (2010). *Education–research interface of climate change adaptation and disaster risk reduction*. AUEDM Report. Kyoto, Japan, 33pp.

Dhar Chakrabarti, P. G. (2010). Integrating disaster risk reduction with climate change adaptation: Recent initiatives in South Asia. In: R. Shaw, J. Pulhin & J. Pereira (Eds), *Climate change adaptation and disaster risk reduction: Issues and challenges*. UK: Emerald.

IPCC. (2007). IPCC Fourth Assessment Report. IPCC (Intergovernmental Panel on Climate Change), Geneva, Switzerland.

ISDR. (2008). *Climate change adaptation and disaster risk reduction*. Briefing note. Geneva, Switzerland, 12pp.

Kameyama, Y., Sari, A., Soejachmoen, M. H., & Kanie, N. (2008). *Climate change in Asia: Perspective of future climate regime* (253pp.). Tokyo, Japan: UNU Press.

Pettengell, C. (2010). *Climate change adaptation: Enabling people living in poverty to adapt*. OXFAM Research Report. London, UK, 48pp.

Sharan, D. (2008). *Financing climate change mitigation and adaptation: Role of regional financing arrangements*. ADB Sustainable Development Working Paper no. 4. Manila, Philippines, 20pp.

Shaw, R., Pulhin, J., & Pereira, J. (2010). *Climate change adaptation and disaster risk reduction: Issues and challenges*. UK: Emerald.

Tjandradewi, B. I., & Marcotullio, P. J. (2009). City-to-city networks: Asian perspectives on key elements and areas for success. *Habitat International*, *33*, 165–172.

UNDP. (2008). Human Development Report 2007/2008: Fighting Climate Change: Human Solidarity in a Divided World. UNDP, NY, USA, 399pp.

Uy, N., & Shaw, R. (2010). Climate change adaptation in ASEAN: Actions and challenges. In: R. Shaw, J. Pulhin & J. Pereira (Eds), *Climate change adaptation and disaster risk reduction: Issues and challenges*. UK: Emerald.

World Bank. (2008). *Development and climate change: A strategic framework for the World Bank group* (39pp.). Washington DC, USA: World Bank.

World Bank. (2010). *Managing climate risk: Integrating adaptation into World Bank group operation*. Available at http://web.worldbank.org/WBSITE/EXTERNAL/COUNTRIES/SOUTHASIAEXT/0,,contentMDK:21469804~menuPK:2246552~pagePK:2865106~piPK:2865128~theSitePK:223547,00.html. Retrieved on July 11, 2010.

WWF. (2009). Mega stress for mega cities: A climate vulnerability ranking of major coastal cities in Asia. WWF Report. Geneva, Switzerland, p. 40.

CHAPTER 2

IMPACTS OF CLIMATE CHANGE INDUCED HAZARDS AND ADAPTATION PROCESSES IN BANGLADESH: AN OVERVIEW

Tawhidul Islam, M. Aminur Rahman and Fuad Mallick

INTRODUCTION

Living with physical hazards is an everyday issue for the people of Bangladesh because different seasons bring different kinds of uncertainties. Sometimes traditional knowledge and practices (Alam, 2007) help them to adjust to these conditions; in some occasions, these hazards turn into catastrophic disasters causing deaths and bringing unbearable damages in different sectors. In addition to the tangible damages incurred due to hazards, in most of the cases, these hazards injure the internal social organizations at different levels (for example, household, community, and institutional) and affect their assets. Thus, the community resilience is greatly weakened, which makes people susceptible to upcoming hazards. Scientists suggest that climate change-induced threats and uncertainties in the forms of erratic rainfall patterns that result in drought conditions and sometimes floods, abnormal foggy conditions, change in the wind direction and the characteristic patterns of seasons, anomalies in the temperature

Climate Change Adaptation and Disaster Risk Reduction: An Asian Perspective
Community, Environment and Disaster Risk Management, Volume 5, 19–36

ISSN: 2040-7262/doi:10.1108/S2040-7262(2010)0000005008

regimes, and the occurrence of cyclones will bring new dimensions to existing situations. These natural hazards, temperature rises, and sea level rise–induced inundations will contribute to the breakdown of the traditional systems of living; they also bring change in topographical factors (for example, flooding), biophysical factors (changes in the crop yields, runoff, risks of the spread of infectious diseases, changes in the vegetation pattern), and socioeconomic factors (per capital income, health, education, population density) (World Bank, 2001).

Based on this background, this chapter reviews the proposed and ongoing works and action plans regarding climate change issues in Bangladesh. The review work aims to understand (1) how climate change dynamics are locally perceived and their likely impacts, (2) the level of preparedness of the government and nongovernment sectors in areas of disaster risk reduction (DRR) and climate change adaptation (CCA). In this regard, Intergovernmental Panel on Climate Change (IPCC) predictions, the local Providing Regional Climates for Impacts Studies (PRECIS) models, and other relevant exercises are evaluated. The review results suggest that temperature has moderately increased for most of the weather stations in Bangladesh, compared to the past records, and rainfall patterns have been found to be more erratic. The chapter then evaluates the action plans, which are proposed in the Bangladesh Climate Change Strategy and Action Plan (BCCSAP) by the National Adaptation Programme of Action (NAPA), to combat climate change uncertainties and to reduce the impacts on communities.

VULNERABILITY TO HAZARDS AND COMMUNITY RESILIENCE: AN ASSESSMENT

Environmental hazards create differential risks to the people. Distribution of risks is again differential in respect to class in the community, because the level of exposure to natural hazards of certain groups of people is determined by the social and economic status to which they belong. In this regard, contemporary disaster management research put more emphasis on community vulnerability assessments (Winser, Blaikie, Cannon, & Davis, 1994) to understand the natural hazards and their related risks more clearly in the way the community perceives them.

Kelly and Adger (2000) explained the nature of the relationship between vulnerability and adaptation in the purview of climate change. They defined vulnerability in terms of "the capacity of individuals and social groups to respond to, that is, to cope with, recover from or adapt to, any external stress placed on their livelihoods and well-being." They put the socio-economic and institutional constraints at the center of vulnerability analysis. Based on the entitlement concept of Sen (1981), Kelly and Adger (2000) mentioned that the availability of resources and the ability of individuals and the community to access and control those resources determines how they can reduce their vulnerability in adverse situations. They identified that the underlying causes and processes are responsible for making the people vulnerable to natural hazard events. In another attempt, Adger, Hughes, Folke, Carpenter, and Rockstorm (2005) claimed that decline of resilience also contributes to social and ecological vulnerability of a community. They also mentioned strong social-ecological systems that incorporate diverse mechanisms for living with, and learning from, change and unexpected shocks and suggested that disaster management requires multilevel governance systems that can enhance the capacity to cope with uncertainty.

Eakin (2005) showed that vulnerability is a multistep phenomenon because globalization and market liberalization, emerging as external forces, contribute in structuring the livelihood pattern of the community, which in turn determines the level of a community's capacity to face disasters. Schroter, Polsky, and Patt (2005) mentioned that people undertake vulnerability assessments in regard to climate change, but the approach of vulnerability assessment itself is not critically examined. They proposed a methodological guideline of vulnerability assessment, setting five major criteria for it: (1) They should have a knowledge base from various disciplines and stakeholder participation, (2) be place-based, (3) consider multiple interacting stresses, (4) examine differential adaptive capacity, and (5) be prospective as well as historical.

However, one of the earlier works to address climate change problems in Bangladesh was undertaken by Warrick, Bhuiya, and Mirza (1994). In five briefing documents, they mainly tried to present the scenarios of the greenhouse effect and climate change within the physical, social, and legal frameworks of the country. They put little emphasis on the community resilience build-up process.

World Bank (2001) also identified climate change–induced problems in coastal and freshwater resources, agriculture, human health, ecosystems, and biodiversity. In their document, World Bank considered that the community resilience/well-being issues should be seen as contingent in the security of these areas.

CLIMATIC CHANGE INDUCED HAZARDS AND THEIR CONSEQUENCES IN BANGLADESH

The presence of uplands in the eastern, northern, and western parts of Bangladesh and the Bay of Bengal in the southern part determines the characteristics of climatic, hydrological, and geomorphological processes of Bangladesh. The occurrence of low-lying floodplain with fertile soil and the availability of irrigation water from a large number of rivers, canals, and creeks strongly support the primary economic activities such as agriculture, fisheries, and livestock rearing. As a result, a large portion of people are heavily dependent on the nature-based resources, which in turn makes the community vulnerable to hazards because the primary economic activities are always at risk to hazards. The fused nature of physical and human interaction sometimes collapses when some of the dynamic physical processes[1] appear beyond their characteristic pattern, for example, in the form of hazards like floods and cyclone. In many instances, these hazards turn into disasters when people find limited opportunities to cope with the situation with their own capacity and resources. It is imperative to mention that the impacts of these physical hazards are sometimes aggravated by man-made shortcomings like unplanned land use change. Thus, the hazards that generally occur in Bangladesh can be seen within a broad spectrum of physical and human processes and events that influence one another.

As mentioned previously, a majority of the people in Bangladesh, both directly and indirectly, depend on primary economic activities based on the natural productive systems of the environment. This primary productivity could continue to occur as a result of the existence of these physical (and climatological) processes. In this context, any alteration in the climatic pattern or physical processes may affect the productive nature to which the communities are accustomed. Experts believe the alteration in the climatic systems happens at a faster pace than the natural rate of human adaptation to cope with the situation. These uncertainties make the community, their institutions, and natural productive systems vulnerable to shocks. Thus, change in the climatic pattern and related hazards such as floods, cyclones, and anomalies in the rainfall pattern concurrently put the natural systems, economic activities of the communities, and infrastructure in a point of insecurity.

Reviewing Local-Level Climate Change Studies

IPCC indicated historical rise in the temperature (Houghton, 2004) and declared that this increase is human induced (IPCC, 2007). At the same time, IPCC (2007) mentioned that the general circulation model (GCM) produced prediction scenarios that should be used at global scales because the resolution of the input data were not high enough to foresee local scale changes. In these circumstances, there was a demand at local levels to examine climate change issues based on local climatic data. As a response, some work has been conducted in Bangladesh to assess local scale climate change and its relevance to natural hazards. A brief review on this work is provided in the following sections to show how climate change issues are locally perceived.

The first major model-based work on local-level change in temperature and rainfall elements of climate had been undertaken by Climate Change Cell (Climate Change Cell (CCC), 2009) of the government of Bangladesh. They used a model called PRECIS (Providing Regional Climates for Impact Studies), developed by the Hadley Centre in the United Kingdom. The model used data from 31 weather stations in Bangladesh, and regression analysis was performed; the model outputs are provided in a 50-kilometers by 50- kilometers (31 miles by 31 miles) grid format. The months of December, January, and February were considered as winter, and March, April, and May were considered the summer season. The following are the major findings of the PRECIS model:

- Rainfall during monsoon and post-monsoon periods will increase, where it will remain close to the historical amount during dry season.
- Rainfall during pre-monsoon will fluctuate in different years.
- Over the country, rainfall will increase 4 percent, 2.3 percent, 6.7 percent in 2030, 2050, 2070, respectively, in reference to the observed baseline period.
- Monthly average maximum temperature will change from -1.2°C to 4.7°C in 2030, from -1.2°C to 2.5°C in 2050, and from -1.2°C to 3.0°C in 2070.
- Maximum temperature will increase during the monsoon period, and it will decrease in other periods.
- The monthly average minimum temperature will increase in all periods and vary from 0.3°C to 2.4°C in 2030, from 0.2°C to 2.3°C in 2050, and from -0.6°C to 3.3°C in 2070.
- The variation of rainfall and temperature (both maximum and minimum) in any location in Bangladesh and in a particular month is larger than the seasonal and annual average.

A second major study was also conducted by CCC (2008), during which it was attempted to characterize changes of Bangladesh climate in the context of agriculture and irrigation. It also considered March, April, and May as summer and November, December, January, and February as winter season. The research showed that annual and seasonal mean temperatures are found to be in increasing trends. The overall trend in mean annual temperature is found +0.10°C and +0.21°C per decade for the years between 1948 and 2007 and between 1980 and 2007, respectively. The study concludes that warming has been more rapid in recent decades. The initial National Communication report prepared by the government to fulfill the submission requirement to United Nations Framework Convention on Climate Change (UNFCCC) and the National Adaptation Programme of Action also mentioned increasing predictions for Bangladesh (Table 1). However, it is interesting to note that the model results do not always comply with other prediction results. For instance, local PRECIS model results differ from Initial National Communication (INC) and IPCC Third Assessment Report (TAR) model results. The following table indicates a significant departure from one assessment to the other.

All of these works mainly provide future-changing scenarios of climatic elements (that is, temperature and rainfall) in Bangladesh, based on

Table 1. Comparative Assessment of Temperature and Rainfall Change Model Predictions for Bangladesh.

Sources/Models	Seasons	Temperature Increase °C (T)$T_{max}/T_{min/Aver}$		Rainfall Change %	
		2030	2050	2030	2050
PRECIS (local model)	Annual	+0.02/+0.37/NA	+0.01/+0.44/NA	+4.0	+2.3
	Winter (DJF)	−0.03/+0.13/NA	−0.05/−0.03/NA	−8.7	−4.7
	Summer (MAM)	+0.16/+0.48/NA	+0.27/+0.52/NA	+4.1	−3.5
	Rainy (JJAS)	+0.23/+0.64/NA	+0.01/+0.74/NA	+3.8	+3.0
	Seasons	2030 (average)	2050 (average)	2030	2050
Initial National Communication (INC) Report to UNFCCC, World Bank	Rainy	+0.7	+1.1	+11	+28
	Winter	+1.3	+1.8	−3	−37
BDNAPA/IPCC TAR	Annual	1.0	1.4	+5	+6
	Winter (DJF)	1.1	1.6	−2	−5
	Rainy (JJA)	0.8	1.1	+6	+8

Note: DJF, December, January, February; MAM, March, April, May; JJAS, June, July, August, September; JJA, June, July, August.

available historical data. They give national-level changing trends applicable for all the regions of the country, which may not provide local scale insights useful for relating that change to different sectors like agriculture or fisheries of that area. In addition, spatially different regions of the country show distinct characteristics influenced by proximities to sea or uplands/hills, hydrological regimes, land cover, and soil conditions that determine the solar albedo, and area-specific heat flux (for example, especially the influence of the urban heat island).

Islam (2010), in this regard, determined a linear trend of change based on regression models and examined the change based on the yearly deviation of values of temperature and rainfall variables from that historical average (at least an average of 30 years of climatic data). The months of December and January were considered as winter season, and April and May were considered as summer. Comparisons are also made for summer average minimum and maximum temperatures. Fig. 1 shows an increase in both the maximum and minimum current temperature for both summer and winter seasons, from base year temperature for all the weather stations of Bangladesh. A significant rise appeared in the winter minimum temperature.

Islam (2010) averaged rainfall occurrences in different months of a year for about a half century, and plotted stations so that both spatial and temporal dimensions of rainfall fluctuations could be determined. The analysis indicates that the weather stations located in the coastal belts and situated in the uplands/hills show a tendency of high rainfall incidence, especially during monsoon period. There is a slight variation in the annual total rainfall occurrence at different weather stations.

IPCC estimates a rise in the sea level, which would be in the range of 15 to 95 centimeters (5.9 to 37.4 inches) by 2100. A 10-centimeter (3.9-inch) sea level rise that would most likely happen before 2030 would inundate 2,500 square kilometers (965 square miles), about 2 percent of the total land area. With the high-end estimates, sea level rise in Bangladesh would inundate about 18 percent of the country by 2100.

INITIATIVES TO ENHANCE "COMMUNITY RESILIENCE" TO TACKLE CLIMATE CHANGE PROBLEMS IN BANGLADESH

Adaptation and mitigation are two measures that are being considered globally to address climate change–induced problems. Bangladesh has little

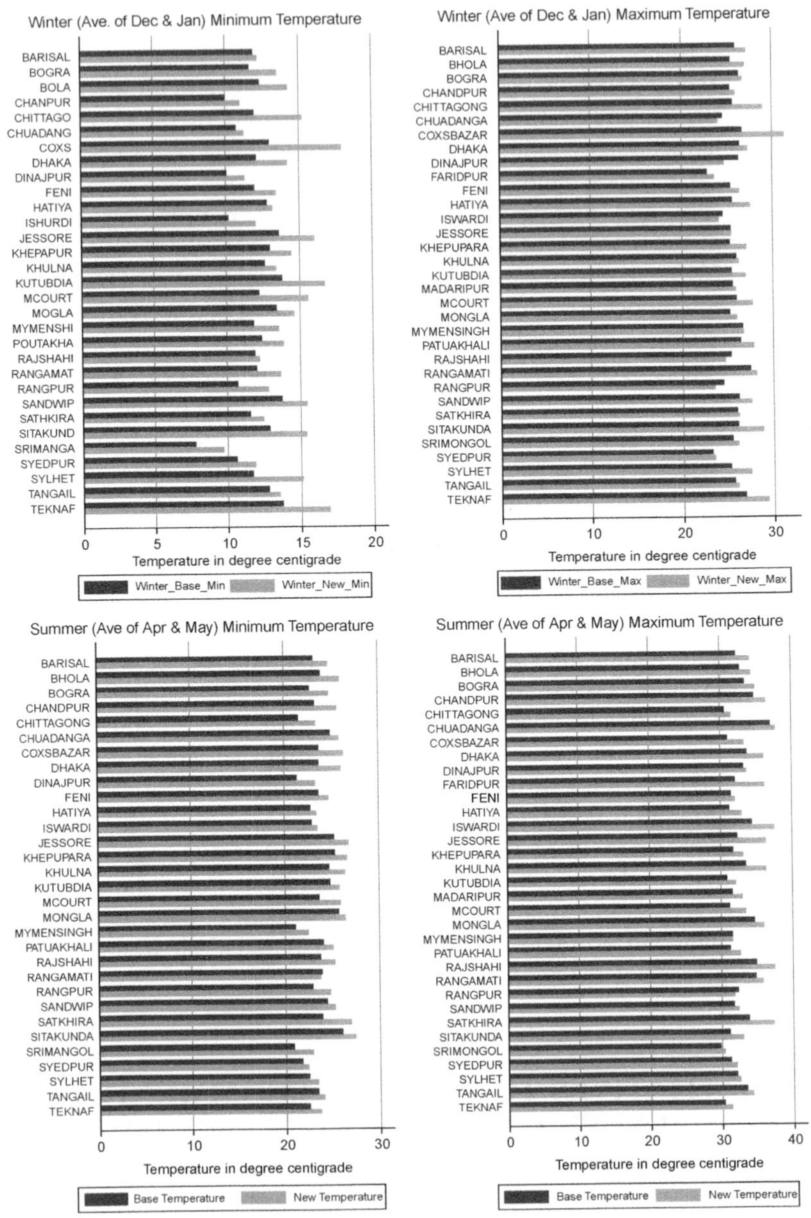

Fig. 1. Comparison of Summer and Winter Temperature Fluctuations for the Different Weather Stations of Bangladesh.

to do with mitigation because its contribution in emitting greenhouse gases that cause global warming is insignificant; rather adaptation is considered more pragmatic for the community. The following discussion focuses on the existing efforts in Bangladesh that aim at enhancing community capacity and how these support improved community resilience to reduce climate change–induced impacts.

The government of Bangladesh produced the National Adaptation Programme of Action (NAPA) (Government of Bangladesh, 2007) to set strategic guidelines for minimizing climate change risks in different sectors. NAPA strongly emphasized developing temperature- and saline-tolerant crop varieties, constructing new flood shelters, and strengthening urban infrastructure and industries. The following suggested interventions, outlined in the NAPA document, are to be undertaken at the government level to reduce climate change vulnerability:

(i) Promoting adaptation to coastal crop agriculture to combat increased salinity.
(ii) Adaptation to agriculture systems in areas prone to enhanced flash flooding in northeast and central region.
(iii) Promoting adaptation to coastal fisheries through culture of salt-tolerant fish species in coastal areas of Bangladesh.
(iv) Adaptation to fisheries in areas prone to enhanced flooding in North East and Central region through adaptive and diversified fish culture practices.
(v) Construction of flood shelter, and information and assistance centre to cope with enhanced recurrent floods in major floodplains.
(vi) Reduction of climate change hazards through coastal afforestation with increased community participation.
(vii) Providing drinking water to coastal communities to combat enhanced salinity due to sea level rise.
(viii) Enhancing resilience of urban infrastructure and industries to cope with the impacts of climate change including floods and cyclone hazards.

In addition to NAPA implementation, the nongovernmental organizations (NGOs) of Bangladesh have been trying to play an important role in reducing climate change–induced problems. Their actions in recovery and rescue of people and property during disasters and postdisaster efforts are significant and appreciable. They have been trying to assist people with different kinds of programs, with the consideration that these interventions may provide backup support to community-level adaptive initiatives coping with natural hazards. Some of the external interventions focus on livelihoods protection; some are aimed at infrastructure (house) building, rehabilitation, and protection; and some of the projects are deployed in reducing psychological shocks and trauma.

INSTITUTIONAL RESPONSE FOR CLIMATE CHANGE ADAPTATION (CCA)

The government of Bangladesh is aware of the country's natural sensitivity to climate variability in general, and climate change in particular. There are several policy responses related to climate change. The indirect responses include addressing the impacts of climate change through programs that reduce vulnerability, by alleviating poverty, generating employment, and diversifying crops. The more direct responses are through disaster risk reductions and management schemes, and through specifically targeting climate change by mainstreaming climate change issues into sectoral plans and national policies.

Bangladesh is signatory to the UNFCCC. In 1992, the government of Bangladesh signed the UNFCCC, and ratified it in 1994. The Ministry of Environment and Forest (MOEF) is responsible for coordinating the UNFCCC process in Bangladesh. A National Climate Change Committee, comprised of members from all relevant government and nongovernment organizations, was constituted in 1994, for policy and guidance and to oversee the implementation of obligations under the UNFCCC process. In addition to the Climate Change Cell (CCC), working under the auspices of Ministry of Environment and Forest, other government institutions that are relevant for climate change include an inter-ministerial committee on climate change, headed by the Ministry of Environment and Forest and with representation from relevant government ministries as well as NGOs and research institutions, and a National Environment Committee to determine environmental policies chaired by the prime minister with representation from Members of Parliament (MPs) as well as government and civil society.

Among the programs, the Participatory Disaster Management Program (PDMP) focuses on disaster management and prevention, and also on adapting to climate change. The emphasis of the program is on preparedness, such as raising awareness, enhancing knowledge and skills of personnel in handling disasters, establishing disaster action plans in the most disaster prone areas, promoting local-level risk reduction measures, and improving early warning systems.

Bangladesh has also taken up a Comprehensive Disaster Management Programme (CDMP) with United Nations Development Programme (UNDP) and other donor assistance. The aim of the program is to put greater emphasis on disaster preparedness and risk reduction. The Local Disaster Risk Reduction Facility (LDRRF), another component of the CDMP, has the task of keeping the CCC of the Department of Environment (DoE) informed. LDDRF aims to improve coordination between development-oriented and disaster management aspects of the government of Bangladesh at the local level.

The government of Bangladesh's structural and organizational preparedness for disaster has been succinctly described in the 2008 Bangladesh Climate Change Strategy and Action Plan (BCCSAP). The government has so far invested over $10 billion (at constant 2007 prices) to make the country more climate resilient and less vulnerable to natural disasters. Flood management embankments, coastal polders, and cyclone shelters have been built, and important lessons have been learned on how to implement such projects successfully in the dynamic hydrological conditions in Bangladesh, ensuring active participation of communities. A comprehensive system of disaster preparedness and management, including *Standing Orders on Disaster*, which details the responsibilities of government officials and others in times of disaster, has also been put in place. The government demonstrated its competence in dealing with disasters in 2007, when the country suffered two serious floods and a severe tropical cyclone (Cyclone SIDR) in the same year.

CCA-DRR Convergence in Combating Climate Change

Climate change adaptation (CCA) is a long-term process and has a connotation with sustainable development and enhancement of equity in resources and benefit sharing; all these are mutually reinforcing. On the other hand, the disaster risk reduction (DRR) approach mainly focuses on immediate needs-based improvements of the systems (for example, social, economic, biophysical), to reduce vulnerabilities of the communities. In other words, DRR aims mainly to enhance predisaster capacity building and to support disaster preparedness plans of the communities. These preparedness plans differ in the type of disasters they face (for example, floods, cyclones). However, both the CCA and DRR approaches have some overlap and, at the same time, have distinct features while considering temporal dimension of climate change.

People in some areas have been adapting to change in the physical contexts since historical times (Islam, 2007b). Therefore, backup supports to strengthen the existing adaptation processes through DRR programs are considered appropriate for some areas, while new interventions may be needed for some areas. However, designing these location- and community-specific adaptation options advocates the need for understanding the sociocultural fabric, the type and level of knowledge and skills of the communities and their institutions, and the physical contexts in which they situate. The Bangladesh Climate Change Strategy and Action Plan (BCCSAP) addresses these issues by capturing both the DRR and CCA options. The following presents a comprehensive list and set plans that BCCSAP gives to promote adaptation in changed conditions:

Theme 1. Food Security, Social Protection, and Health-Specific Programs

- P 1: Institutional capacity development
- P 2: Climate-resilient crop variety development
- P 3: Adaptation to agricultural drought
- P 4: Adaptation in fisheries sector
- P 5: Adaptation in livestock sector
- P 6: Adaptation in health sector
- P 7: Water and sanitation program
- P 8: Livelihood protection in ecologically vulnerable areas
- P 9: Attention to special vulnerable groups

Theme 2. Comprehensive Disaster Management–Specific Programs

- P 1: Flood forecasting and early warning
- P 2: Cyclone storm surge warning
- P 3: Awareness raising and public education toward climate resilience
- P 4: Risk management against loss of income and property

Theme 3. Infrastructure-Specific Programs

- P 1: Repair and maintenance of flood embankments
- P 2: Repair and maintenance of cyclone shelters
- P 3: Repair of coastal polders
- P 4: Improvement of urban drainage
- P 5: Adaptation against floods
- P 6: Adaptation against future cyclones
- P 7: River training

Theme 4. Research and Knowledge Management-Specific Programs

- P 1: Establish research/knowledge management and training center on climate change
- P 2: Climate change modeling
- P 3: Adaptation against sea level rise (SLR)
- P 4: Monitoring ecosystem and biodiversity changes
- P 5: Macroeconomic/sectoral economic impacts

Theme 5. Mitigation and Low Carbon Development-Specific Programs

- P 1: Energy efficiency in production and consumption of energy
- P 2: Gas exploration and reservoir management
- P 3: Development of coal mines and coal-fired power stations
- P 4: Renewable energy development
- P 5: Lower emission from agricultural lands
- P 6: Urban waste management
- P 7: Afforestation and reforestation program

Theme 6. Capacity Building and Institutional Strengthening–Specific Programs

- P 1: Revision of sectoral policies for climate resilience
- P 2: Mainstreaming climate change
- P 3: Human resource development
- P 4: Strengthening institutional capacity
- P 5: Mainstreaming climate change in media

Although the government of Bangladesh has prepared its action plan for implementation, donor communities and NGOs have proposed certain CCA and DRR options for floods and freshwater resources. These adaptation options, in substantive terms, are not very far or different from those identified in BCCSAP. As a matter of fact, some intervention options are quite similar. The following lists some of the options relevant to floods, in six categories:

1. Structural flood and erosion control measures, including river training
   - Adaptation to increased morphological dynamics by ensuring protection against riverbank erosion
   - Full flood-protection embankments
   - Controlling of flooding in combination with compartmentalization
   - Design for major infrastructure such as barrages and river training works in the main rivers

2. Drainage, storage, irrigation, and other efficient water use measures
   - Ensuring additional inflows from upstream
   - Increased storage of water
   - Channel restoration to reduce waterlogging through dredging and river training
   - Improved irrigation efficiency
   - Promotion of optimal use of both surface and ground waters
3. Participatory/community management of the Operation and Maintenance (O&M) of the structures
   - Arrangement for (participatory) operation and maintenance of water resources infrastructure
   - Involving local communities in maintaining flood-protection embankments
4. Flood management, including forecasting and warning, evacuation to flood shelter, raising mounds as flood shelter, and the like
   - Improved flood warning and forecasting
   - Evacuation of flood-vulnerable people and their valuables
   - Landfills and flood shelters (elevated land) and flood refuge areas
5. Crop diversification to reduce or save crop damage or loss
   - Crop diversification
   - Development of drought-tolerant crop varieties
   - Faster growing or stronger and less flood-sensitive crops or varieties
6. Land use control in flood-vulnerable areas and flood insurance
   - Discouraging future development in high-risk areas
   - Expanding access to credit and crop insurance

Currently, both the CCA and DRR approaches are combined, and government is trying to mainstream this fused approach through sectoral policies and programs specially designed for disaster mitigation.

AREAS TO ADDRESS FOR FURTHER IMPROVING THE COMMUNITY RESILIENCE IN BANGLADESH

Developing long-term community resilience depends on equitable distribution of resources and wealth among the people because these are the determining factors for them to take decisions independently and act pragmatically. Similarly, fair distribution of services like health and education are crucial and determine the sustained capacity of the community to stand strong in most of the conditions. But in Bangladesh, equitable resource allocation and the asset-building policies rarely favor the poor (Afsar, 2005); similar conditions persist in the service sector as well. In this regard, it might be mentioned that land reform is still left unresolved; unused public land (called *khas* land locally) distribution is still skewed ;to the powerful in society (Barakat, Zaman, & Raihan (2001)). Forests are destroyed when dominant groups in the society grab forestlands. In some cases, forestlands are leased out by the state authorities (Islam, 2007a) to the powerful groups. Similar kinds of problems can be seen in the wetland resources management systems. The current wetland leasing system favors the local and urban elites. Thus, the marginalization of certain groups of people as a result of exclusion from resources and services causes poverty and many different kinds of problems that, sometimes, create a very new set of problems that are fundamentally different from the original ones. However, it is imperative to mention that depriving common people from the natural resources (that support provisional, supportive, and regulatory services) and excluding them from equitable sharing of wealth and service-related provisions seriously hampers the development of strong community resilience.

CONCLUSION

Climate change will not necessarily give rise to any new hazards in Bangladesh; rather, the known hazards will have added dimensions leading toward worse scenarios, and the intensity and frequency of common hazards will increase. Thus, climate change may cause the breakdown of the morphological processes, biophysical systems, and socioeconomic conditions. Breakdown of these very foundations on which strength of the community depends might have deep negative impacts that may undermine local and collective initiatives to tackle climate change risks. In this context, the relief and recovery approach of disaster management in Bangladesh has been replaced by a holistic approach incorporating the processes of hazard identification and mitigation, community preparedness, and integrated response efforts (IWM, 2009). Efforts are being implemented that target the strengthening of preparedness capacities and enhancing community resilience under the disaster risk reduction approach. In addition, NGOs, donors, and government departments are undertaking specific programs and action plans to address the climate change–induced hazards as called climate change adaptation approach.

However, a critical assessment may reveal that the NAPA and BCCSAP documents failed to foresee the climate change–related problems as multi-stepped problems, as Eakin (2005) suggested. In this regard, it is noteworthy that the *Hyogo Framework for Action (2005–2015)* for disaster management urged the respective governments to create an enabling environment so that the community is more resilient to combat disasters. But these two guidelines paid less attention to outlining how the community will be resilient with their resources (for example health, natural resources), knowledge (education, skills), and freedom to make the adaptation as a means of autonomous or reactive adaptation. NAPA suggested including climate change issues in secondary- and tertiary-level education, but little focus has been given to how local knowledge could be mined, documented, and disseminated for inclusion in different layers of decisions.

Lastly, it should be mentioned that it would remain a big challenge for Bangladesh to achieve success in disaster risk reduction (DRR) and climate change adaptation (CCA) if human-induced hazards are not reduced. For instance, some of the disasters (sometimes termed "natural disasters") in Bangladesh are the results of combined outcomes of many different human interventions and the effects of climate change. For example, all the water inundation events may not be attributed to climate change; rather, unplanned development activities are sometimes responsible for flood hazards (Hughes, Adnan, & Dalal-Clayton, 1994). Similarly, shrimp cultivation, withdrawal of water by the upstream neighboring countries, and sea level rise all may collectively contribute to salinity problems in the coastal areas of Bangladesh. In this respect, the different origins of certain events that have identical consequential signs (for example, floods or waterlogging) may need to be identified properly to ascertain the contribution climate change would have on these events. This distinction may help to design the right CCA approach for the country to deal with major and minor factors responsible for making the communities susceptible to adversities.

NOTE

1. Effects of north-south continental level (terrestrial to oceanic) gradient of wind pressures, monsoon wind systems, the influence of the jet stream stretched from Southeast Asia to northern Africa, waters from glacier melt in the Himalayas, fluctuations in the terrestrial and sea surface temperature all contribute and make up the physical processes of Bangladesh.

REFERENCES

Adger, W. N., Hughes, T. P., Folke, C., Carpenter, S. R., & Rockstorm, J. (2005). Social-ecological resilience to coastal disasters. *Science, 309*(5737), 1036–1039.

Afsar, R. (2005). Migration and rural livelihoods. In: K. A. Toufique & C. Turton (Eds), *Hands not lands: How livelihoods are changing in rural Bangladesh*. Dhaka: Kazi Ali Toufique (BIDS).

Alam, K. (2007). *Drowning sand and the holy banana tree. The tale of people with disability and their neighbors coping with Sharbanasha floods in the Brahmaputra-Jamuna Chars of Bangladesh*. Dhaka: Handicap International.

Barakat, A., Zaman, S. U., & Raihan, S. (2001). *Political economy of khas land in Bangladesh*. Dhaka: Association for Land Reform and Development (ALRD).

Climate Change Cell (CCC). (2008). *Characterizing long-term changes of Bangladesh climate in context of agriculture and irrigation*. Climate Change Cell, Department of Environment, Ministry of Environment and Forests, Component 4b, Comprehensive Disaster Management Programme, Ministry of Food and Disaster Management, Bangladesh.

Climate Change Cell (CCC). (2009). *Generation of PRECIS scenarios for Bangladesh (Validation and Parameterization)*. Climate Change Cell, Department of Environment, Ministry of Environment and Forests, Component 4b, Comprehensive Disaster Management Programme, Ministry of Food and Disaster Management, Bangladesh.

Eakin, H. (2005). Institutional change, climate risk, and rural vulnerability: Cases from Central Mexico. *World Development, 33*(11), 1923–1938.

Government of Bangladesh (2007). *National Adaptation Programme of Action (NAPA)*. Final Report, November 2005.

Houghton, J. (2004). *Global warming*. Cambridge: Cambridge University Press.

Hughes, R., Adnan, S., & Dalal-Clayton, B. (1994). *Floodplains or flood plans? A review of approaches to water management in Bangladesh*. London: IIED and RAS.

IPCC. (2007). *Working group II. Contribution to the intergovernmental panel on climate change fourth assessment report climate change 2007: Climate change impacts, adaptation, and vulnerability*. Version for policy makers. Institute of Water Modelling, Dhaka, Bangladesh: Intergovernmental Panel on Climate Change.

Islam, S. T. (2007a). Deforestation in Bangladesh: Language and power in action. *Narrative Inquiry, 17*(1), 141–156.

Islam, S. T. (2007b). Indigenous floating cultivation: A sustainable agricultural practice in the wetlands of Bangladesh. *Development in Practice, 17*(1), 130–136.

Islam, T. (2010). *Climate change in Bangladesh: A closer look into temperature and rainfall data*. Dhaka: University Publishers Limited.

IWM (2009). *Assistance to climate change adaptation and disaster risk reduction in Bangladesh: Support to national flood forecasting and warning services in Bangladesh*. Report supported by Danish Government, Institute of Water Modelling, Dhaka, Bangladesh.

Kelly, P. M., & Adger, W. N. (2000). Theory and practice in assessing vulnerability to climate change and facilitating adaptation. *Climate Change, 47*(4), 325–352.

Schroter, D., Polsky, C., & Patt, A. G. (2005). Assessing vulnerabilities to the effects of global change: An eight-step approach. *Mitigation and Adaptation Strategies for Global Change, 10*(4), 573–596.

Sen, A. (1981). *Poverty and famine: An essay on entitlement and deprivation*. Oxford: Clarendon Press.

Warrick, R. A., Bhuiya, A. H., & Mirza, M. Q. (1994). *The greenhouse effect and climate change*. Dhaka: Bangladesh Unnayan Parishad.

Winser, B., Blaikie, P., Cannon, T., & Davis, I. (1994). *At risk: Natural hazards, people's vulnerability and disasters*. London: Routledge.

World Bank. (2001). *Bangladesh: Climate change and sustainable development*. Dhaka: World Bank Office.

CHAPTER 3

OVERVIEW OF DROUGHT RISK REDUCTION APPROACHES IN BANGLADESH

Umma Habiba, Yukiko Takeuchi and Rajib Shaw

INTRODUCTION

Many people as well as the government in Bangladesh perceive floods and cyclones as recurrent environmental hazards in the country. They also view that these two hazards are the main contributors to crop loss in the country. But, in reality, droughts afflict the country at least as frequently as do major floods and cyclones, averaging about once in 2.5 years (Adnan, 1993, p. 1; Erickson, 1993, p. 5; Hossain 1990, p. 33). In some years, droughts not only cause a greater damage to crops than floods or cyclones, but they also generally affect more farmers across a wider area (Paul, 1995). If not institutionally and economically tackled, the consequences tend to have a far-reaching effect on the given society, and the socioeconomic problems would assume a chronic pattern.

The international community considers Bangladesh as the country most threatened by disasters. During the period between 1991 and 2000, Bangladesh has suffered from 93 large-scale natural disasters that killed 0.2 million people and caused loss of properties valued at about $59 billion in the agricultural and infrastructure sector (Climate Change Cell, 2009). According to the forecasts of IPCC's Fourth Assessment Report regarding

Climate Change Adaptation and Disaster Risk Reduction: An Asian Perspective
Community, Environment and Disaster Risk Management, Volume 5, 37–58

ISSN: 2040-7262/doi:10.1108/S2040-7262(2010)0000005009

climate change in Bangladesh, average temperature has increased by 1°C for May and by 0.5°C in November. According to the report, it is also stated that Bangladesh would experience heavier monsoons and that the melting of Himalayan glaciers will cause higher river flows and severe floods. Rainfall will become heavier and more erratic while droughts will increase in frequency.

In Bangladesh, drought is a recurrent phenomenon, afflicting the country at least as frequently as major floods and cyclones (Paul, 1998). Drought mostly affects the country in pre-monsoon and post-monsoon periods (Banglapedia, 2006). Drought damaged about 218 million tons of rice between 1973 and 1987 in Bangladesh (Climate Change Cell, 2009). Depending on the impact on human habitat and environment, drought in Bangladesh can be classified as follows (Mirza & Pal, 1992; Das, 1997):

1. Agricultural drought: Shortage of moisture in the soil for crop growth;
2. Hydrological drought: Falling of the surface and subsurface water levels and decreases in stream flow, ultimately affecting soil moisture; and
3. Economic drought: A condition adversely affecting the economy at large.

During the last 50 years, Bangladesh suffered from drought conditions about 20 times. The more severe drought-affected years were 1951, 1961, 1973, 1975, 1978, 1979, 1981, 1982, 1992, 1994, 1995, and 2000. The 1973 drought was one of the severest in the present century and was responsible for the local famine in northern Bangladesh in 1974. In the 1975 drought, 47 percent of the country was affected, causing suffering to about 53 percent (Task Force Report 1990; Adnan, 1993). The consecutive drought of 1978 and 1979 directly affected 42 percent of cultivated land and reduced rice production by an estimated 2 million tons (Brammer, 1987). The losses due to drought in 1982 were more than double the losses caused by floods in the same year. But the most persistent drought was in 1989 (Kafiluddin, 1991). Again, the 1997 drought caused a reduction of around 1 million tons of food grain, of which about 0.6 million tons were transplanted aman. Although droughts were not continuous, they did affect the low rainfall zones of the country.

According to the report of the National Drought Mitigation Center (2006), Bangladesh has already shown an increased frequency of droughts in recent years. Droughts are increasingly being reported in Rajshahi, Chapai-Nawabganj, Naogaon, Rangpur, Bogra, Pabna, Dinajpur, and Kustia regions. Almost every year, all areas are affected by drought, but the northwestern part of the country, however, is considered to be the most drought-prone. These regions are relatively dry, receiving much lower rainfall compared to the rest of the country (Paul, 1998). However, drought can hit both drought-prone and non-drought areas (Murshid, 1987, p. 38; Paul, 1995). The aim of this chapter is to disclose the overall situation of drought, including its causes, impacts, risk mitigation, and why drought-resilience approach is necessary in Bangladesh context.

DISASTER RISK IN BANGLADESH

Bangladesh is one of the most disaster-prone countries in the world (Shahid & Behrawan, 2008). Almost every year, the country experiences disasters of one kind or another, such as tropical cyclones, storm surges, coastal erosion, floods, northwesters, tornadoes, and droughts, causing heavy loss of life and property and jeopardizing development activities (Ali, 1996). High spatial and temporal climatic variability, extreme events, high population density, high incidence of poverty and social inequity, poor institutional capacity, inadequate financial resources, and poor infrastructure have made Bangladesh highly vulnerable to disaster (Ahmed, 2004). The common disasters of Bangladesh are briefly discussed here, and drought is mentioned in the following parts of the chapter.

Bangladesh has its own history of major disasters. The cyclone of 1970 and subsequent flood took away the lives of over 300 thousand people and damaged the equivalent of about US$2.5 billion in crops and property; the catastrophic cyclone of 1991 killed over 120,000; and the 2004 floods that inundated over 34 percent of the country resulted in 747 deaths. The cyclones of 1876, 1919, 1961, 1963, 1965, 1970, 1985, 1988, 1991, 1994, 1995, 1997, 2007, and 2009 were also of severe nature. In 2007, Cyclone Sidr made landfall on southern Bangladesh, causing over 2,000 deaths and severe damage (the cyclone death toll mounted to 2,388); in 2009, Cyclone Aila caused 330 deaths, made 1 million people homeless, and left damages that totaled $40.7 million. In Bangladesh, an estimated 20 million people were at risk of postdisaster diseases due to Aila.

Floods have wreaked havoc in Bangladesh throughout history. Annually, floods inundate an average of 15 percent of the total geographical area. The severe floods of 1822, 1854, 1922, 1955, 1966, 1974, 1987, 1988, 1998, 2002, 2004, and 2007 are worth mentioning. The flood of 1988 during August and September inundated an 89,000 km^2. area, hitting 52 districts of the country and causing the loss of 1,517 human lives. The 1998 flood in Bangladesh, with an unprecedented duration of 65 days, inundated 53 districts covering about a 100,000 km^2 area, and took 918 lives. The 2004 flood in Bangladesh inundated 40 districts and took 747 human lives. But, the most recent flood occurred in 2007. According to government statistics, 298 people died, and a total of 10,211,780 people were badly affected by it.

Annually, about 8,700 ha of land are eroded as a result of riverbank erosion, displacing about 200,000 people. Earthquakes are considered to be an emerging risk for the country, with the capital Dhaka and other major cities such as Sylhet and Chittagong being extremely vulnerable. The earthquakes of 1869, 1885, 1897, 1918, 1930, 1934, 1959, 1997, and 1999 are noteworthy.

Bangladesh can be considered to be in a state of extreme starvation, in which the population of an area can suffer due to scarce food supply. The famines of 1866, 1896, 1943, and 1974 are noteworthy.

Road accidents, ferry tragedies, and building collapses are the major human-induced disasters in the country. About 70 million of its people will be affected due to climate change (UN Human Development Report 2007/08), and several sectors like water resources, coastal resources, agriculture, health, livelihoods, food security, habitat, and infrastructure are threatened by climate change. Cyclones and floods pose the greatest risk to Bangladesh at the country level. At the regional level, the southeast is particularly vulnerable to cyclones, droughts, earthquakes, and floods. Generally, the western part of the country and the hilly areas in Chittagong are more prone to drought.

DROUGHT RISK IN BANGLADESH

Drought is a normal part of climate, an extreme climatic event often described as a natural hazard (Wilhite, 2000). A broad definition of drought is a deficiency of precipitation over an extended period of time, usually a season or more, which results in a water shortage for some activity, group, or environmental sectors. In the context of Bangladesh, Brammer (1987) defines drought as a period when supply of moisture in the soil is less than that which is required for satisfactory crop growth during a season when crops are normally grown. This definition resembles one provided by Heathcote (1974), who defines drought as a "shortage of water harmful to man's agricultural activities. It occurs as an interaction between agricultural activity (i.e., demand) and natural events (i.e., supply), which results in a water volume or quality inadequate for plant and/or animal needs."

According to the definition, drought is different from other disasters in three aspects (Wilhite, 2000). First, it is a "creeping phenomenon," making its onset difficult to determine. Its effects accumulate slowly over a considerable period of time, and may linger for years even after the termination of the event. Second, the absence of a precise, common definition of drought adds to confusion about its occurrence and severity. Third, damage due to drought does not normally involve damage to infrastructure (unlike flooding, earthquakes, and so on).

Fig. 1 represents the maps of kharif and rabi crop areas of Bangladesh that are drought-prone. Droughts mainly occur in areas with higher evaporation than precipitation. Weather is not the only cause of drought. Human activities contribute to the effects of a drought. In Bangladesh, however, droughts are related to climate variability and nonvariability of surface water resources. More precisely, Bangladesh faces drought due to rainfall.

Rainfall

Bangladesh extends from 20°34′ N to 26°38′ N latitude and from 88°01′ E to 92°41′ E longitude. Climatically, the country belongs to subtropical regions where monsoon weather prevails throughout the year. The average annual rainfall of the country varies from 1,329 mm in the northwest to 4,338 mm in the northeast (Shahid, Chen, & Hazarika, 2005). Fig. 2 shows the annual average total rainfall of drought-prone areas compared to the annual average total rainfall of Bangladesh. At the national level, about 80 percent of the annual rainfall occurs between July and October, with almost no rainfall during winter (October to March) (Hossain, Islam, & Saha, 1987).

During the dry season (of 7 months) in some regions, the evapotranspiration exceeds the amount of rainfall by an amount of more than 0.5 times the rainfall. The water deficit exists in the dry months of the year, as the demand is higher than the total rainfall. During monsoon, the available rainfall may fulfill the crop water demand, but during the dry season, the deficit is too large to meet the demand. So to meet the crop water demand, alternative sources need to be investigated.

Again, based upon General Circulation Models (GCM) using Model for Assessment of Greenhouse-gas Induced Climate Change (MAGICC)/ Scenario Generator (SCENGEN), Agarwala, Ota, Ahmed, Smith, and Van Aalst (2003) gave a report that the future climate change projections show increased rainfall during monsoon season and declining rainfall in winter months (Table 1). The monsoon season rainfall is projected to increase, the rainfall availability may increase significantly causing more intense rainfall and/or longer dry spells.

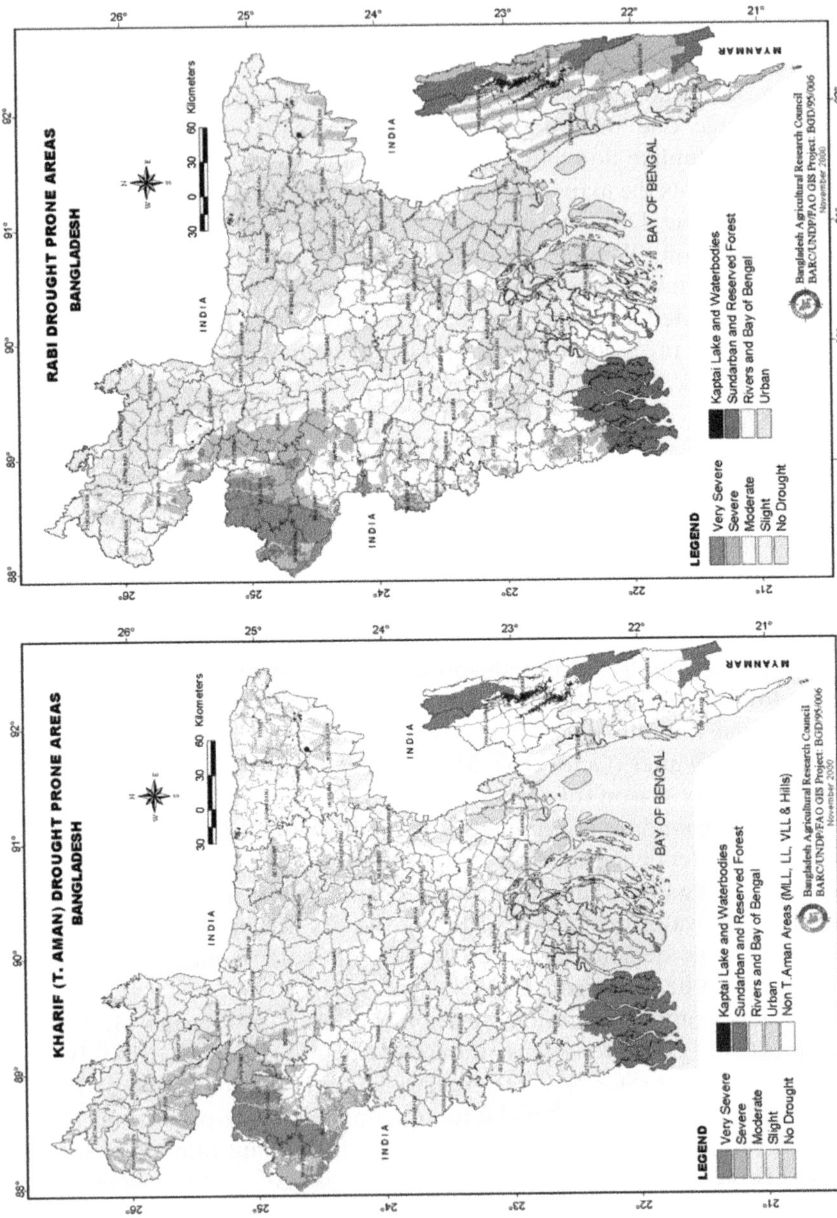

Fig. 1. Kharif and Rabi Drought-prone Areas of Bangladesh. *Source:* Bangladesh Agricultural Research Council (BARC).

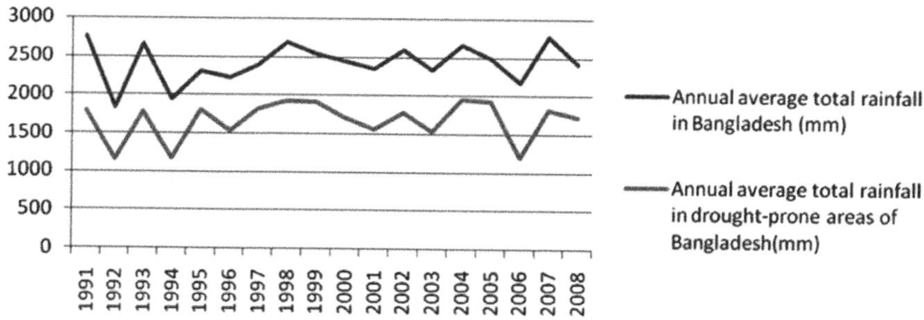

Fig. 2. Annual Average Total Rainfall of Drought-prone Areas, 1991–2008.

Table 1. Climate Change Scenarios for Bangladesh, NAPA.

Year	Mean Temperature Change (°C)			Mean Precipitation (%)			Sea Level Rise		
	Annual	DJF	JJA	Annual	DJF	JJA	IPCC (Upper Range)	SMRC	NAPA
2030	1.0	1.0	0.8	5	−2	6	14	18	14
2050	1.4	1.6	1.1	6	−5	8	32	30	32
2100	2.4	2.7	1.9	10	−10	12	88	60	88

Note: DJF, December, January, February; JJA, June, July, August; NAPA, National Adaptation Platform of Action; SAARC Meteorological Research Center (SMRC).
Source: Adopted from IPCC 2001, OECD Report 2003; cited in NAPA.

Temperature

Annual and seasonal temperatures are found to have increasing trends in Bangladesh in general. The magnitudes of the trends are dependent on the period of analysis of the available data. The overall trend in mean annual temperatures is found to be +0.10° and +0.21° per decade (equivalent to +1.03°C and +2.14°C per century) for data periods between 1948 and 2007 and between 1980 and 2007, respectively (Table 2). It thus appears that the warming has been more rapid in the recent decades. Average winter season (December, January, and February) maximum and minimum temperatures show respectively a decreasing and an increasing trend annually at 0.001°C and 0.016°C (Rahman & Alam, 2003). SMRC (2000)

Table 2. Seasonal and Annual Mean Temperature in Bangladesh.

Season	Trend of Mean Temperature (°C per Century) in Bangladesh	
	For data period of 1948–2007	For data period of 1980–2007
Winter (Nov.–Feb.)	+1.67	+1.33
Summer (Mar.–May)	+0.26	+2.15
Monsoon (June–Oct.)	+1.05	+2.44
Annual (Jan.–Dec.)	+1.03	+2.14

Source: Climate Change Cell (2009).

mentioned that the monthly and annual mean maximum and minimum temperature in Bangladesh for a period of 30 years revealed a statistically significant increasing trend for annual mean temperature, a slightly decreasing trend for annual mean temperature and slightly increasing trend of annual mean temperature. The average temperature ranges from 7.2°C to 12.8°C during winter and 23.9°C to 31.1°C during summer. According to the recent reports, the temperature is generally increasing in the monsoon season with maximum and minimum temperatures exhibiting an increasing trend at the rate of 0.05°C and 0.03°C each year (Huq, Rahman, Konate, Sokona, & Reid, 2003). Conversely, average winter maximum and minimum temperatures are exhibiting a decline and a thriving trend annually at the rate of 0.001°C to 0.016°C (Huq et al., 2003).

Groundwater Level

Groundwater is the most important water supply source in Bangladesh. The depth of aquifers varies from zero to 54 m below ground surface. Groundwater is mainly extracted by installing wells for the development of water supply systems. Sufficient water replenishment takes place during monsoon season in normal years. However, the continuous high pumping of groundwater leads to overexploitation and groundwater depletion (Fig. 3). Recent statistics show that the availability of groundwater is around 21 billion cubic meters (Bangladesh Bureau of Statistics (BBS), 2005).

But groundwater is extensively used for domestic and agricultural purposes, and groundwater levels begin to fall in response to rapid drainage of surface water. In Bangladesh, 33.34 percent (4.7 million hectares) of total cropped area (14.17 million hectares) has been brought under irrigation. Groundwater marked a 70 percent contribution in the total irrigated area

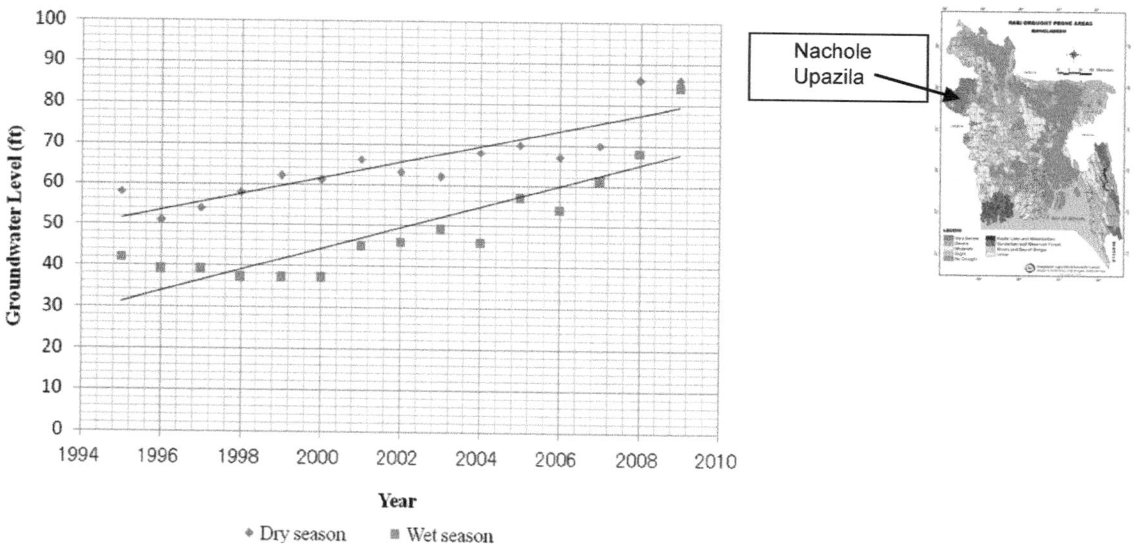

Fig. 3. Groundwater Depletion in Nachole Upazila in Chapai-Nawabganj District. *Source:* Barind Multipurpose Development Authority (2009).

while surface water provided the rest between 2004 and 2005. Excessive use of groundwater used for irrigation purposes causes groundwater depletion that is not completely replenished during drought years.

DROUGHT IMPACTS

Impacts of drought are diverse and often ripple through the economy. They are often referred to as direct or indirect, or they are assigned an order of propagation (that is, first-, or second-order) (see Kates, Ausubel, & Berberian, 1985). The first-order impact of drought is observed in the form of a decrease in food production, and the second-order impact is decreased employment and income.

Socioeconomic Impact

Barlow, Cullen, Lyon, and Wilhelmi (2006) found that drought is the highest natural hazard that negatively impacts human livelihood. People face tremendous problems in their daily lives due to reduced food production, abnormal increases in food grain prices, and nonavailability of jobs. Small farmers and landless labors especially suffer, compared with large farmers. Food consumption includes a household's ability to meet its food needs on a sustainable basis.

Drought also stresses the supply of drinking water and irrigation. Vegetables and many other foods are in short supply during that period. People start to consume wild plants, tubers, and leaves not normally eaten (Jallow 1995, p. 35). As a consequence, calorie consumption in households is lower than that of households not exposed to this situation. Besides, other symptoms like widespread malnutrition among the population increase the risk of dengue, malaria, diarrhea, and so on.

At this stage, human distress can cause people to sell and/or mortgage their land and livestock, and to sell their belongings because of the need for cash to buy food at a high price. At the same time, perhaps they take a loan from a bank or another NGOs to meet the family expenses and cover crop cultivation expenditures. This can lead to indebtedness, asset depletion, and the erosion of investment capacity. In drought time, people of that area defer purchases of clothing and luxury items. Although they may avoid migration, seasonal and urban migration does occur in drought-prone areas to cope with drought. Fishermen change their occupation; in the dry period, farmers even change their occupations and engage in different work like pretty business, and so on. Drought occasionally creates water conflicts among the different users for crops and fish.

In Bangladesh, agriculture is the main economic activity, providing about 62 percent of total employment, even though its contribution to the gross domestic product (GDP) has sharply declined from around 26 percent in 1997 to about 20 percent in 2006 (Bangladesh Bureau of Statistics [BBS], 2006; MoA, 2007). About 65 percent of the rural population of Bangladesh depends on agriculture for its livelihood, and this sector feeds about 150 million people. But drought adversely affects crop production in Bangladesh. It affects different rice crops, including high-yielding varieties of boro, aus, and aman (rice varieties in local language). Bangladesh has reported that every year, 0.45 million hectares of land is affected by very severe drought during the Rabi season while 0.40 million hectares and 0.34 million hectares are affected during the pre-kharif and kharif seasons, respectively (Table 3). For example, the drought of 1994–95 in the northwestern districts of the country led to a shortfall of rice production of about 3.5 million tons (Rahman & Biswas, 1995, p. 7). Drought is usually followed by famine, resulting in loss of life and poor agricultural production (Karim, 1994). Between 1996 and 1997, with a production target of 8.7 million metric ton of aman paddy, a total of 4.9 million hectares was cultivated. But drought conditions damaged about 2.6 million hectares of paddy land. Of this, 2.2 million hectares were in the 16 districts of northwest Bangladesh (Table 4). The Bangladesh Disaster Preparedness Forum (BDPF) compiled the damages to land, people, and crops for the 1996–97 drought conditions in the country as follows:

Table 3. Impacts of Annual Drought in Bangladesh.

Classification of Drought	Area Affected (Million Hectare)		
	Rabi season (November–February)	Pre-Kharif season (March–June)	Kharif season (July–October)
Very severe	0.446	0.403	0.344
Severe	1.71	1.15	0.74
Moderate	2.95	4.76	3.17
Slight	4.21	4.09	2.9
No drought	3.17	2.09	0.68
No T-aman	–	–	4.71

Source: Climate Change Cell (2009).

Table 4. Damage Due to Drought Condition.

Damaged Areas/Districts	Affected Land (Million Hectare)	Affected Farmers/ Population (Million)	Crop/ Resources Damaged (US$ '000)
Rangpur, Bogra, Dinajpur, Thakurgaon, Panchagarh, Kurigram, Nilphamari, Gaibandha, Lalmonirhat, Serajgonj, Naogaon, Natore, Rajshahi, etc. (Total, 16 districts)	2.20	6.49	67.65
Barisal, Bhola, Barguna, Jhalakathi, Patuakhali	0.06	1.75	18.38
Noakhali, Lakshmipur, Feni	0.17	5.07	53.38
Sunamganj	0.04	1.17	12.35
Chakaria and Lama	0.02	0.78	5
Barind Area	0.04	1.12	12.35

Source: BDPF, 1998.

Jute also suffered severely due to drought. Farmers are not interested in cultivating jute due to the lack of water for retting. Its production is drastically reduced from the past. It should be noted that such as, to other crops (all rabi crops, sugarcane, tobacco, wheat, and so on) as well as perennial agricultural resources, such as bamboo, betel nut, various fruits, like, litchi, mango, jackfruit, banana, and so on, are considered, the loss will be substantially much higher, even amounting to double the loss of rice production.

Out of total fish production in Bangladesh, the total marine catch was 4,97,573 tons, total; inland catch was 1,060,181 tons; and total production from aquaculture was 1,005,542 tons (DoF, 2009). Collectively, fisheries and aquaculture play a major role in nutrition, employment, and foreign exchange earnings, with about 12 million people associated with the fisheries sector, of which 1.4 million people rely exclusively on the activities related to fisheries (Shah, 2003). Drought also disrupts fish cultivation and fishing activities, because seasonal variations have diverse implications for fishing, hatchery operations, table fish production, and the livelihoods of a wide range of people directly and indirectly involved with fisheries and aquaculture (Haque, 2007). Due to natural and man-made causes, aquatic biodiversity (especially species of fish and other aquatic organisms in open water) has been declining sharply.

The livestock population of Bangladesh includes 22.3 million cattle and buffalo, 14.6 million sheep and goats, and 126.7 million poultry. Lack of grazing facilities and decreasing food constrain the mass rearing of cattle and goats during drought period. Livestock are affected by air temperature, humidity, wind speed, and thermal radiation, which influence their growth, milk production, reproduction, health, and well-being. Sometimes, cattle and poultry suffer from heat stroke, even affected by different diseases like black quatrain, anthrax, and so on.

Environmental Impacts

The occurrence of fluctuating rainfall and drought in this world is not new but the duration has been increasing alarmingly (Islam et al., 2008). The environmental change has been increasing in Bangladesh and causes damage to plant and animal species, and increased incidence of the diseases that bring pathogens and parasites. Drought dries out the natural water bodies and thus causes loss of wild and cultural stocks. It also reduces water levels in reservoirs, lakes, and ponds, increases salt concentration, water temperature, affects air and water quality and degrades landscape quality and causes soil erosion. It also causes loss of biodiversity and extinction of some fish species, and diminishes the reproductive capacity of cattle and poultry.

DROUGHT RISK MITIGATION APPROACH

At a Physical Level

In the event of drought, the government of Bangladesh has undertaken an irrigation project by installing deep tube wells in the region to increase agricultural productivity. The Barind Multipurpose Development Authority (BMDA) is actively involved in irrigating the northern district. Irrigation is necessary for many types of crops, especially for boro rice cultivation. Records show that the irrigation coverage was less than 2 million hectares in the early 1980s, and took almost two decades to reach about 5 million hectares. It also reveals that the growth rate was a bit higher in the last decade (starting from 1994) than the earlier decade (starting from 1982). An ongoing drought forces farmers to irrigate lands for crop cultivation. Even in the last year, farmers used irrigation for typically rain-fed aman crops.

At an Institutional Level

The Ministry of Food and Disaster Management (MoFDM) has the mandate to coordinate all disaster management activities within the country. The government of Bangladesh has taken a number of significant steps during the last few years to build up institutional arrangements from a national level to the union levels for effective and systematic disaster management facilities mitigating the sufferings of disaster victims. To maintain proper coordination among the concerned ministries, departments, line agencies, local government body (LGD), and community people, and also to ensure their proper functioning to mitigate sufferings of the people, the government of Bangladesh has formulated a set of mechanisms for council and communities from national down to the grassroots level (Fig. 4). To increase resilience at all levels, from the national to community level, and to reduce damage and losses from natural disasters and the impacts of climate change, the government of Bangladesh has launched the Comprehensive Disaster Management Program (CDMP), which started field implementation in 2004.

Among other things, the CDMP also addresses the risk associated with the climate variability and change, including livelihood adaptation to climate change. The government of Bangladesh approved the CDMP to increase resilience at all levels and reduce damage and losses from natural disasters. The government also undertook relief measures by providing drinking water, food grains, and food subsidies to special groups and through food-for-work programs. The rural work program of the government provides employment to the population affected by drought and helps mitigate the drought severity.

The Disaster Management Bureau (DMB) coordinates drought relief work with local governments. DMB also has activities in human resource development, database and information services, and documentation on disaster management. Barind Multipurpose Development Authority (BMDA) and different NGOs like BRAC, PROSHIKA, CARITAS, TMSS, and TRINOMUL are actively involved in drought-prone areas.

To meet the increased demand for food, the government and NGOs have initiated modern agricultural technology and practices such as increased irrigation coverage, and growing high-yielding variety (HYV) seeds and drought-tolerant varieties. The Department of Agriculture Extension (DAE) also provides farmers in drought-prone areas with some training programs on supplementary irrigation for T. Aman and rabi crops. Bangladesh Agricultural Research Institution (BARI), Bangladesh Rice Research Institute (BRRI), and DAE distributed drought-tolerant rice, wheat, barley, sweet potato, and mung bean varieties to drought-prone farmers. Even in the recent year, the

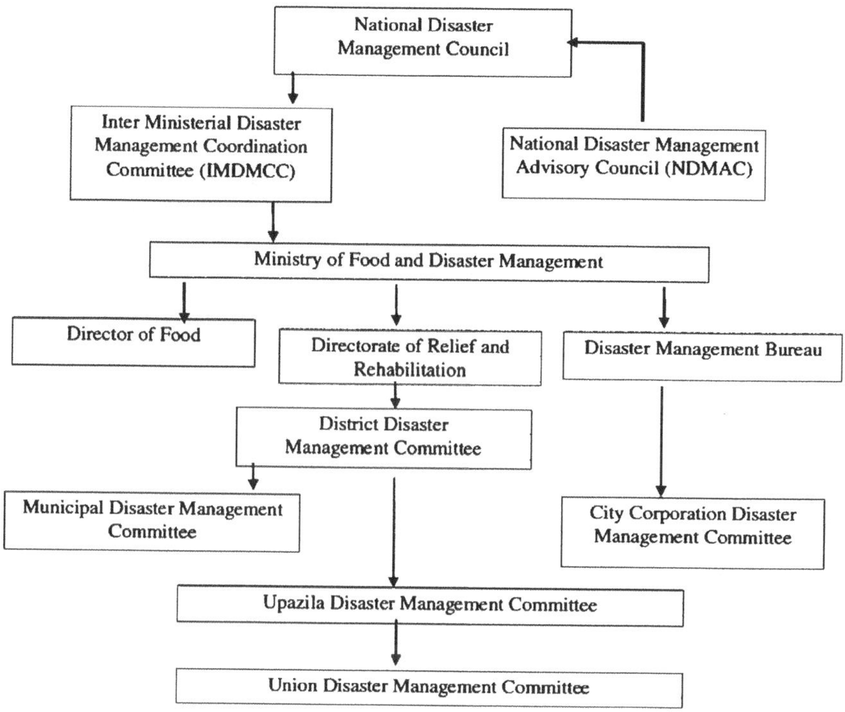

Fig. 4. Disaster Management Systems in Bangladesh. *Source:* Selvaraju, Subbiah, Baas, and Juergens (2006).

government offered free electricity for millions of farmers to pump underground water after a severe lack of monsoon rains.

Early Warning System

The Bangladesh government could not yet develop powerful institutions like Western Europe and America that are able to forecast a complex climate related disaster named el Niño, but they established institutions at the national level such as the Space Research and Remote Sensing Organization (SPARRSO), the Bangladesh Meteorological Department (BMD), and the Bangladesh Water Development Board (BWDB), for monitoring and forecasting disasters. To integrate climate risk into development plans and processes, the government, through its Department of Environment (DOE), has recently launched the preparation of the Climate Change Policy and

Action Plan for Bangladesh. The government established the Climate Change Cell in the DOE under the Ministry of Environment and Forest and has developed a network of 34 "focal points" in different government agencies, research, and other organizations.

At the national level, BMD generates weather and climate information relevant to drought risk management. It also creates public awareness on the basis of severity and risks associated with projects that will better prepare at-risk communities and public officials to mitigate the consequences (Fig. 5). Mass media plays an important role in disseminating weather and climatic information to farmers, farmers' groups, and community associations at local the level.

At an Indigenous Level

To minimize the adverse effects of drought, the households practice various adjustments at the household and community levels. At the household level, people intend to reduce the effects of drought hazard by using two types of drought-mitigating techniques. These techniques are referred to as agricultural and nonagricultural adjustment. People practice some agricultural adjustments, such as resowing crops, to compensate for the reduction in the crop area, and others, such as applying irrigation water, to increase crop yield (Brammer 1987, p. 24–25). As well, pond digging is done to facilitate the judicious use of water for domestic and agricultural purposes, changing cropping pattern during drought time.

Because of higher price of food grains during drought period, people need additional cash to buy crops for consumption. For this reason, they generally practice nonagricultural adjustments. People may either sell and/or mortgage their land and livestock, and sell their belongings to earn additional cash. At the community level, friends, neighbors, relatives, and affluent members of the *samaj* may help the drought victims by providing cash, loans, food, and clothes.

NEED FOR AN INTEGRATED APPROACH

Because drought is a natural disaster that onsets slowly, it starts unnoticed and develops cumulatively. It is one of the natural hazards that hampers lives and properties. Drought affects almost all areas of Bangladesh, especially the northwest region. Agriculture and the other primary livelihood activities and human health are badly affected by drought.

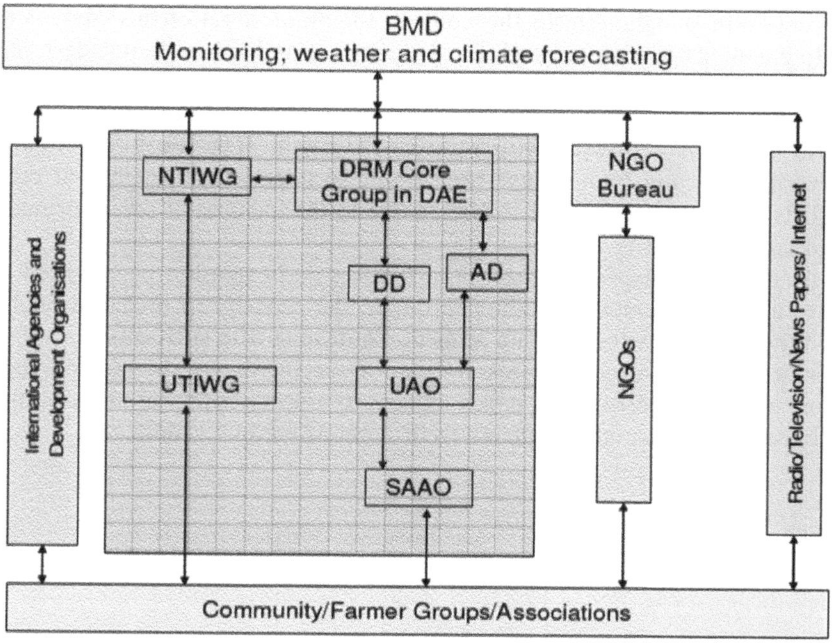

Fig. 5. Institutional Setup for Generation and Application of Weather and Climate Forecast Products for Drought Risk Management in Bangladesh. *Note:* AD, Additional Director; BMD, Bangladesh Meteorological Department; DAE, Department of Agricultural Extension; DRM, disaster risk management; DD, Deputy Director; NTIWG, National-Level Technical Implementation Working Groups; SAAO, Sub-Assistant Agricultural Officer; UAO, Upazila Agricultural Officer; UTIWG, Upazila Technical Implementation Working Groups. *Source:* Selvaraju et al. (2006).

Again, drought is intimately related with food security, which is critical for the poor who spend a greater percentage of their income on food and therefore suffer more when prices increase.

In this paper, we considered a number of studies addressing the impact of droughts on agriculture (Jabbar, Chaudhury, & Huda, 1982); Karim, Ibrahim, Iqbal, & Ahmed, 1990; Jabbar, 1990; Saleh, Mazid, & Bhuiyan, 2000; Mazid et al., 2005), food production (Ahmed & Bernard, 1989; Ericksen, Ahmad, & Chowdhury, 1993), land degradation (Rasheed, 1998; Karim and Iqbal, 2001; Government of Bangladesh, 2005), economy (Ericksen et al., 1993; World Bank Bangladesh, 2000), and society (Ericksen et al., 1993; Paul, 1998). But none of these studies highlighted drought risk reduction in Bangladesh. In this context, the main target of this study is to help create a holistic approach for drought risk reduction. Bangladesh has achieved remarkable success in disaster management (early warning and postdisaster management) related to floods and tropical cyclones, but very little has been addressed with regard to drought. The local community, governments, and NGOs have been trying to cope with drought but their efforts have not been effective enough due to the increase of severe drought and lack of capacity.

As agriculture-based country, both farmers and extension people need to have a clear understanding about drought, the capacity to predict drought, and the know-how for building resilience in dealing with such situations. For this, a mass awareness is necessary, and a specific diagnostic methodology must be developed to allow policy makers to assess whether a policy option enhances or undermines resilience to drought at various scales.

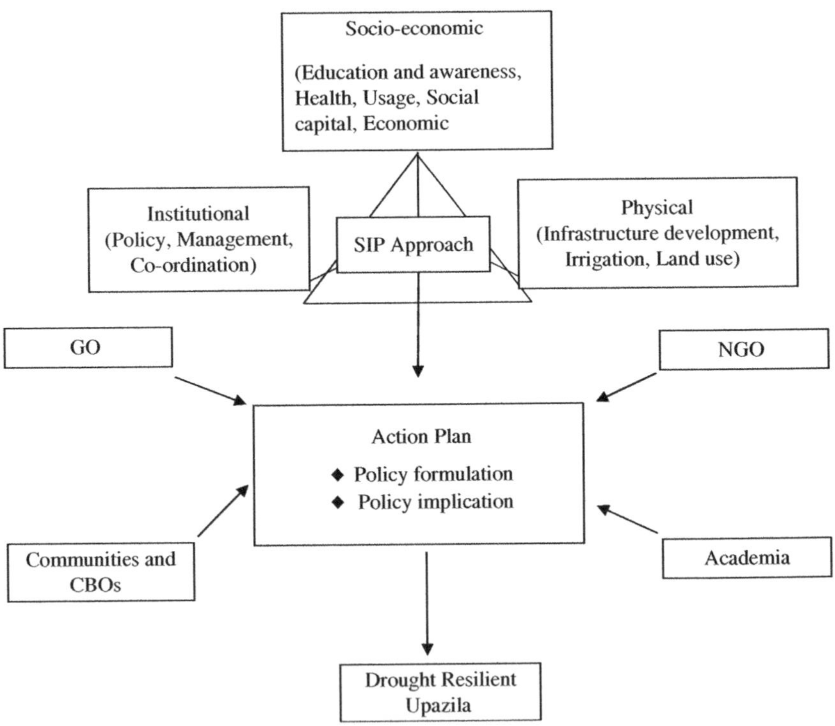

Fig. 6. Model of Drought-resilient Upazila Developed through SIP Approach.

The SIP (socioeconomic, institutional, and physical) approach is a study by which the existing level of different socioeconomic, institutional, and physical conditions of a targeted area can be measured for how the people are coping with drought. Based on the three dimensions of the SIP model (socioeconomic, institutional, and physical), a methodological approach is developed for determining the drought resilience of that area (Fig. 6). For an empirical application of the SIP approach, a field survey of the targeted area must be conducted. The field studies, coupled with the existing relevant secondary information, help planners to understand the problems and recognize potentials within areas for which enhancing resilience to drought is of the utmost importance. This also offers information by which we can measure the overall existing level of drought resilience of that area. Thus, the initiative helps to determine the strength and weakness of different sectors of various dimensions of drought resilience, so that government and different organizations can prioritize the sector for policy implication, can provide inputs for policy formulation, and finally can help to minimize the drought risk.

ACKNOWLEDGMENTS

The first author thanks the JASSO and GCOE-ARS projects for financial support.

REFERENCES

Adnan, S. (1993). *Living without floods: Lessons from the drought of 1992*. Dhaka: Research and Advisory Services.

Agarwala, S., Ota, T., Ahmed, A. U., Smith, J., & Van Aalst, M. (2003). *Development and climate change in Bangladesh: Focus on coastal flooding and the Sundarbans*. France: OECOD.

Ahmed A. U. (2004). *Adaptation to climate change in Bangladesh: learning by doing*. UNFCCC Workshop on Adaptation, Bonn, Germany, June 18, 2004.

Ahmed, R., & Bernard, A. (1989). *Rice price fluctuation and an approach to price stabilization in Bangladesh*. Washington: International Food Policy Research Institute.

Ali, A. (1996). Vulnerability of Bangladesh to climate change and sea level rise through tropical cyclones and storm surges. *Water Air Soil Pollution, 92*(1–2), 171–179.

Bangladesh Bureau of Statistics (BBS). (2005). *Compendium of environment statistics of Bangladesh*. Dhaka: Ministry of Planning, Government of Bangladesh.

Bangladesh Bureau of Statistics (BBS). (2006). *Agriculture sample survey of Bangladesh 2005, National, Vol. 1*. Dhaka: Government of Bangladesh.

Banglapedia. (2006). Drought in Bangladesh. Available at http://banglapedia.search.com.bd/HT/D_0284.htm

Barind Multipurpose Development Authority. (2009). *Barind Multipurpose Development Authority*. Bangladesh: Ministry of Agriculture.

Barlow, M., Cullen, H., Lyon, B., & Wilhelmi, O. (2006). Drought disaster in Asia. In: J. Arnold, R. Chen, U. Deichmann, M. Dilley, A Lerner Lam, R. Pullen, and Z. Trohanis (Eds.), *Natural disaster hotspots case studies*. World bank disaster risk management series, no. 6. Available at http://www.fao.org/ag/agn/agns/files/HLCLClimateChange andFoodSafety.Pdf.

BDPF. (1998). *Disaster Report, 1997*. Bangladesh Disaster Preparedness Forum, Dhaka.

Brammer, H. (1987). Drought in Bangladesh: Lessons for planners and administrators. *Disasters, 11*(1).

Climate Change Cell. (2009). *Component 4B of Comprehensive Disaster Management Programme*, Department of Environment. Ministry of Environment and Forests (MoEF). Bangladesh.

Das, S. C. (1997). *Drought in Bangladesh: Its impact and Mitigation*, 19th Indian Geographical Congress, National Association of Geographers India (NAGI). Thiruvananthapiram, Kerala, India, December 29–31.

DoF. (2009). *Fisheries statistical yearbook of Bangladesh (2007–2008)* (p. 42). Bangladesh: Department of Fisheries.

Ericksen, N. J., Ahmad, Q. K., & Chowdhury, A. R. (1993). *Socio-economic implications of climate change for Bangladesh*. Dhaka: Bangladesh Unnayan Parishad.

Haque, M. M. (2007). *Decentralised fish seed network in Northwest Bangladesh: Impacts on rural livelihoods*. Ph.D. thesis, Institute of Aquaculture, University of Stirling, Scotland, UK, p. 491.

Heathcote, R. L. (1974). Drought in South Australia. In: G. F. White (Ed.), *Natural hazards: Local, national, global*. New York: Oxford University Press.

Hossain, M. (1990). Natural calamities, instability in production and food policy in Bangladesh. *The Bangladesh Development Studies, 18*(1), 33–54.

Hossain, M., Islam, A. T. M. A., & Saha, S. K. (1987). *Floods in Bangladesh. Recurrent disaster and people's survival* (pp. 3–17). Dhaka, Bangladesh: University Research Centre.

Huq, S., Rahman, A., Konate, M., Sokona, Y., & Reid, H. (2003). *Mainstreaming adaptation to climate change in least developed countries (LDCs)*. Nottingham, UK: Russel Press.

IPCC. (2001). *Climate change 2001: Impacts, adaptation and vulnerability*. UK: Cambridge University Press.

Jabbar, M. A. (1990). Causes and effects of drought/aridity in Bangladesh using remote sensing technology. In *Proceedings of ESCAP workshop on remote sensing technology in application to desertification/vegetation type mapping*, Tehran, August 1990.

Jabbar M. A., Chaudhury M. U., & Huda M. H. Q. (1982). Causes and effects of drought/aridity in northwest Bangladesh. In *Proceedings of the first thematic conference on remote sensing of arid and semi-arid lands*, Cairo, Egypt, January 1982.

Jallow, S. S. (1995). Identification of and response to drought by local communities in Fulladu West district, the Gambia. *Singapore Journal of Tropical Geography, 16*(1), 22–41.

Kafiluddin, A. K. M. (1991). *Disaster preparedness for Bangladesh floods and other natural calamities* (p. 251). Dhaka, Bangladesh: Padma Printers & Colour Limited.

Karim, N. (1994). Disasters in Bangladesh. *Natural Hazards, 11*, 247–258.

Karim Z., Ibrahim A., Iqbal A., & Ahmed M. (1990). *Drought in Bangladesh agriculture and irrigation schedule for major crops*. Bangladesh Agricultural Research Council (BARC), Publication no. 34, Dhaka, Bangladesh.

Karim, Z., & Iqbal, M. A. (2001). *Impact of land degradation in Bangladesh: Changing scenario in agricultural land use*. Dhaka, Bangladesh: Bangladesh Agricultural Research Center (BARC).

Kates, R. W., Ausubel, J. H., & Berberian, M. (1985). *Climate impact assessment*. New York: John Wiley.

Mazid, M. A., Mortimer, M. A., Riches, C. R., Orr, A., Karmaker, B., Ali, A., Jabbar, M. A., & Wade, L. J. (2005). Rice establishment in drought-prone areas of Bangladesh. In: K. Toriyama, K. L. Heong & B. Hardy (Eds), *Rice is life: scientific perspective for the 21st century*. Manila, Philippines: International Rice Research Institute.

Ministry of Agriculture. (2007). *Handbooks of agriculture statistics*. Dhaka, Bangladesh: Policy and Planning Support Unit, Ministry of Agriculture. Available at http://www.moa.gov.bd/sattisctics/statistics.htm

Mirza, A. M. K., & Pal, S. (1992). *Natural disaster and Bangladesh environment*. Dhaka: CESR.

Murshid, K. A. S. (1987). Weather, new technology and instability in foodgrain production in Bangladesh. *The Bangladesh Development Studies, 15*(1), 31–56.

National Drought Mitigation Center. (2006). *What is drought? Understanding and defining drought*. Cited in Drought Diagnosis and Diga. Available at http://www.drought.unl.edu/whatis/concept.htm

Paul, B. K. (1995). *Farmers' and public responses to the 1994–95 drought in Bangladesh: A case study*. Manhattan: Kansas State University. KS 66506.

Paul, B. K. (1998). Coping mechanisms practiced by drought victims (1994/95) in North Bengal, Bangladesh. *Applied Geography, 18*, 355–373.

Rahman, A., & Alam, M. (2003). *Mainstreaming adaptation to climate change in least developed countries (LDC)*. Working Paper no. 2: Bangladesh Country Study. IIED, London.

Rahman, A., & Biswas, P. R. (1995). Devours resources. *Dhaka Courier, 11*(42), 7–8.

Rasheed, K. B. S. (1998). Status of land resource use and desertification, drought and land degradation in Bangladesh: Obstacles and effective policy options and measures for sustainable use of land resources. In *Proceedings of the national awareness seminar on combating land degradation/desertification in Bangladesh*, April 1998, Dhaka.

Saleh, A. F. M., Mazid, M. A., & Bhuiyan, S. I. (2000). Agrohydrologic and drought-risk analyses of rainfed cultivation in northwest Bangladesh. In: T. P. Tuong, S. P. Kam, L. J. Wade, S. Pandey, B. A. M. Bouman & B. Hardy (Eds), *Characterizing and understanding rainfed environments* (pp. 233–244). Manila, Philippines: International Rice Research Institute.

Selvaraju, R., Subbiah, A. R., Baas, S., & Juergens, I. (2006). Livelihood adaptation to climate variability and change in drought-prone areas of Bangladesh. In: *Improved adaptive capacity to climate change for sustainable livelihoods in the agriculture sector – DP9/1-BGD/01/004/01/99*. Rome: Asian Disaster Preparedness Center, Food and Agriculture Organization of the United Nations.

Shah, M. S. (2003). Human resource development activities in fisheries sector. In: *Fish Fortnight Compendium 2003* (pp. 57–59). Dhaka, Bangladesh: Department of Fisheries, Ministry of Fisheries and Livestock, Bangladesh.

Shahid, S., & Behrawan, H. (2008). Drought risk assessment in the western part of Bangladesh. *Natural Hazards, 46*(3), 391–413.

Shahid, S., Chen, X., & Hazarika, M. K. (2005). *Assessment aridity of Bangladesh using geographic information center*. Dhaka: Bangladesh.

SMRC. (2000). Recent climate changes in Bangladesh (SMRC Publication no. 4). SAARC Meteorological Research Centre, Dhaka, September 2000.

Wilhite, D. A. (2000). Drought as a natural hazard: Concepts and definitions. In: D. A. Wilhite (Ed.), *Drought: A global assessment* (Vol. 1, Chapter 1, pp. 3–18). London, UK: Routledge Publishers.

World Bank Bangladesh. (2000). *Bangladesh agriculture in the 21st century*. Dhaka, Bangladesh: World Bank Bangladesh Publication.

CHAPTER 4

CLIMATE CHANGE ADAPTATION AND DISASTER RISK REDUCTION IN CAMBODIA

Huy Nguyen, Rajib Shaw and Prabhakar SVRK

INTRODUCTION

The kingdom of Cambodia is located in mainland Southeast Asia, between latitudes 10° and 15′ N and longitudes 102° and 108′ E. Cambodia covers an area of 181,035 km^2 and is divided into 21 provinces. It is bordered to the north by Thailand and Laos, to the east and south by Vietnam, and to the south and southwest by the Gulf of Thailand. Most of Cambodia's land is relatively flat with vast tracts of land given over to rice production. Other areas of Cambodia are mountainous, including the Dangrek, Cardamom, and Elephant mountain ranges (Kingdom of Cambodia, 2001).

As a tropical country, Cambodia is bathed in almost year-round sunshine and has a high average temperature. There are two distinct seasons, the dry and the monsoon. The monsoon lasts from May to October, with southwesterly winds ushering in the clouds that bring 75 percent to 80 percent of the annual rainfall, often in spectacular intense bursts for an hour at a time with fantastic lightening displays. The dry season runs from November to April, averaging temperatures from 27°C to 40°C. The coolest and most comfortable months for those from cooler climates are between October and January.

Climate Change Adaptation and Disaster Risk Reduction: An Asian Perspective
Community, Environment and Disaster Risk Management, Volume 5, 59–79

ISSN: 2040-7262/doi:10.1108/S2040-7262(2010)0000005010

Cambodia is known as a least developed, agrarian country facing many difficulties of social and economic development. Over 80 percent of the population live in rural areas, with 52 percent in the central plains and 30 percent around Tonle Sap Lake. The country also has 435 km of coastline and low elevation in the central plains (Kingdom of Cambodia, 2001).

More than one-third of Cambodians live below the poverty line, struggling to survive on less than US$1 a day. Cambodia is highly vulnerable to the impacts of natural disasters like floods, drought, hurricanes, and landslides, which are increasing in frequency and intensity due to climate change. Cambodia has the fifth highest flood risk out of 162 countries, and floods affect 1.8 million people every year (Plan UK, 2005).

DISASTER IMPACTS AND DISASTER RISK-REDUCTION MECHANISMS

Disaster Impacts

The main natural disasters in Cambodia are flood and drought; very few typhoons occur in the country but, when they do, they bring very serious damages such as the Typhoon Ketsana in 2009. About 80 percent of Cambodia's territory lies within the Mekong River, which is known to have large fluctuations of water levels between the dry and wet seasons. This has caused a cycle of droughts and floods almost every year, damaging agricultural production and livelihoods and constraining Cambodian development and poverty alleviation. It is estimated that floods kill about 100 people annually and cause agricultural losses of $100 to 170 million each year (UNISDR, 2006). The risk of floods and droughts, moreover, is likely to increase in the future with the impact of climate change. Flood and drought risk management is a key priority for poverty alleviation and future development of the provinces (O'Brien, 2001).

Currently, Cambodia is particularly prone to annual river flooding during the monsoon season, while other phenomena also frequently occur such as tropical storms, droughts, forest fires, and so on. Those hazards are major factors that contribute to the reduction of the pace of sustained economic development in this country. Many Cambodian communities, mainly those situated along the two major watersheds – Tonle Sap and Mekong Rivers, have proven to be extremely vulnerable to the effects of these natural hazards. Localized flooding caused by monsoon thunderstorms is a serious threat as they periodically sweep the country. These natural phenomena are both a curse and a blessing as the farmers depend on the annual rainfall for crop production and have developed an extensive water management system to trap and store water to be utilized during periods of drought. In this way, many parts of the country are capable of harvesting a primary rice crop and a secondary harvest of vegetables or other cash-producing commodities. But in consecutive extreme flood years, high-water floods wash away dams, dikes, and distribution structures; destroy crops and livestock; damage homes, temples, schools, clinics, roads, and other community infrastructure; and even cause loss of human life. This increasingly leads Cambodia to food shortages, loss of economic output, health contamination, and consequently hunger and poverty.

Every year, floods of varying intensity affect Cambodia, especially along the two major natural watersheds. Both the Mekong River and Tonle Sap dominate almost all of Cambodia's paddies and forestlands. The Mekong River bisects the eastern third of the country north to south, flowing out of Laos through Cambodia and into Vietnam with its 500 km length. Annual monsoon rains swell the Mekong, causing the Tonle Sap to reverse it course, flooding Tonle Sap Lake and affecting the northwest region of the country.

Cambodia has experienced severe floods in the years of 1961, 1966, 1978, 1984, and 1991. In 1997, 1998, 2000, and 2002, many parts of the country experienced quite severe droughts, causing total crop failure in many places (Nguyen et al., 2009), but the data on the damages were not available for some drought years making it difficult to assess the damage cost from drought. Table 1 indicates the disaster events and their damages between 1996 and 2009.

Disaster Risk-reduction Mechanisms

In Cambodia, the general policy for flood mitigation and protection is to accelerate flood mitigation programs in areas where flooding and waterlogging have serious effects on the social and economic development of the area. The key governmental institutions have been established and examined as the following: The National Committee for Disaster Management (NCDM) has the responsibility for disaster preparedness, response, and mitigation. The prime minister is the president of the NCDM, with membership comprising all Ministers, as well as representatives of the Royal Cambodian Armed Forces, Cambodian Red Cross, and Civil Aviation Authority. At the provincial level, there is a committee for disaster management (Provincial Committee for Disaster Management [PCDM]), chaired by the respective governor, and

Table 1. History of Disaster in Cambodia, 1996–2002.

Year	Disaster	Damaged and Affected
1996	Severe Mekong flood	In the 1996 floods, continuous heavy rainfall in China, Laos, Vietnam, and Cambodia inundated the Mekong River, affected 1.3 million Cambodians, with 600,000 hectares of crops and 50,000 homes damaged or destroyed; 13 provinces were affected
1999	Mekong flood and cyclones	37,527 people of 10 provinces were affected, and 17,732 hectares of rice crop and 491 houses were destroyed
2000	Severe flood caused by Mekong River	3,448,629 people were affected, 768 houses were damaged, and 347 lives were lost
2001	Severe flood caused by Mekong River	429,698 families, equivalent to 2,121,952 people were affected; people killed: 62 (70% were children); houses destroyed: 2,251 houses
2002	Mekong River flood and drought	Drought: – People affected: 442,419 families (2,017,340 individuals) Flood: – People affected: 296,234 families (1,439,964 individuals) 1,082 houses destroyed. Deaths: 29
2005	Drought	600,000 people affected
2009	Typhoon Ketsana	Impacts in 10 provinces, 20 people killed, 48 people injured, and 5,693 people evacuated

Source: MAFF, 2000; NCDM, 2009; IFRC, 2009.

including district governors and representatives of the provincial offices of various ministries, of the communes, and of the local office of the armed forces. Under the PCDM, the District Committee for Disaster Management (DCDM), and the Commune Committee for Disaster Management (CCDM) have been established.

The World Bank study on *Natural Disaster Risk Management in Cambodia: Reducing Vulnerability* (2007) assessed that the NCDM's capacity is still seriously limited due to resource constraints. Although the NCDM has reasonable capacity for dealing with immediate relief and response for serious floods, as a committee with no standing budget, it is limited in its ability to coordinate key line ministries, local government, and communities for broader disaster risk management efforts. Floods and droughts in Cambodia are often localized and could be better handled by the affected provinces. In addition, the Strategic National Action Plan for Disaster Risk Reduction (SNAP-DRR) in Cambodia (2008–2013) has identified the strengthening of the NCDM as a critical priority, and the strengthening of the PCDM as the one of the first priorities, while acknowledging the continued need for technical support and resource allocation from the national government/NCDM. A recent United Nations Disaster Assessment and Coordination Team (UNDAC) report (March 2009) assessed that the DCDM and CCDM have some capacities to deal with relatively small-scale floods. Regarding structural measures for flood and drought management, it is under the responsibility of the Ministry of Water Resources and Meteorology (MOWRAM); the MOWRAM has a provincial office called Provincial Department of Water Resources and Meteorology (PDWRAM). It appears that the MOWRAM would make plans and implement the large-scale infrastructure investments whereas the PDWRAM is delegated to plan and implement smaller ones, particularly at the commune level. The MOWRAM is also responsible for hydrological and meteorological data collection and analysis. Although the director of the PDWRAM is appointed by the MOWRAM, most staffs are recruited by the PDWRAM, and the coordination between the MOWRAM and PDWRAM is not necessarily smooth; the resources allocation and priority investment are often determined by the MOWRAM at the central level.

National Strategy

The country's National Poverty Reduction Strategy (NPRS) explicitly identifies natural disasters, particularly floods and droughts, as critical factors that have increased and continue to increase socioeconomic vulnerabilities of the rural poor, including placing a disproportionate burden of coping on women. The National Strategic Development Plan (NSDP) for 2006–2010, which synthesizes the goals and objectives of Rectangular Strategy, NPRS, and the Cambodia Millennium Development Goals (CMDG), does not directly address disaster risk reduction but incorporates it in the areas of social welfare, water resource management, agriculture, and rural development. Within these priority sectors, the disaster risk-reduction activities identified include the protection of rural areas from natural hazards, enabling communities for disaster preparedness and risk reduction, and reducing the vulnerability of the poor to the external shocks including natural hazards. Cambodia launched its *Strategic National Action Plan 2008–2013 for Disaster Risk Reduction (SNAP-DRR)* in March 2009. Prepared by the NCDM and the Ministry of Planning, this strategy identifies six key disaster risk reduction (DRR) components with four levels of priority (critical, first, second, and third). The critical priorities are to mainstream the DRR into the government's development planning, formulate a national disaster risks management (DRM) policy and legislation, and strengthen national DRR coordination mechanism. The first level of priorities is aimed at integrating the DRR elements in sectoral policies and investment planning. The priority actions include "a matrix approach"; while integrating the DRR elements in the key ministries (such as the Ministry of Land Management, Urban Planning, and Construction, the Ministry of Water Resources, the Ministry of Agriculture, Forestry and Fisheries, the Ministry of Rural Development, and the Ministry of Health), PCDM, DCDM, and CCDM are proposed to be strengthened to facilitate the implementation of the DRR activities on the ground.

The SNAP-DRR proposes alignment with the National Adaptation Program of Action to Climate Change (NAPA) toward identifying and supporting common activities and projects. The NAPA was approved in 2006 by the government. Both the SNAP-DRR and the NAPA seek to address community vulnerability to hazards. Despite a number of key government policies and pronouncements recognizing the importance of disaster risk reduction, the actual practice of government institutions and local governments remains focused primarily on responding after the occurrence of a disaster event. Although some government ministries are already implementing disaster risk-reduction activities and projects, their efforts are largely uncoordinated. Also, potential synergies with other ministries, local government, international organizations, nongovernment organizations, and civil society are not fully realized. The Strategic National Action Plan facilitates the identification of disaster risk-reduction activities of the various government ministries and agencies and consolidates them into a single comprehensive program of action.

Institutional, Legal, and Regulatory Framework

Cambodia has established a disaster management committee at the national, provincial, district, and commune levels (respectively NCDM, PCDM, DCDM, and CCDM). The NCDM was established in 1995, through a subdecree signed by the prime minister. The NCDM has a permanent secretariat, which has the following five departments: (1) Administration and Finance, (2) Information and Relations, (3) Emergency Response and Rehabilitation, (4) Preparedness and Training, and (5) Search and Rescue (Fig. 1).

At the village level, there are pressing concerns about the provision of safe water, sanitation, and road access. Some provinces have developed a response plan for annual flooding with support of NCDM; however, the plan has been prepared through a top-down approach, and villagers are often not aware of the plan. It should also be noted that the PCDM, DCDM, and CCDM do not meet outside the annual flood season, as these committees are mainly focusing on postdisaster response, rather than risk reduction. It is also feared that these committees are not ready for major flood events. As indicated previously, the MOWRAM is the key to addressing flood and drought risks in the country and has a provincial department called PDWRAM. Investment decisions are mainly made by the MOWRAM, as most of the donors provide the external support through the MOWRAM; MOWRAM tends to support the large infrastructure projects, putting less emphasis on the smaller structural measures at the commune levels. Although the PDWRAM has a good understanding on such commune-level investment, it has little resources at its discretion to implement such investments on its own (Eng Rinbol, 2009).

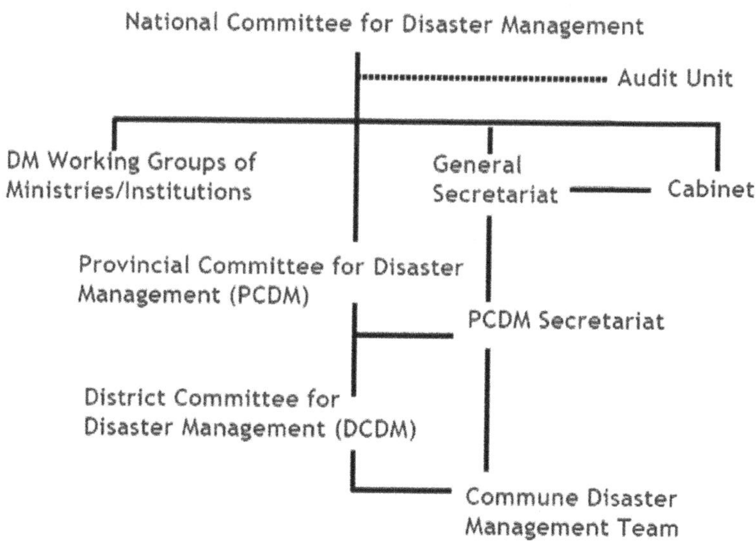

Fig. 1. NCDM Organizational Structure. *Source:* Ross Sovann (2006).

CLIMATE CHANGE IMPACTS AND CLIMATE CHANGE ADAPTATION MECHANISMS

Climate Change Impacts

Impact of Global Warming on Cambodia's Climate

The global warming scenarios used the models named SRESA2 and SRESB1. SRESA2 describes a very heterogeneous world. The underlying theme is self-reliance and preservation of local identities. Fertility patterns across regions converge very slowly, which results in continuously increasing global population. Economic development is primarily regionally oriented, and per capita economic growth and technological change are more fragmented and slow. SRESB1 describes a convergent world with the same global population that peaks in mid-century and declines thereafter, rapid change in economic structures toward a service and information economy, with reduction in material intensity, and the introduction of clean and resource-efficient technology. With these characteristics, the SRESA2 will lead to higher future greenhouse gas (GHG) emissions, whereas SRESB1 will lead to lower future GHG emissions. The global temperature and sea level rise under these two scenarios is presented in Table 2.

The impact of global warming on Cambodia's climate was assessed using two General Circulation Models (GCM), that is, The Center for Climate System Research of Japan (CCSR) and Australia's Commonwealth Scientific and Industrial Research Organization (CSIRO). From validation

Table 2. Temperature and Sea Level Rise Relative to 1990.

Year	2000	2025	2050	2100
Global temperature (°C)				
SRESA2: Best guess	0.2	0.5	1.2	2.9
: Range	0.15–0.25	0.3–0.7	0.8–1.6	2.0–4.1
SRESB1: Best guess	0.2	0.7	1.1	1.9
: Range	0.15–0.25	0.5–0.9	0.7–1.6	1.2–2.7
Sea level (cm)				
SRESA2: Best guess	2	10	21	60
: Range	0–4.0	4.0–20	9.0–41	25–112
SRESB1: Best guess	2	10	21	48
: Range	0–4.0	4.0–22	9.0–42	18–85

Source: McSweeney, New, and Lizcano (2008).

of this study, it was indicated that the two GCM models used in this analysis (CCSR and CSIRO) were not very suitable for use in Cambodia because the two models were developed for use in Japan and Australia, which are very different geographical regions. The deviation of GCM models from the observed rainfall data was very significant. The deviation of monthly wet season rainfall could reach 794 mm, especially during the rainy season. Therefore, correction factors were developed and used in the subsequent analysis.

Under the SRESA2 scenario and using the CSIRO model, it was estimated that the temperature in Cambodia would increase by up to 2.0°C, whereas it would increase up to 2.5°C using CCSR model. Under SRESB1, the increase was much smaller, that is, 1.35°C under CSIRO and 1.60°C under CCSR. Rainfall in Cambodian would also increase from the current condition. The magnitude increase also varies with location, time, the GCM model, and emission scenario. Lowland areas seem to be more affected than high land areas. Under SRESA2, annual rainfall in 2100 would increase between 3 percent and 35 percent from the current rainfall depending on location, whereas the increase is smaller under SRESB1.

Impact of Climate Change on Sectors

Climate change is expected to affect Cambodia significantly as it is one of the most vulnerable countries in the region. At stake are recent gains in the fight against poverty, hunger and disease, and the lives and livelihoods of the majority of Cambodians. Climate change will lead to higher temperatures, changes in precipitation patterns, rising sea levels, and distribution and frequency of weather-related disasters, posing risks for already vulnerable agriculture, food, and water supplies.

Cambodia's vulnerability to climate change is linked to its characteristics as a post-civil war, least developed, agrarian country with 80 percent of the population living in rural areas, as well as its low elevations on the central plain. Climatic events such as floods and droughts are already recognized as some of the main contributors to poverty. Sea level rise is one problem caused by climate change. A study conducted by the Ministry of Environment indicated that if sea level rises by up to 1 m, about 0.4 percent (4,444 ha) of the area would be under water (Table 3).

The impact of climate change on sectors was assessed using analog and modeling approaches. The analog approach used past experiences to assess the possible impact of climate change in the future, whereas the modeling approach used a statistical model to express the relationship between the variability of climate and sectors' outputs.

Table 3. Area of Land to be Lost in Case of Sea Level Rise by 1 m in Koh Kong Province.

No.	Description	Area Under Water (ha)
1	City/Town	279.6
2	Village	77.9
3	Swimming Beach	10.4
4	Marsh	32.5
5	Forest	181.8
6	Secondary Forest	3.5
7	Shrub	12.7
8	Mangrove	3,114.3
9	Grassland	301.2
10	Rice Paddy	29.6
11	Crops	9.8
12	Shrimp Farm	345.2
13	Others	46.0
	Total	4,444.1

Source: Ministry of Environment, 2001.

Agriculture

Agriculture remains a significant part of the Cambodian economy and accounts for about two-thirds of employment (MAFF, 2001). The agriculture sector remains highly relevant to Cambodian economic growth in the medium term, not only because of its ongoing contribution to employment but also because other recent growth sectors including the garment industry, construction and tourism, have contracted in recent times (ACIAR). The agricultural sector contributes around 31 percent of the gross domestic product (GDP) and engages 84 percent of the population. The high dependence on agriculture, with 80 percent of farmers growing rice (60 percent for subsistence) but with only 7 percent of the crop area being irrigated, makes this important sector extremely vulnerable to any change in rainfall patterns (MAFF, 2001). The lack of processing capacity and dependence on a single rice cropping cycle means food security is highly climate dependent. The combination of high poverty levels and great dependence on agriculture is the main mechanism attributed to the country's extreme vulnerability to climatic events. Between 1998 and 2002, floods accounted for 70 percent of production losses of rice – the single major agricultural crop of the country – while drought accounted for 20 percent of losses. Climatic variations are anticipated to further increase the severity and frequency of flood and drought events.

Based on field observations, flooding mostly occurred due to the increase in water levels of the Mekong River and Tonle Sap Lake (ADPC, 2007). These two water bodies link each other. The Mekong River starts from the Tibet Plateau. The increase of water level in the Mekong is closely related with the rainfall throughout the Mekong basin. This might explain why the flood occurrence was not always related with Cambodian rainfall. For future study, the relationship between water flow in the Mekong River and Tonle Sap Lake and rainfall variability in the entire basin as well as the relationship between El Niño/La Niña-Southern Oscillation (ENSO) and rainfall variability in the basin need to be investigated. By increasing our understanding of these issues, the timing and the intensity of flood and drought events in rice-growing areas of Cambodia can be better estimated. Some studies indicated that intensity and the frequency of ENSO events under changing climate might increase.

This means that with changing climate, the yield of wet season rice would increase while that of dry season rice would remain the same or decrease. However, there is a chance that, under changing climate, rice yield in some provinces would be more variable than current conditions due to the increase in flood frequency and intensity, in particular in rice-growing areas surrounding Tonle Sap Lake and Mekong River.

Rice production in Battambang, Prey Veng, and Takeo in 2000 exceeded the demand of the corresponding provinces, whereas at Kampong Cham there was a deficit of about 100,000 tons. Under changing climate, rice production in the four provinces is over the demand if the government could increase rice productivity by about 1 ton/ha per 25 years from the current productivity. In 2025, rice production from these four provinces can meet 56 percent of the national demand and, in 2050, it could increase to 67 percent under the SRESA2 scenario and 65 percent under the SRESB1 scenario. In 2100, under the SRESA2 model, contribution from these four provinces to the total national demand would be almost the same as in 2050, whereas it would be lower under the SRESB1 model (McSweeney, New, & Lizcano, 2008).

Forestry

According to the Holdridge Classification System, under current climatic conditions, Cambodia's forest is dominated by dry forest (60 percent), followed by wet forest (20 percent) and moist forest (20 percent). Under changing climate, the area of wet forest would decrease while moist forest would increase and dry forest would remain the same (Ministry of Environment, 2001). This change indicates that forest productivity and biodiversity might also change. The high rate of deforestation may accelerate the loss of forest biodiversity and reduce forest productivity due to the increase in human activities in the forest areas. The magnitude of change in the forest area due to climate change depends on GCM and emission scenarios used in the analysis. In this study, the pattern of change of forest type due to climate change was quite similar in both the CCSR and CSIRO models. Further study on the impact of climate change on forest biodiversity and forest productivity needs to be carried out. Based on studies of other neighboring countries, there is a possibility that forest productivity of Cambodian forest may change and some of the species may disappear in a changing climate. However, this study will require more data in terms of quantity and quality, to determine if this is the main problem the country will face.

Coastal Zone and Fishery

The coastal zone of Cambodia consists of three provinces (Kompong Som, Kampot, Koh Kong, and one autonomous city Kep). The total area covered by these provinces and the autonomous city is about 17,237 km^2. In one research study conducted by the UNFCCC and the Ministry of Environment in 2002, the Koh Kong province has been assessed because this province covers over 64 percent of the coastal zone (11,160 km^2) and is the most vulnerable to the impact of sea level rise according to a preliminary analysis of the impacts of a 1 m sea level rise on the Cambodian coastal zone (Ministry of Environment, 2002). This is due to the fact that most areas along the coastline in this province are low-lying. Table 2 shows that the sea level under the two scenarios would rise by about 0.5 m. If the sea level rise were only 0.5 m, only small areas would be under water. However, if the global effort to reduce GHG emissions is not carried out (SRESA2 scenario), rainfall might increase under changing climate and the water flow from the river may also increase; as a result, the area close to the riversides might be exposed to frequent floods. Under the SRESA2 scenario, the rainfall in the main four river basins (Stung Metoek, Stung Russei Chrum, Stung Sala Munthun, and Stung Chhay Areng) of the Koh Kong province would increase between 2 percent and 15 percent, and this would be translated to the increase in water flow of the four river basins between 2 and 10 m^3/s. Under the SRESB1 model, the rainfall might decrease between 2 percent and 5 percent, and this might not have significant impact on the river flow.

Cambodia's fisheries are also highly vulnerable to climatic variations. The Tonle Sap fishery alone accounts for a significant 7 percent of the GDP, and as a sector contributes between 9 and 12 percent of the GDP and substantially to incomes, jobs, and food security. Women's livelihoods are particularly at risk in this sector due to their significant post-harvest participation. Cambodia's fishery sector is almost all capture fisheries with very limited aquaculture, thereby making the sector highly vulnerable to flow changes in the Mekong (notably, the flood pulse), particularly if exacerbated by hydropower development.

Human Health

The impact of climate change on human health is not well understood. However, it was reported that climate change would have indirect and direct impacts on human health. The direct impacts include exposure to thermal extremes. Frequency and/or intensity of extreme weather events may increase, and this could result in death, injuries, psychological disorders, and damage to public health infrastructure. The indirect impacts include the disturbances of ecological ecosystems that cause changes in geographical range and the incident of vector-borne diseases, infectious diseases, malnutrition, and hunger, which in turn disturb child growth and development. Furthermore, sea level rises may force population displacement and cause damage to infrastructure. This will lead to increased risk of infectious diseases and psychological disorders (Ministry of Environment, 2005). Other factors for welfare such as safe water supply, good sanitary facility, and hygiene education should also strongly influence the number of cases. Under a changing climate, it was found that the average number of malaria cases in most of provinces in Cambodia would consistently decrease under the SRESB1 scenario, ranging from −1 to −62 percent. In contrast, under the SRESA2 scenario, the number of malaria cases would gradually increase, ranging from −1 to 16 percent. For future studies, it is recommended that the model to estimate malaria cases be improved using longer historical data and covering most of the sensitive areas. Continued observation and good database management would be a priority activity for facilitating the studies.

Climate Change Adaptation Mechanisms

Climate change has become a real challenge for all countries throughout the world. The Royal Government of Cambodia (RGC) clearly recognizes this issue and is fully committed to the global efforts to address climate change, both at the national and international levels. Therefore, Cambodia ratified the United Nations Framework Convention on Climate Change (UNFCCC) in 1995 and acceded to its Kyoto Protocol in 2002.

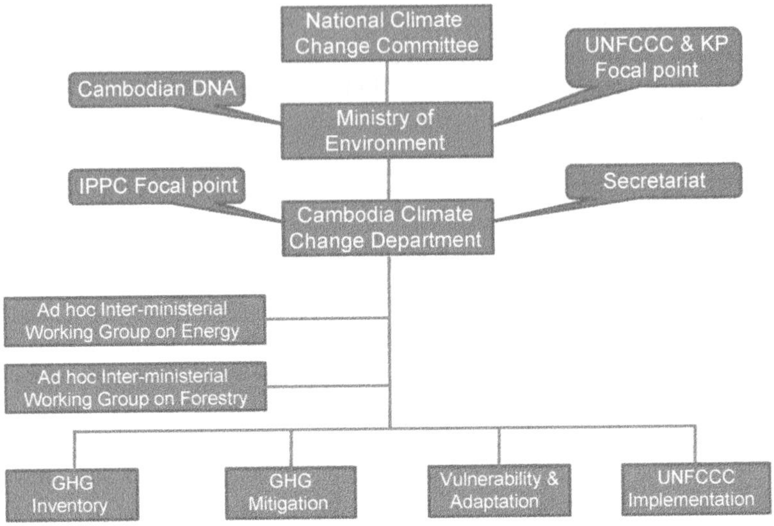

Fig. 2. Climate Change Institution in Cambodia. *Source:* Navann, 2009.

As a least developed agrarian country, Cambodia is highly vulnerable to climate change, the more so as it has low adaptive capacity to changing climate conditions. In recent years, there were more frequent and severe floods and droughts, which have resulted in a significant number of fatalities and considerable economic losses. For this reason, the RGC has recently taken the first steps in developing a Cambodian National Adaptation Program of Action to Climate Change (NAPA). The NAPA is directed by the National Climate Change Committee, as shown in Fig. 2. The main goal of the Cambodian NAPA is to provide a framework to guide the coordination and implementation of adaptation initiatives through a participatory approach, and to build synergies with other relevant environment and development programs. Cambodia's NAPA presents priority projects to address the urgent and immediate needs and concerns of people at the grassroots level for adaptation to the adverse effects of climate change in key sectors such as agriculture, water resources, coastal zone, and human health.

Agriculture

The impact of climate change on rice production in Cambodia would not be substantial if the government could meet the existing 1999–2010 Agriculture Development Plan. Studies in the four provinces indicated that between 2025 and 2100, the rice production may exceed demand, if rice productivity were increased by about 1 ton/ha per every 25 years from the current productivity. Adaptation options that have been identified for increasing rice productivity under changing climate are the following:

- Improvement of genetic or development of new high-yielding varieties;
- Improvement of crop management and cultural practices;
- Development of capacity to adapt to current extreme climate such as development of early warning system to extreme climate, development maps showing the provinces of rice-growing areas prone to flood and drought;
- Development of irrigation facilities in many parts of lowland areas;
- Increasing planting index in suitable areas; and
- Diversification of foods.

Forestry

In forestry sectors, there are at least three options for reducing the impact of climate change, namely:

Forest Plantation Establishment

Promotion of forest plantation establishment is important to relieve pressure on the natural forests. The optimal use of unproductive land for forest plantation establishment should be encouraged. Tree species used in the designated areas should match with socioeconomic and biophysical conditions of the areas as well as global markets. Therefore, maps of land quality index for tree plantations should be verified or established.

Conservation of Protected Areas

There is a need for establishing appropriate legal and policy frameworks, protected area management plans, and an effective monitoring system. Strengthening law enforcement and community participation in protected area management is also critical. Programs for protecting critical wildlife habitats and for the expansion of species and forest communities should also be enhanced, in the likely affected areas in particular. Programs to rehabilitate the protected forests also need to be promoted through enhanced natural regeneration techniques using native and exotic tree species.

Improvement of Forest Resource Management

The common goal of forest management is to achieve sustainable management through the utilization of forest resources in sustainable ways.

Partially, this can be done through the promotion of improved silvicultural systems and techniques to forest concession holders, such as techniques to reduce the impact of logging.

Coastal Zone

Considering the possible impact of sea level rise on the coastal zone is important for the government, as it would have a very significant impact on the country. Possible measures for consideration include:

- Developing a national strategic response to sea level rise for the coastal areas;
- Investigating further potential impacts of sea level rise on biogeophysical, socioeconomy, marine resources, fresh water, infrastructure, human settlements, and agricultural production;
- Formulating a comprehensive adjustment and mitigation policy for sea level rise in the context of integrated coastal zone management;
- Developing computer-based information systems covering the results of surveys, assessments, and observations to minimize the impact of sea level rise resulting from climate change;
- Increasing public awareness on the effect of sea level rise on Cambodia's coast;
- Identifying potential donors, either multilateral or bilateral sources to assist the country in adaptation to sea level rise; and Establishing cooperation frameworks, training, technology transfer, surveillance of climate change in case of sea level rise, and the sharing of experiences to assist the government in establishing preparedness response to climate change.

Human Health

For the health sector, control measures being introduced to reduce malaria cases include early diagnosis and treatment of the diseases, having communities living in high-risk areas utilize parathyroid-treated mosquito nets to control the vectors, strengthening program management and supervisory practices, and increasing funds for the provision of mosquito nets and insecticides. In addition, they include introducing and promoting information, education, and communication (IEC) programs for health such as increasing community understanding on the most critical causes of disease – watercourses and containers where mosquitoes' breed and personal habits, which are conducive to the attack of parasitic and infectious organisms. These programs should also promote and expand upon low-cost preventative methods such as improving personal hygiene and surrounding environments by destroying the insect breeding sites, using bed nets, and so on. The educational materials should be simple, as the level of education in rural communities is rather low.

CHALLENGES AND OPPORTUNITIES FOR DISASTER RISK REDUCTION AND CLIMATE CHANGE ADAPTATION IN CAMBODIA

The biggest challenge for Cambodia for disaster risk reduction and climate change adaptation is to link NCDM, PCDM, MOWRAM, and PDWRAM to establish an effective institutional framework for addressing the disaster risks and climate change impacts in a comprehensive manner through both structural and nonstructural measures. Therefore, ideally, the NCDM would be responsible for overall coordination and guidance for the disaster management, whereas the implementation responsibilities would be transferred to the PCDM to enable a multisectoral approach; the MOWRAM would also support the large-scale infrastructure intervention that may cover more than one province, whereas the PDWRAM would have adequate resources to carry out priority structural measures, particularly at the commune levels in collaboration with the PCDM, DCDM, and CCDM.

The actions of the PCDM on flood early warning, weather forecasts, and prediction of the extent and severity of the flooding were incorrect and not timely, and as a consequence, the level of preparedness was inadequate. Flood prediction was only available for a one-day forewarning and only at a national and provincial level. Indeed, from a preparedness perspective, the length of forewarning was not adequate. Additionally, although television broadcasts provide situational updates on the damages of floods, no public awareness about their consequences and what local actions need to be taken before and during the flood were provided.

The absence of partnership agreement and implementation of guidelines for PCDM collaborative action prohibited the effectiveness of coordination and cooperation between the PCDM, NGOs, and international organizations. One of the most important lessons learned from the flood of 2000 is that there is an urgent necessity for improving interagency or interorganizational coordination. As a starting point, if there is recognition that future-damaging disasters will happen, the NCDM must be assisted to improve its capacity, system, and procedures in damage and needs assessment and reporting. Such information is important if coordination is to be achieved. At the moment, there are areas that need to be improved, as various organizations and stakeholders cited to the Assessment Team.

Through several DRR programs implemented in Cambodia, one of the important lessons learned is that Community Based Disaster Risk Management (CBDRM) methodologies do work in communities and need to be continually used as a core concept in all future DRR programs. Results of DRR projects revealed a need for a broader information sharing and partnership focus, with an emphasis on having good linkages with the government sector. Training and capacity building is effective at the community level. However, this can only be so true with the support from the NCDM, the PCDM, and the DCDM. The cooperation and coordination between the three pillars of disaster management in the country – national, provincial, and district – is critical to successful DRR programs. As in the case of the Takeo province's CBDRM program, villagers learned what to do in disaster preparedness and mitigation. With effective government support at all levels, this experience can happen in all of Cambodia's rural and most vulnerable communities.

Future Outlook Statement

A key role in the establishment of a national platform for DRR in Cambodia would have to be played by the NCDM. At the level of a "national platform," this would have to be based on a revitalized and strengthened NCDM playing a wider risk reduction role by establishing a suitable additional mechanism to bring together all stakeholders to have a shared vision and a concerted programmatic approach to implementing DRR in the country. Here, UN agencies, donors, and NGOs would be key associate agencies to be formally brought into the mechanism. For development and implementation of a national program on DRR, the NCDM 2001 strategy would still remain as a framework within which to initiate a programmatic approach to DRR. There is need for a revitalized and revised national strategy, together with solid procurements of energy and resources. This would be a useful first step for the success of future DRR in the country.

Some key recommendations to consider are:

- Ensuring community participation and government linkages, including communications with authorities and DM focal points.

- Focusing projects more on sustainable livelihoods and strengthening capacity.
- Mainstreaming DRR into national and local development planning rather than "stand-alone" DRR projects.
- Accessing more resources and longer funding cycles to enable DRR initiatives to be fully integrated into development plans.
- Addressing food security issues for rural households when constructing DRR projects at a community level, as this is a primary concern for most rural households.

CONCLUSION

Cambodia is a country vulnerable to climate variability and climate change. Flood and drought are two major climate hazards commonly experienced in Cambodia. Their impacts include the yearly destruction of infrastructures, properties, crops, livestock, and losses of lives. The adaptive capacity to flood and drought is poorly developed in Cambodia. With the support of international donors, the Royal Government of Cambodia has formulated long-term strategic sectoral plans and programs. However, the implementation of these programs relies almost exclusively on external funding sources. Poor governance in all sectors and across all management levels is a major barrier to achieving sustainable development. This includes overlap or conflict of mandates among agencies, lack of coordination and information sharing across sectors and agencies, limited number of professional staff, and low commitment of government staff. Even though Cambodia is vulnerable to climate hazards and future climate change, current policies have not thoroughly considered global warming issues. The Ministry of Environment has developed a national action plan on climate change; however, this plan only focuses on mitigation aspects.

Planned adaptation to future climate change should be based on current individual, community, and institutional behavior that, in part, have been developed as a response to current climate. However, common practices local people use to adapt to existing climate hazards have not received enough attention from government and international agencies. These practices have been implemented sporadically due to the lack of financial support. Delays in developing and enhancing local communities' capacity in responding to climate hazards may increase their vulnerability, and lead to increased costs at a later stage.

For Cambodia to optimize the use of global funding for climate-related issues, there is a crucial need for policy makers to quickly increase their understanding of climate change as well as their capacity to identify eligible and appropriate programs. National policies and regulations consider climate change aspects, and assist local communities in increasing their adaptive capacity to climate variability.

ACKNOWLEDGMENT

The work published in this chapter was supported by the Ministry of Education, Culture, Sports, Science, and Technology (MEXT) of Japan through doctoral fellowship to the primary author at Kyoto University; Japanese Society for Promotion of Science (JSPS) project fund; and Graduate School of Global Environmental Studies (GSGES) project fund through grant research to the second author.

REFERENCES

ADPC. (2007). *Safer communities, case study 4: Sustaining the flood preparedness and emergency management system in Cambodia*. Bangkok, Thailand: ADPC.

Eng, Rinbol. (2009). Cambodia: Mainstreaming flood and drought risk mitigation in East Mekong Delta. Presentation at the Expert Group Meeting on Innovative Strategies towards Flood Resilient Cities in Asia-Pacific, July 21–23, 2009, Bangkok.

IFRC. (2009). *Typhoon Ketsana in Cambodia*. Phnom Penh, Cambodia: International Federation of Red Cross and Red Crescent Societies report.

Kingdom of Cambodia. (2001). *Draft second five-year socioeconomic development plan 2001–2005*. Phnom penh, Cambodia: Ministry of Planning.

MAFF. (2000). *Agricultural statistics. Department of Planning, Statistics and International Cooperation*. Phnom penh, Cambodia : Ministry of Agriculture, Forestry and Fisheries.

MAFF. (2001). *Summary of Agricultural Activities*. Report no. 358/038. Ministry of Agriculture, Forestry and Fisheries, Phnom penh, Cambodia.

McSweeney, C., New, M., & Lizcano, G. (2008). *Cambodia climate change profile*. Oxford University.

Ministry of Environment. (2001). *Vulnerability and Adaptation Assessment to Climate Change in Cambodia*, Phnom Penh.

Ministry of Environment. (2002). *Initial National Communication to UNFCCC*. Phnom penh, Cambodia.

Ministry of Environment. (2005). *Analysis of policies to address climate change impacts in Cambodia (Final draft)*. Phnom penh, Cambodia.

Navann, O. (2009). Overview of climate change activities in Cambodia, kick-off workshop approach to reducing deforestation and associated greenhouse gas emissions in

Cambodia under a post-2012 international climate change policy framework. Ministry of Environment, March 2009.
NCDM. (2009). Cambodia: National progress report on the implementation of the Hyogo Framework for Action (2007–2009). Phnom Penh, Cambodia.
O'Brien, N. (2001). Risk mitigation and disaster management among rural communities in Cambodia. Presentation at the Expert Group Meeting on innovative strategies towards flood resilient cities in Asia-Pacific July 21–23, 2009, Bangkok.
Plan UK. (2005). Cambodia disaster risk reduction. A project proposal, Plan UK. London, UK.
Sovann, R. (2006). Cambodia disaster information management and database. Presentation at the workshop to improve the compilation of reliable data on disaster occurrence and impact, April 2–4, 2006, Bangkok, Thailand.
UNISDR. (2006). *Cambodia disaster statistic*. Bangkok, Thailand: UNISDR Asia-Pacific region.

CHAPTER 5

INTEGRATED CLIMATE CHANGE ADAPTATION IN CHILIKA LAGOON FISHERIES, INDIA

Shimpei Iwasaki and Rajib Shaw

INTRODUCTION

Over the past decades, changes in climate have been commonly observed in many parts of the world. It is apparent that changes in temperature and rainfall and resulting increases in frequency and intensity of flood and drought events have affected ecological and social systems on the earth. According to the Fourth Assessment Report of the Intergovernmental Panel on Climate Change, warming atmosphere of the climate system is now unequivocal (IPCC, 2007). Climate change poses significant risks to the livelihoods, culture, and health of millions of people (Barnett, 2003). Ecological and climatic disasters – hurricanes, tornadoes, draught, flooding, landslides – are becoming more frequent, resulting in devastation to family and communities, especially the poor living in precarious environments (Ogata & Sen, 2003). Related to this, the number of climate-led disaster events and affected populations has been increasing during the last decades according to the data from EM-DAT (2010).

Climate Change Adaptation and Disaster Risk Reduction: An Asian Perspective
Community, Environment and Disaster Risk Management, Volume 5, 81–103

ISSN: 2040-7262/doi:10.1108/S2040-7262(2010)0000005011

Particularly, lagoon areas that are targeted in this chapter are considered some of the places most vulnerable to climate variability. A lagoon is an area of shallow water separated from the sea by barriers such as low sandy dunes. The barriers are constantly eroded by waves and wind, requiring continuous sediment deposition to maintain the lagoon environment. In other words, lagoon areas are formed and maintained through sediment transport process (Anthony et al., 2009). The geographical location is highly exposed to such climatic and environmental factors as sea level rise, increased level of inundation and storm flooding, seawater intrusion, coastal erosion, and water pollution. Lagoons are physically subject to various influences from not only the coastal and brackish environment but also the adjacent marine and terrestrial areas, including the watersheds. Water quantity and quality conditions in a lagoon are influenced by the rate at which the lagoon loses or gains water from evaporation, precipitation, groundwater input, surface runoff, and exchange with the sea (*ibid.*). In particular, the extent of salinity level and patterns of hydrological cycle play significant roles on maintaining diverse and productive resource conditions in the lagoon environment. These factors are largely dependent on changes in wind and weather as well as the volume of fresh water from upstream and tidal variability at intervals. If a lagoon environment loses balance of water influx and efflux from the watersheds or the sea combined with climate variability, there is a great possibility for change in the whole situation, which can collapse fertile ecosystems and cause subsequent natural resource degradation plaguing human lives, livelihoods, and dignity. Table 1 indicates the classification of climate change impacts in the lagoon environment, according to a category that the UNISDR (2009) presented.

In this chapter, we therefore focus on impacts and adaptive countermeasures of climate change in the local context of the lagoon environment. To ensure the objective, this chapter introduces a case study of the Chilika Lagoon, where the people are largely dependent on fishery resources. The research is divided into four parts. First, we delineate research scope and methods to explore climate change impacts and related countermeasures in local context. Second, we describe the profile of the field study site (Chilika Lagoon), and then illustrate the trend of climate variability and related people's perception. Third, we analyze livelihood conditions in fishing communities of Chilika Lagoon and identify vulnerability to climate changes. Fourth, the research highlights existing fishery adaptive countermeasures against changes in climate and draws implications about a need for integrated climate change adaptation (ICCA) in Chilika Lagoon fisheries.

Table 1. Classification of Climate Change Impacts in the Lagoon Environment.

Hazard Type Related to Climate Change	Physical Forms or Events	Major Root Causes
Biological hazard	– Outbreaks of epidemic diseases and infections – Weed invasion (shrinkage of water cover area)	– Increase of temperature – Salinity change – Sedimentation
Geological hazard	– Sedimentation/water shallowness – Coastal earthquake/ Tsunami – Inundation/surge – Seawater intrusion – Coastal erosion	– Landslides, rockslides, surface collapses, and debris or mud flows – Closure of sea mouth – Sea level rise
Hydrometeorological hazard	– Salinity change – Cyclones (also known as typhoons and hurricanes) – Floods – Droughts – Heat waves/cold spells	– Loss or gain of water from evaporation, precipitation, groundwater input, surface runoff, and exchange with the sea – Land use change – Sedimentation – Change of temperature

Source: Modified from UNISDR (2009).

LINKING RURAL LIVELIHOODS WITH CLIMATE CHANGE ADAPTATION

Although climate change adaptation has been discussed over the past several years through organizational and adaptive responses, not enough attention has been paid to focusing on adaptation at the community level and to integrating adaptation methods into a policy perspective (Shaw, 2007). Specially, particular adaptation measures faces some difficulties in scaling up from certain community or area (Van Aalst, Cannon, & Burton, 2008). Identification of adaptive capacity is context-specific and varies from community to community, among social groups and individuals, and over time (Smit & Wandel, 2006). Local experiences may provide effective information for planning the future in certain communities, but the lessons learned from the case studies are not necessarily applicable to other communities or regions. Thus, community-based adaptation to climate change requires the recognition of multiple stimuli beyond those related to climate, including political, cultural, economic, institutional, and technological forces (*ibid.*).

In rural areas, people depend on rather climate-sensitive occupations such as fishing, farming, and forestry, which are closely linked to people's lifestyles. In particular, small-scale fishermen that are targeted in this article are subject to strong restrictions from nature's forces so that, for instance, they tend to adapt to migratory fish distribution on the basis of climate-related seasonality and fish ecology. In other words, their practices must adjust to these biological factors, causing the fishermen to separate their living area from other communities. This linkage between climate and practice means that climate change adaptation measures can be commonly scaled up from the sampling communities to the larger archetypal livelihood region of which it is a part. Given that the impacts of climate change cover multifaceted aspects of rural livelihoods, there is an urgent challenge to understand the broad relevance to livelihood contexts and their linkage with climate change.

With regard to these discussions, the term *vulnerability* is crucial in exploring the impacts and countermeasures of climate change. Vulnerability can be portrayed in the degree to which a system is susceptible or unable to cope with negative effects of natural or man-made variability and extremes. It is usually characterized as forms of the characteristics, extent, and frequency of exposures and sensitivity, and people's capacity to adapt to these hazards (see Fig. 1). In the discipline of climate change, Bernett and Adger (2007, p. 641) argued, "the vulnerability of people to climate change depends on the extent to which they are dependent on natural resources and ecosystem services, the extent to which the resources and services they rely on are sensitive to climate change, and their capacity to adapt to changes in these resources and services." There is a need to understand the degree of exposures, sensitivity, and adaptive capacity. In particular, adaptive capacity is a critical key in building people's resilience to climate change, given that the climate is becoming more variable and creating various risks. Hence, the term *adaptation* that takes into account uncertainty including climate change and does experimentation and learning in an iterative process was elaborated in a case study of Chilika Lagoon fisheries in this chapter.

With these recognitions, fieldwork was carried out in four fishing communities of Chilika Lagoon, where two communities are located on an inland and two are coastal areas along the Bay of Bengal (see Fig. 2).

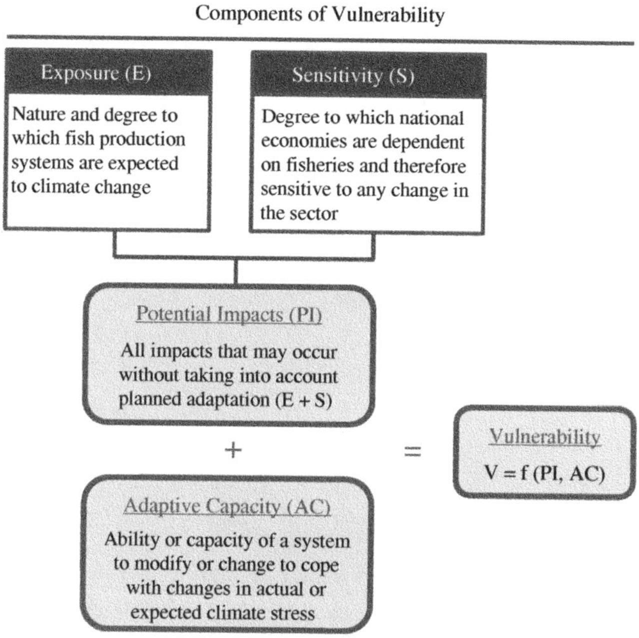

Fig. 1. Components of Vulnerability. *Source:* Revised from FAO, 2009.

The research used both qualitative and quantitative data, including structural questionnaires, participatory rural appraisal (PRA), key informant interviews, and secondary data. In this chapter, the primary data from structural questionnaires were collected from 25 households in each fishing community, bringing the total to 100 household samples. The questionnaires used a multiple-choice format and were compiled into six sections: general household information, climate risk perception, economic, social, and environmental impacts of climate change, and coping mechanisms. Further, Venn diagrams, problem trees, and seasonality maps were applied as a PRA exercise for identification of vulnerability. In addition to the livelihood assessment in four sample communities, greater emphasis was made on implementing integrated lagoon watershed management in the catchment of Chilika Lagoon. Throughout these studies, semistructured interviews with key informants (government officers, village leaders, NGOs, researchers) were carried out to validate and complement the information.

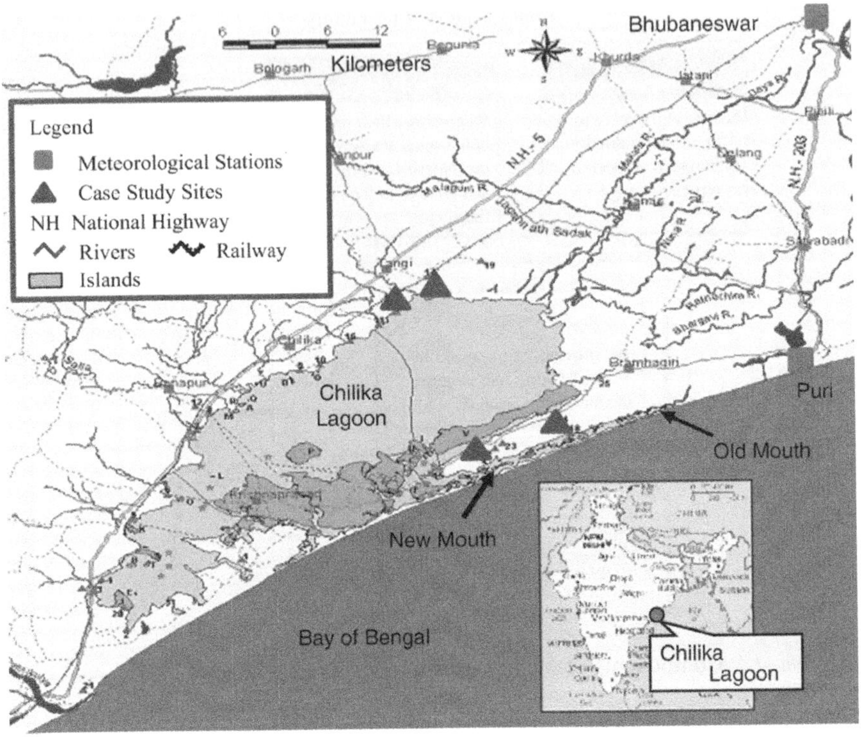

Fig. 2. Map of Chilika Lagoon. *Source:* Revised from CDA data.

TREND OF CLIMATE VARIABILITY AND PEOPLE'S PERCEPTION

Profile of Chilika Lagoon

Chilika Lagoon is the largest lagoon in India, and is situated between 19°28′ and 19°54′ north latitude and 85°05′ and 85°38′ east longitudes (see Fig. 2). The lagoon extends from the southwest corner of Puri and Khurda districts to the adjoining Ganjam district of Orissa state. The average lagoon area is 1,055 km^2, which increases to 1,165 km^2 from July to December and shrinks to 906 km^2 from December to July (Pattnaik, 2002). Chilika Lagoon becomes less saline during the rainy season due to floodwaters from 52 rivers and rivulets. It becomes more saline during the dry season as the supply of flood water is cut off when the south wind begins to blow and saline waters enter from the Bay of Bengal at high tide (Patro, 2001).

The water environment in Chilika Lagoon is a unique assemblage of marine, brackish, and freshwater ecosystem with estuarine characteristics. This combination has endowed a highly productive ecosystem and a valuable biodiversity, including endangered species like the Irrawaddy dolphin (*Orcaella brevirostris*). These valuable characteristics allowed the Chilika Lagoon to be listed as a wetland of international importance under the Ramsar Convention and become the first Ramsar site of India in 1981. There are 127 reported fishing communities in Chilika Lagoon. The rich ecosystem of Chilika Lagoon enables more than 200,000 fishers and thousands of local people who are engaged in allied fishery business activities to support their livelihoods (CDA, 2005a). In regard to fishing occupation, traditional fishermen include seven subcaste groups: Keuta, Niari, Kartia, Kandara, Gokha, Tiara, and Nolia (Mitra & Mahapatra, 1957). Most of them belong to Schedule Caste (SC), and their societal status is quite low as they belong to the poorest of society.

Trend of Climate Change in Chilika Lagoon

Chilika Lagoon, which is a part of Orissa state, India, is considered to be a climate-sensitive area in India. Although limited quantitative data on climate change effects in Chilika Lagoon are available, floods, droughts, heat waves, and sometimes cyclones are common features at the state level. The socioeconomic condition of the people in Orissa state is strongly influenced by various natural forces (Gupta & Sharma, 2000).

Temperature

Chilika Lagoon is typically tropical with average annual maximum and minimum temperatures of 39.9°C and 14°C, respectively (CDA, 2010). Changes in temperature, even small changes in water temperature, are expected to exert strong pressure upon fish ecology (WWF, 2005). Temperature variations also affect people's health, undermining their capacity for operating the fisheries. According to the statistical fixed-point observations in two meteorological stations (Bhubaneswar and Puri), it seems that the apparent warming trend has not been seen in Chilika Lagoon (Fig. 3), as opposed to the world's expectation (IPCC, 2007). However, it

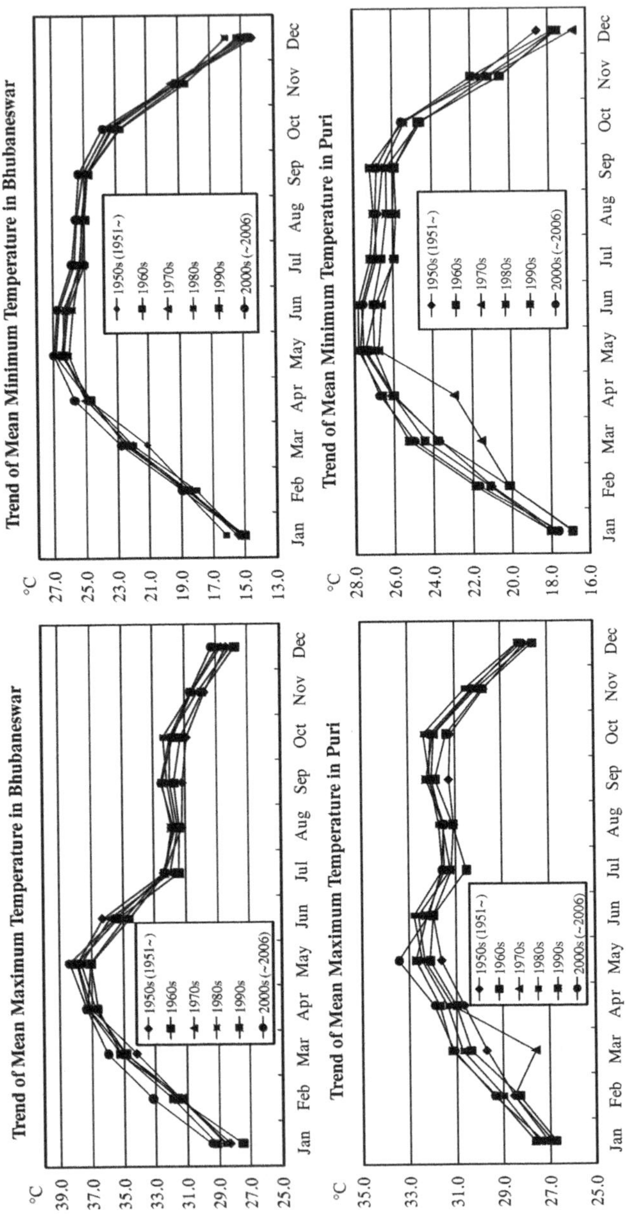

Fig. 3. Trend of Mean Highest and Minimum Temperature in Bhubaneswar and Puri.

seems that there is a slight warming trend of mean maximum and minimum temperature, especially in the hottest month.

On the other hand, some districts of Orissa state experienced unusual severe high temperatures in 1998, where 954 people died, the worst tolls in all of India (IDR team, 2000). As well, in 2004, strong sunstroke prevailed in 16 districts of Orissa state, from April to June, causing the deaths of nine people in the Chilika Lagoon area (SRORD, 2005).

Cyclones

Chilika Lagoon, which covers Khurda, Puri, and Ganjam districts, is classified as a "very high damage risk zone (50 m/s)" in terms of wind and cyclones. The lagoon is situated along the Bay of Bengal, which is one of the six major cyclone-prone regions of the world. The annual cyclone season along the Bay of Bengal normally occurs in the months of April to May and October to November. Increasing warm sea temperatures (approximately 26.5°C) in the Bay of Bengal induce cyclones to become more intense as they get close to the coastal areas of India (Gupta & Sharma, 2000). On an average, every year, four to five cyclones strike the coastal regions of India (Shanmugasundaram, Arunachalam, Gomathinayagam, & Lakshnaman, 2000). Coastal districts of Orissa state have been hit by 11 severe cyclones and 55 cyclone storms, with a probable maximum storm surge height between 3.2 and 5.5 m in the last 120 years (Gupta & Sharma, 2000).

Importantly, in 1999, Orissa state experienced a "super cyclone," its worst catastrophic cyclone in 100 years. The super cyclone swept the entire Orissa coastal area, affecting 15,681,072 people, 14,586 villages, causing 9,893 deaths, and damaging 1,661,683 houses (*ibid.*). Fortunately, Chilika Lagoon avoided much of the possible loss of human life, though the cyclone caused tremendous damage to fishing equipment and houses in and around the lagoon. Cyclones have a short-term as well as long-term impact on fishery livelihoods in terms of damages to vital infrastructures (schools, hospitals, road networks), and create less access to medical supports, water use, and official loan provision. Those losses were evident in the Chilika Lagoon after the 1999 super cyclone, which led to traumatic perceptions for fishing communities.

Floods and Droughts

In addition to cyclones, floods are common in the Chilika Lagoon. The majority of the annual rainfall occurs from June to November (see Fig. 4).

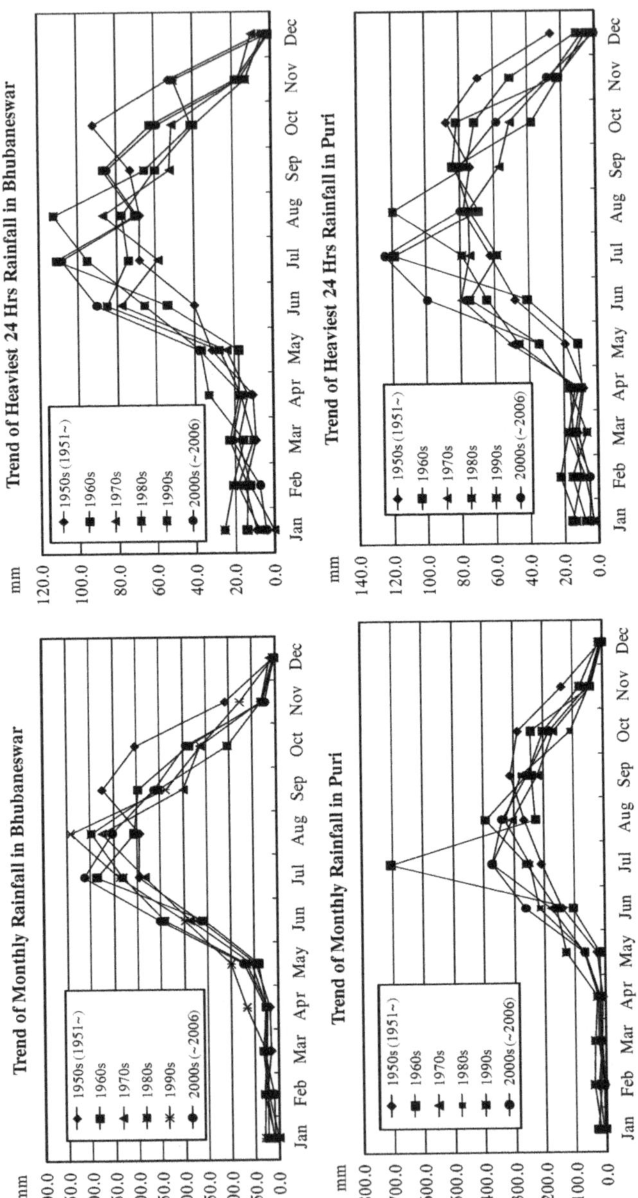

Fig. 4. Trend of Monthly and Heaviest 24 Hours Rainfall in Bhubaneswar and Puri.

The southwest monsoon brings much rainfall from June to September, while the northeast monsoon brings some rain in October and November. The highly concentrated rainfall in July and August provoked flooding of the major streams. Chilika Lagoon receives fresh water from 52 channels, especially two principal streams (that is, Daya and Bhargavi Rivers), which are subbranches of the Mahanadi River system that is prone to be affected by extreme floods.

From 1834 to 2007, 76 flood events were recorded in Orissa state (OSDMA, 2007). Although flood control efforts have been completed by the Orissa state government (for example, construction of dams and weirs), 11 floods were recorded during the past quarter century (1988 to 2007), while 32 floods were recorded during the century from 1834 to 1933. The report indicates that vulnerability has been slightly increasing for decades, both in terms of frequency and intensity of floods. Importantly, flood events have been continuously recorded in Orissa state since 2003. In terms of flood, the fishermen tend to suffer losses or damages of fishing boats and materials. There is a need to promote a coordinated flood response in which people move their boats and nets to a secure location.

Droughts are also of concern for the people's livelihoods. It is likely that average monthly rainfall near Chilika Lagoon is significantly decreased except during the monsoon season (see Fig. 3). In the last decade (1995 to 2004), there have been six moderate and severe droughts in various parts of Orissa state (SRORD, 2004, 2005). Although fishing communities are less vulnerable to droughts than agrarian communities, the former are faced with some difficulties in having an access to adequate water for drinking, cooking, and bathing, and available water is often of low quality. The water scarcity and pollution combined with anthropogenic pressures endanger the health of the people and interfere with every aspect of life.

Risk Perception of Climate Change

Although various exposures to climate changes in Chilika Lagoon have been reviewed through secondary data and research papers, people's perception is also of importance in considering their behavior and responsive actions (Grothmann & Patt, 2005). According to the results of the questionnaire survey, many people, irrespective of inland versus coastal areas, perceived the phenomenon of climate change in the form of increasing drought (65 percent) and high temperatures (78 percent) in Chilika Lagoon. Likewise, most of them (72 percent) believe that a catastrophic cyclone

Table 2. Ranking of Risk Perception of Climatic Change Hazards.

	Cyclones	Floods	Drought	High Temperature
Inland	1st (3.7)	2nd (3.5)	4th (3.0)	3rd (3.1)
Coast	1st (3.2)	2nd (2.8)	4th (2.6)	3rd (2.7)
Total	1st (3.5)	2nd (3.1)	4th (2.8)	3rd (2.9)

Note: The ranking is based on the total numbers of risk perception. Value in brackets shows the average number of risk perception.

like the 1999 super cyclone will occur in Chilika Lagoon in the near future. On the other hand, inland versus coastal people have different perceptions of climate change, especially in the form of floods. Seventy percent of people in inland areas tend to feel that flood frequency and intensity are the same as they used to be. However, 62 percent of those who live in coastal areas feel the variability has increased.

As for climate exposures and sensitivity, there is a consensus about the risk perception between the two areas (see Table 2). Most of the people ranked cyclones as the most prevalent climate hazard. Each climate hazard (cyclones, floods, droughts, and high temperatures) was regarded as risky or most risky. In this respect, it is worth noting that the people in inland areas seemed to be more aware of climate variability and extreme events than the people in coastal areas. According to the informant interviews, differences can be related to the super cyclone of 1999. The super cyclone affected Chilika Lagoon, though the cyclone did not strike there directly. In contrast, fishermen in inland areas incurred heavy losses from the cyclone because they went to fish in Chilika Lagoon. Before the cyclone made landfall, the government issued numerous early warnings. Despite that, people in inland areas did not take the warnings seriously. They did not believe that the hazardous cyclone would strike Chilika Lagoon, nor did they believe the predictions of damage to human health and boats. On the other hand, coastal communities bore the brunt of the cyclone to a great extent because the fishermen chose not to fish in those areas. Responses between people in these two areas were reflected in their perception of vulnerability; the vulnerable people in coastal areas were identified as disabled people, followed by women, fishers, and children. In contrast, people in inland areas who were most vulnerable were fishers, followed by women, children, and disabled people. Also, due to the proximity to seawater, the coastal communities may have more experience with related impacts rather than the inland communities. The different experiences between the two areas likely changed their perceptions of risk due to climatic change hazards.

Livelihood Impacts

This section aims to assess livelihood conditions in fishing communities and link them with climate change in local context. Specially, the research focuses on fishing activities in terms of environmental, economic, human, and physical aspects, addressing both the negative and positive.

Environmental Impacts

Chilika Lagoon has experienced severe environmental degradation, primarily from siltation, since the 1980s. In combination with anthropogenic pressures such as deforestation, livestock overgrazing, and industrialization, seasonal climate forces such as cyclones and floods brought a lot of silt into the lagoon. Silt promotes agglutination and precipitation reactions when it flows into salty water, leading to silt accumulation in the lagoon area, especially near the sea mouth. The silt accumulation reduces the water spread area and hinders the exchange of water between the sea and river. The closure of the sea mouth affects the salinity level in the lagoon and prevents exchange of migratory fish species, which account for nearly 80 percent of lagoon fisheries catch (Pattnaik, 2005). Siltation into the lagoon also encourages prolific growth of freshwater invasive species. The area of Chilika Lagoon dominated by invasive plants increased from 20 km^2 in 1972 to 685 km^2 in May 2000 (CDA, 2005a). These changes reduce the area of fishing grounds in Chilika Lagoon. That, in turn, leads to loss of income that can render fishers more vulnerable to climate change. Moreover, the weed invasion presents physical difficulties for boat navigation, further undermining people's adaptive capacity to climate change.

Economic Impacts

The livelihoods of fishing communities largely depend on the fisheries from Chilika Lagoon; the impacts against changes in climate render them more to the brink. Most of the people (97 percent) have no secondary occupation, so fishing is a sole income source for them. On the whole, members of fishing boats in Chilika Lagoon are limited to family or neighbors, and their profits are equally distributed among them. Household incomes per month in inland and coastal areas are Indian Rupee (INR) 1,190 (approximately US$30) and INR 1,332 (approximately $34) in rainy season, versus INR 2,294 (approximately $58) and INR 2,564 (approximately $65) in the dry season, respectively.[1] The income is not sufficient to cover their living expenses, especially during the rainy season, because they cannot often go fishing due to extreme rainfall and winds. In fact, 90 percent of respondents borrowed money from fish merchants at an average of INR 29,483, with no interest. The loan provisions are addressed by purchasing or repairing fishing gears, instead of informal fish trade promises at lower price. As a result, the dependency pushes households to the brink and can lead to nearly defunct fishery cooperative societies in terms of fish marketing (Iwasaki & Shaw, 2008). Due to a lack of money, around 50 percent of households have housing that is highly vulnerable to climate hazards. On this account, expenditures for food and fishing gears were regarded as the highest priorities whereas electric products and savings were identified as the lowest ones.

Human Impacts

Changes in climate may endanger health conditions. According to problem tree analysis on climate change, all fishing communities commonly identified disease issues affected by climate change. In particular, the local people were vulnerable to increasing temperatures and water pollution. Due to increasing temperatures, the local people often developed eye inflammation (48 percent). Likewise, water pollution affected their health in various ways. In all targeted communities, there is no tap water available. Water supply depended upon three types of well waters (pipe well, tube well, and dug well) and ponds. In response to climate changes and anthropogenic pressures, these drinking water sources have become scarce and muddy. Both quantitative and qualitative changes in water resources increased people's exposure to various diseases such as cholera (17 percent), diarrhea (26 percent), and viruses (35 percent). On the other hand, there are no sanitary facilities in fishing communities so that once heavy rainfall occurs, people are forced to move to public spaces or even within villages for toilets, a situation susceptible to social disputes among villagers or among neighbor villagers. Given that they have less access to medical support due to remote distance and lack of money, their low health resistance may undermine their fishing activities and livelihoods.

Related to education, around 25 percent of people over 15 years (N = 410) had no educational experience whereas only around 5 percent could enter college or university. Most of the people (59 percent) dropped out before high school. On this account, only 55 percent of people can read and write their native language (Oriya), resulting in various constraints on alternative jobs, health care, and formal credit taking that are largely linked to enhancement of other capital assets. Lower education limits alternative jobs except fishing and related industries so that there is an increasing trend of fishing population in Chilika Lagoon. It creates high competition over limited fishery resources and triggers disputes among fishers. More importantly, children who have started fishing can be commonly observed in Chilika Lagoon. They are more vulnerable to climate variability and extreme events than the adults, but there is no choice for them but to engage in the fisheries. The root cause for that was mainly attributed to lack of money, but also linked to affairs of formal schooling. According to the problem tree analysis, some people in coastal areas tend to be reluctant to go to high school because of caste discrimination. High schools are located in nonfishing villages where the sense of caste discrimination remains, leading to one or more obstacles for entering high school.

Physical Impacts

Natural calamities in combination with climate change often result in a great degree of devastation to basic infrastructures and physical goods, as well as a large number of casualties. The physical environment surrounding fishing communities is often very poor. The fishermen are forced to make use of unreliable boat jetties attached only to raised ground. Villagers stress that the catastrophic cyclones and floods damage physical capital, including infrastructures (e.g., community roads and fish landing centers) and physical goods (e.g., boats, nets, and houses). Once such climate hazards occur, community roads are easily damaged by fallen trees. At the same time, poor housing conditions are affected to a great extent, and unsanitary environments cause various diseases.

Among physical capital assets, villagers identified fishing boats as the most seriously vulnerable to climate change (76 percent). Forty-eight percent reported loss of boats. The loss leads to suspension of fishing activities, resulting in danger to family maintenance. Some of them may attempt to apply for loan financing from fish merchants.

Table 3. Existing Adaptive Strategies for Climate Change Impacts in Chilika Lagoon Fisheries.

Indicator	Major Problems Related to Climate Change	Root Causes	Adaptive Strategies
Environmental impacts	– Decrease of fish species – Decrease of fishing grounds – Boat navigation trouble	– Soil erosion (deforestation from upstream) – Closure of sea mouth – Proliferation of freshwater weeds	– Hydrological interventions – Involvement of stakeholders for watershed management
Economic impacts	– Lack of money – Indebtedness	– Decrease of fish species – Exploitative marketing structure	– Strategic loan provision for fishing equipments
Human impacts	– Dropouts from schooling	– Less fishery income – Caste discrimination in schooling	– Eradication of racial discrimination – Enhancing disaster awareness
Physical impacts	– Poor infrastructure – Loss or damages of fishing equipments	– Lack of political power – Lack of unity for disaster prevention	– Promotion of disaster insurance scheme

RESPONDING TO CLIMATE CHANGE IMPACTS

On the basis of the previous discussions, existing adaptive strategies for climate change in Chilika Lagoon fisheries were presented in Table 3. The strategies can be compiled into two components with due consideration of positive and negative aspects: climate-resilient fishing communities and integrated lagoon watershed management.

ENHANCING CLIMATE-RESILIENT FISHING COMMUNITIES

Practicing fisheries are very prone to be physically affected by various types of climate variability and subsequent environmental problems. Tremendous natural hazards that frequently occur in Chilika Lagoon cause serious damages to fishing communities. In particular, people such as the aged, the disabled, women, and children are vulnerable. Despite the early ages, for instance, children forced to start fishing are being exposed to climate hazards such as strong sunshine and cold winds. There is no choice for them but to engage in the fisheries; the entrance to schooling is disallowed mainly due to lack of financial resources. In this respect, the fishermen previously had opportunities for loan finance from private and government banks, though they failed to repay their loan amount to the banking institutions partly due to the prevailing psychology of "loan waiver"; debtors could easily escape the duty of loan repayment from the lenders while continuing to lie about who were debtors. The failures induced a high dependency on fish merchants (Samal & Meher, 2003), which undermined the fishers' capacity to adapt to climate change in terms of proper marketing activities and saving. In this respect, it is interesting to note an innovative approach developed by South Indian Federation of Fishermen Society (SIFFS). SIFFS provided loan finance from banks to a large number of fishing communities in southern India after appropriate capacity assessment. Financing decisions were made on not only capacity assessment but also on the introduction of automatic deduction of the loan repayment from payment through cooperative fish marketing. SIFFS encouraged the fishermen to join the cooperative marketing activities for economic improvement as well as for a solution of the "loan waiver" psychology, after due consultation by NGOs and researchers. These efforts may offer one way to improve their livelihoods and make sure that their children are able to go to school and stay in a safe environment. In addition, another effort needs to be made to eradicate caste discrimination in formal schooling. Discrimination induces school children to join the fisheries in spite of that lifestyle being physically vulnerable to climate hazards. Therefore, a strategic loan financial scheme for fishing gears and education expense, and deliberate discussions on casteism among various stakeholders is of high importance for climate change adaptations.

In addition, their physical capital is so low that it will be easily damaged or lost when disastrous events happen in Chilika Lagoon. On this account, promotion of natural calamity insurance supported by Department of Fisheries and Animal Resources Development has been implemented. The insurance scheme enables all people, including the poor, to mitigate sudden shocks from climate variability and extreme events. Taking into account the increased intensity and frequency of natural disasters in Orissa state, the initiatives will play a greater role in responding to the direct and indirect effects of climate change in the long period. There is, of course, negative aspects to be considered. People in inland areas did not take the government's early warnings seriously when the super cyclone of 1999 made landfall. It led to tremendous damages to humans and physical aspects. Furthermore, irrespective of inland versus coastal areas, many people do not tend to adjust themselves to the changes in climate, even though extreme catastrophic climate events such as cyclones and floods attack their private and public goods, including fishing materials. Hence, enhancing disaster awareness and the sense of unity among the local people is recommended to mitigate the impacts of climate change.

TOWARD INTEGRATED LAGOON WATERSHED MANAGEMENT

In the debates between fishery livelihood and climate change, the fishery resource is the core point to be addressed. The resource is the principal form of livelihood for survival and affects lives in various ways. However, the coastal zone that includes Chilika Lagoon is considered to be a vulnerable water ecosystem. It is affected by the confluence of various aspects from upstream to sea. In the case study of Chilika Lagoon, siltation (soil erosion) is identified as among the most serious environmental problems, in combination with climate variability such as extreme floods and cyclones. Exposure to silt accumulation reduces the water spread area and hinders the exchange of water between the sea and river, resulting in decreased salinity levels and subsequent prolific growth of freshwater invasive species. As a result, the Fish landing quantities in Chilika Lagoon rapidly decreased until 2000, thereby leading to the poorest people not being effectively able to adapt to climate variability and extreme events. Furthermore, the weed invasion obstructs passages from boat jetties to fishing grounds, sometimes leading to boat crew clashes and the subsequent disputes among fishers.

In an effort to maintain (and preferably improve) fish stocks, there is a need to integrate fishery resources into a watershed perspective. In this respect, Chilika Development Authority (CDA), which was established in 1991, has conducted a successful hydrological intervention (sea mouth management) in the short run, while adopting participatory micro-watershed management at the watershed level in the long run.

In the hydrological intervention, CDA implemented a successful wetland restoration project in September 2000. The authority enforced innovative hydrological interventions (opening of a new mouth and dredging water channels) in collaboration with research institutions, NGOs, and the local

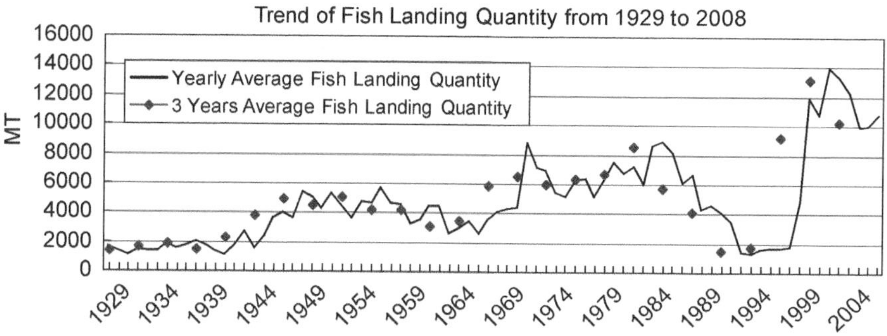

Fig 5. Trend of Fish Landing Quantity from 1929 to 2008. *Source:* Revised from DFGO, 1970; DFGO and CDA, 2005; CDA data.

people. The hydrological interventions resulted in the dramatic improvement of the natural and social lagoon environment (CDA, 2005a). Fish production remarkably improved (see Fig. 5). At maximum, the amounts of fish landings were recorded to be around 14,053 metric tons in 2003–2004, compared to around 1,600 metric tons prior to the interventions. Related to this, the government estimated that fisher family income increased around ten times in 2003–2004, compared to before interventions in 2000 (DFGO and CDA, 2005).

It should, of course, be kept in mind that the technical interventions could be just temporal end-of-pipe measures to cope with siltation. Seasonal climate forces such as cyclones and floods continuously bring a lot of silt into the lagoon. Consequently, there has been a gradual decrease of fish landing quantities after the hydrological implementations (see Fig. 5). In response to this situation, CDA then adopted a concept of participatory micro-watershed management in 2001, which seeks to involve the local people in the project, to mitigate the impacts of siltation caused by extreme climate events combined with anthropogenic pressures from upstream. The idea behind watershed management was largely linked to a new initiative implemented by an externally aided Western Orissa Rural Livelihoods Project (WORLP), funded by the DFID in 2000. Unlike the past approach in Orissa state, the project aims to focus more on livelihood development at a community level, in addition to area development for watershed governance, by introducing an additional component called "watershed plus" (Reddy, 2009). Through watershed plus, exclusive funds for the poorest people in project sites are allocated for their livelihoods and well-being. The objective is meant to improve livelihood for activities that are not land-based (the landless, the poorest, and so on), as well as land-based activities (land owners). Even in cases where project implementations have been satisfactory, these achievements have not been sustained in terms of operation and maintenance of assets created and resources developed due to inadequate participation by concerned communities and user groups (CDA, 2005b). The recognition led to a creation of institutional arrangement at watershed level where watershed associations are registered under the Societies Registration Act, 1860, thus giving the groups community-based and community-owned institutions (Reddy, 2009). Under the circumstances, CDA adopted the watershed management that takes into account involvement of all concerned villagers, including the poor, and created watershed association at a bottom-up approach for integrated soil and water conservation in a sustainable manner. These changes at the watershed level are expected to play an important role in reducing silt accumulation in the lagoon, leading to sustainable lagoon fisheries. In other words, practices regarded as integrated lagoon watershed management with community participation help develop a pathway for strengthening adaptive capacity to climate change as well as sustaining fishery resources. There is a further need to build multilevel collaborations and partnerships such as governments, NGOs, researchers, and local people, including targeting upstream areas as CDA did.

CONCLUSIONS

Regional changes in climate have been observed in many parts of the world. They pose significant risks to people's livelihoods, especially in rural areas where most of them depend on rather climate-sensitive resources. Given that such occupations are subject to strong restriction from nature's forces, combined with climate variability, they are largely required to adjust their livelihoods to these climate factors.

It is important to note that the impacts of climate change cover not just occupational activities but also multifaceted aspects of rural livelihoods that determine the extent of their capacity to adapt to changes in climate. The range of expected fishery strategies covers all encompassing aspects including fishing, marketing, schooling, social cohesion, and environmental and disaster awareness. Increase of fish species, for instance, meets an important requirement of capability for climate-resilient communities but is not a necessary precondition for such communities. Dominance of fish marketing by fish merchants undermines fishermen's capacity to adapt to climate change, and access to fishing grounds also affects them to a great extent. These vulnerabilities can trigger school dropouts and cause younger people to enter the fishing industry, plaguing human lives, livelihoods, and dignity. In this way, climate change does not occur independently of other processes impacting fishing communities, so that there is an urgent challenge in linking the change with developmental processes (Thomas & Twyman, 2005). Each vulnerability factor among their livelihood assets influences the extent of their capacity to adapt to climate change. Therefore, a holistic approach, so-called integrated climate change adaptation (ICCA), that takes into account the multifaceted aspects of rural livelihoods needs to be developed.

With regard to this, expected fishery strategies for ICCA in Chilika Lagoon were presented in Table 3. On the whole, enhancing climate-resilient fishing communities is crucial for adapting to changes in climate variability. The livelihoods of fishermen that encompass numerous patterns of internal relationships among the stakeholders and external effects of fishery governance, market fluctuations, immigration, development of science and technology, and so on, are indirectly tied to their resilience against climate hazards. Apart from the view of capacity building, the perspective of integrated lagoon watershed management plays a role commensurate with dynamic and complex characters of so-called "eco-tone" management. It is important to note that a community-based approach undertaken in the watershed development programs is considered effective to mainstream lagoon fisheries management into a comprehensive approach to ICCA. Such expected strategies undertaken by Chilika Development Authority and associated institutions, including community-based watershed associations, have just begun, so further research is called for understanding the process of these attempts toward integrated climate change adaptation in Chilika Lagoon fisheries.

NOTE

1. An exchange rate of US$1 = INR 39.25 is used throughout this study.

ACKNOWLEDGMENTS

The authors acknowledge the support of the Japan Society for the Promotion of Science (JSPS) grant. The first author is extremely grateful to Ajit Kumar Pattnaik, Durga Prasad Dash, Sudarshan Panda, and Krupasindhu Bhatta for their wholehearted support offered.

REFERENCES

Anthony, A., Atwood, J., August, P., Byron, C., Cobb, S., Foster, C., Fry, C., Gold, A., Hagos, K., Heffner, L., Kellogg, O. D., Lellis-Dibble, K., Opaluch, J. J., Oviatt, C., Pfeiffer-Herbert, A., Rohr, N., Smith, L., Smythe, T., Swift, J., & Vinhateiro, N. (2009). Coastal lagoons and climate change: Ecological and social ramifications in U.S. Atlantic and Gulf Coast ecosystems. *Ecology and Society*, *14*(1), 8Available at http://www.ecologyandsociety.org/vol14/iss1/art8 (accessed on September 30, 2009).

Barnett, J. (2003). Security and climate change. *Global Environmental Change*, *13*, 7–17.

Bernett, J., & Adger, N. W. (2007). Climate change, human security and violent conflict. *Political Geography*, *26*, 639–655.

CDA. (2005a). *Achievement report 2005*. Bhubaneswar, India: Chilika Development Authority.

CDA. (2005b). *A brief resume of the participatory watershed management project in the Dengai Pahad micro watershed – Chilika catchment*. Bhubaneswar, India: Chilika Development Authority.

CDA. (2010). *About Chilika: Climate*. Available at http://www.chilika.com/climate.htm (accessed on October 10, 2010).

DFGO. (1970). *Tha Chilka Lake*. Cuttack, India: Directorate of Fisheries, Government of Orissa.

DFGO, CDA. (Eds). (2005). *Collection and estimation of fish, prawn and crab landings statistics in the Chilika Lagoon, Annual Report – 2003–04*. A collaborative programme of Department of Fisheries, Government of Orissa and Chilika Development Authority, Bhubaneswar, India.

EM-DAT: The OFDA/CRED international disaster database, Université catholique de Louvain, Brussels. (2010). Trends and relationships for the period 1900–2005. Available at http://www.emdat.be/Database/Trends/trends.html (accessed on April 1, 2010).

FAO. (2009). *The state of world fisheries and aquaculture 2008*. Rome, Italy: Food and Agriculture Organization.

Grothmann, T., & Patt, A. (2005). Adaptive capacity and human cognition: The process of individual adaptation to climate change. *Global Environmental Change*, *15*, 199–213.

Gupta, M. C., & Sharma, V. K. (2000). *Orissa super cyclone 99* (p. 122). Delhi, India: National Centre for Disaster Management Publication.

IDR team. (2000). Heat and cold wave. In: S. Parasuraman & P. V. Unnikrishnan (Eds), *India disasters report towards a policy initiative*. New Delhi, India: Oxford University Press.

IPCC. (2007). Climate change 2007: The physical science basis: Summary for policymakers. Available at http://news.bbc.co.uk/2/shared/bsp/hi/pdfs/02_02_07_climatereport.pdf (accessed on September 30, 2008).

Iwasaki, S., & Shaw, R. (2008). Fishery resource management in Chilika Lagoon: A study on coastal conservation in the eastern coast of India. *Journal of Coastal Conservation*, *12*, 43–52.

Mitra, N. G., & Mahapatra, P. (1957). *Bulletin on the development on the Chilika Lake of Orissa survey report on the fishing industry*. Cuttack, India: Orissa Government Press.

Ogata, S., & Sen, A. (2003). *Human security now, commission of human security*. New York: Commission on Human Security.

OSDMA. (2007). *History of floods in Orissa*. Bhubaneswar, India: Orissa State Disaster Mitigation Authority.

Patro, N.S. (2001). New challenges to Chilika. In: Indian Environmental Society (Ed.), *Proceedings of the 5th Asia pacific NGOs environmental conference Apnec-5*, September 22–25, 2000, Agra, India.

Pattnaik, K.A. (2002). Chilika Lake: An overview. In: Chilika Development Authority and Department of Water Resources (Orissa) (Eds.), *Proceedings of the international workshop in sustainable development of Chilika Lagoon*, December 12–14, 1998, Bhubaneswar, India.

Pattnaik, K. A. (2005). *Impact of unauthorised shrimp culture on fishery resources of Chilika Lagoon*. Bhubaneswar, India: Chilika Development Authority.

Reddy, B. G. (2009). Institutional delivery for watershed management in Orissa. In: S. K. S. Premchander & P. Reid (Eds), *Finding pathways: Social inclusion in rural development*. Bangalore, India: Books for Change.

Samal, C. K., & Meher, S. (2003). Cooperative societies of fishermen of Chilika Lake: Problems and prospects. *Journal of Rural Development, 22*(4), 517–525.

Shanmugasundaram, J., Arunachalam, S., Gomathinayagam, S., & Lakshnaman, P. (2000). Cyclone damage to buildings and structures: A case study. *Journal of Wind Engineering, 84*, 369–380.

Shaw, R. (2007). Community-based climate change adaptation in Vietnam: Inter-linkages of environment, disaster, and human security. In: S. Sonak (Ed.), *Multiple dimension of global environmental changes*. New Delhi, India: TERI Publication.

Smit, B., & Wandel, J. (2006). Adaptation, adaptive capacity and vulnerability. *Global Environmental Change, 16*, 282–292.

SRORD. (2004). *Annual report on natural calamities 2003–2004*. Bhubaneswar, India: Special Relief Commissioner, Revenue Department.

SRORD. (2005). *Annual report on natural calamities 2004–2005*. Bhubaneswar, India: Special Relief Commissioner, Revenue Department.

Thomas, S. G. D., & Twyman, C. (2005). Equity and justice in climate change adaptation amongst natural resource-dependent societies. *Global Environmental Change, 15*, 115–124.

United Nations International Strategy for Disaster Reduction (UNISDR). (2009). 2009 UNISDR terminology on disaster risk reduction. Geneva: The United Nations International Strategy for Disaster Reduction.

Van Aalst, K. M., Cannon, T., & Burton, I. (2008). Community-level adaptation to climate change: The potential role of participatory community risk assessment. *Global Environmental Change, 18*, 165–179.

WWF. (2005) Are we putting our fishy in hot water? Available at http://assets.panda.org/downloads/fisherie_web_final.pdf (accessed on September 30, 2009).

CHAPTER 6

CLIMATE CHANGE ADAPTATION AND DISASTER RISK REDUCTION EXPERIENCES IN JAPAN

Yukiko Takeuchi and Rajib Shaw

INTRODUCTION

Japan is an island arc that sits in the monsoon region, and is under the influence of warm and moist air masses in summer and cool air masses in winter. The moisture that is taken in the lower leaves of the air masses over the sea is poured on the country by typhoons in summer, by snowfall in winter, by the "Bai-u Front" (in Japanese) in June and July, and by depressions and fronts in all seasons. Owing to Japan's slender shape and complicated landform, aerial differences in climate are great. Japan is located on the eastern edge of the monsoonal region of Asia, and its climate varies according to seasonal and regional conditions. Typically, heavy rains occur in various parts of the country, both during the rainy season in June and July and during the typhoon season from August to October. This precipitation is predominantly in the form of locally specific temporary downpours. In winter, the northern part of the country usually receives heavy snowfall that causes prolonged floods in spring from the melting of snow. The average amount of precipitation is 1,800 mm (70 inches) a year. This is two or three times the amount received in other areas of the same latitude. In the southern Pacific coast areas, rainfall amounts to 4,000 mm

Climate Change Adaptation and Disaster Risk Reduction: An Asian Perspective
Community, Environment and Disaster Risk Management, Volume 5, 105–125

ISSN: 2040-7262/doi:10.1108/S2040-7262(2010)0000005012

(160 inches). Precipitation in Tokyo is twice as much as other large cities in western countries. Some 50–60% of the annual precipitation in the Pacific coast of Japan is concentrated from June to October. Artificial changes in natural environments are rapid and large, accompanying the great increase in economic activity and exploitation (Nakano, Kadomura, Mizutani, Okuda, & Sekiguchi, 1974). Although the country's 10% of land area is flood prone, about 50% of the population lives in floodplains and almost 75% of the property is concentrated in the floodplains (JWF, 2006).

Owing to these meteorological characteristics, flood disasters caused by heavy rains occur frequently in Japan. Seventy percent of the land is mountainous, with generally high relief. Chains of spinal mountains, some reaching 3,000 m (1,000 feet) above sea level, run through the center of the narrow and long country. Consequently, the rivers are generally short with steep gradient. Erosion and devastation in the mountain areas are very rapid. Rivers have flash flooding with high concentrated peak discharges soon after an intense rainfall. The remaining 30% of the land is mostly alluvial plains where housing, farming, and industries are densely concentrated, consequently increasing the vulnerability to flood disasters. From a socioeconomic perspective, Japan has achieved rapid socioeconomic development since the end of World War II: Its population has increased by 50% (from 83 million in 1950 to 127 million in 2000), and the gross domestic product has increased around tenfold (from ¥48 trillion in 1955 to ¥487 trillion in 2000, equivalent to US$44 trillion). The majority of the population lives in densely populated areas in downstream alluvial plains, forming megacities such as Tokyo and Osaka. Highly valued assets are also concentrated in these urbanized areas. Therefore, Japan inevitably suffers serious socioeconomic damage when flood disasters occur.

DISASTER RISK REDUCTION APPROACHES IN JAPAN

The immense damage Typhoon Isewan caused in 1959 was a turning point for disaster management that led to the planning and preparation of a comprehensive disaster management system, and in 1961, the Disaster Countermeasures Basic Act was passed. Therefore, the disaster management system has been improved and strengthened following the occurrence of large natural disasters and accidents. The Disaster Countermeasures Basic Act is the basis for disaster management in Japan, and the expectation for effective disaster management is that the government, the local government, and designated public corporations are expected to work out disaster management plans and carry them out appropriately. Following are the main contents of the act:

1. Definition of jurisdictions and responsibility for disaster management
2. Disaster management system
3. Disaster management plan
4. Disaster preparedness
5. Disaster emergency response
6. Disaster recovery
7. Financial measures
8. State of emergency

The Central Disaster Management Council was established for the purpose of promoting comprehensive countermeasures in which the prime minister takes the chair and other ministers of state are members. Following are the duties of the Central Disaster Management Council:

1. Prepare and promote implementation of the basic disaster management plan and draft the earthquake disaster management plan.
2. Prepare and promote implementation of the urgent measures plan for major disasters.
3. Deliberate important matters pertinent to disaster management according to requests from the prime minster and/or minister of state.
4. Offer opinions regarding important matters to disaster management to the prime minister and minister of state for disaster management.

Central Disaster Management Council's members include the prime minister, all cabinet ministers, chief of the designated public corporations (governor of the Bank of Japan, president of Japan Red Cross Society, president of Japan Broadcasting Corporation, and president of Nippon Telegraph and Telephone Corporation), and four people of experience or academic standing.

Every year, natural disasters affect people's lives and damage properties. Until the 1950s, numerous large-scale earthquakes and large typhoons claimed the lives of more than 1,000 people; for example, Typhoon Isewan claimed the lives of more than 5,000 people in a single event. However, due to the progress of countermeasures, such as the promotion of national land conservation projects, improvements in weather forecasting technologies, and completion of the disaster management system, the number of deaths and missing people is declining. However, in 1995, more than 6,400 lives were lost in the Great Hanshin-Awaji earthquake, and there is concern that an enormous ocean trench earthquake will probably occur in the Tokai region. Therefore, the menacing threat of natural disasters still lingers (Cabinet Office Government of Japan, 2002).

CLIMATE CHANGE IMPACTS TO NATURAL SYSTEMS IN JAPAN

The Japanese government established meteorological observatory offices in some parts of Japan, from the 1870s prior to the establishment of the Law of Meteorological Observation in 1886. Other meteorological observatory offices were established toward the end of the 19th century. These offices collect data on climate change in Japan. At present, there are 160 meteorological observatory stations. In addition, there are 1,300 stations with Automated Meteorological Data Acquisition System (AMeDAS), which automatically observes temperature and rainfall. The sections titled "Climate Change Impacts to Natural Systems in Japan" and "Climate Change Impacts on Japanese Society" explain the impact of climate change in Japan, based on the report of the Ministry of Education, Culture, Sports, Science, and Technology (MEXT et al., 2009).

Temperature

The temperature in Japan has been gradually increasing, with an average of 1.1°C increase between 1898 and 2008. Increased frequencies of high temperatures occurred in the 1990s (see Fig. 1). As a consequence of rising temperatures, the number of summer days increased, whereas the number of winter days decreased. Based on data predicted by computer simulation, the temperature in Japan will continue to increase, leading to average temperatures exceeding the annual world average and resulting in longer summer days and shorter winter days.

Rainfall

The fluctuations in annual precipitation are large but the increase or decrease shows no clear trend (see Fig. 2). Rainfall patterns of 100 mm and 200 mm per day indicate an increasing trend that may be attributed to the effects of global warming. Comparing the number of days of rainfall over 100 mm/day in the 30 years before the 20th century and the number of days

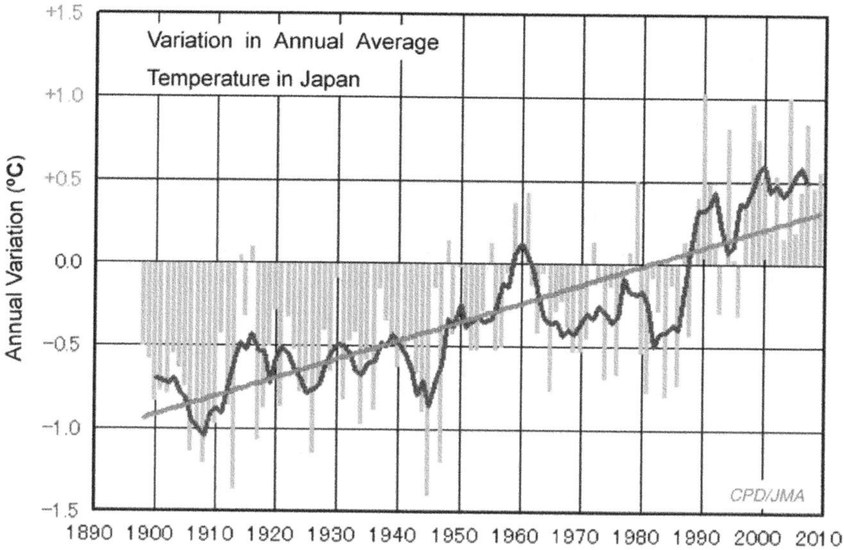

Fig. 1. Variation in Annual Average Temperature in Japan (1898–2008). *Source:* Data from Japan Meteorological Agency.

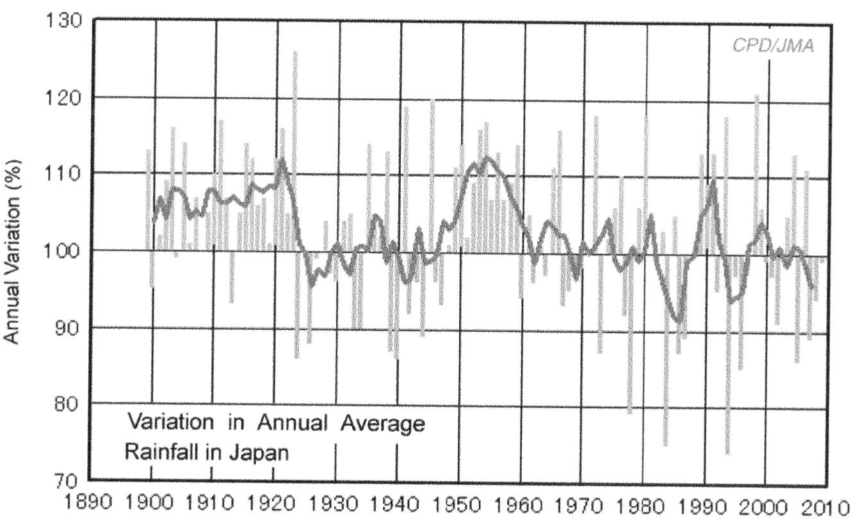

Fig. 2. Variation in Annual Average Rainfall in Japan (1898–2008). *Source:* Data from Japan Meteorological Agency.

of rainfall over 200 mm/day in the 30 years after the 20th century, an increase of 1.2 and 1.4 is observed in the number of days of rainfall over 100 mm/day and the number of days of rainfall over 200 mm/day, respectively. The IPCC AR4 reported that the frequency of heavy rainfall increased in the latter half of the 20th century. This observed increase in the days of heavy rainfall in Japan parallels with that of the IPCC forecast.

The amount of heavy rainfall at the end of the 21st century is also predicted to exceed that of the 20th century. This prediction considers some uncertainty and would need to be elaborated in the simulation results.

Typhoon

Typhoons often trigger flood and landslide events. In 2004, ten typhoons entered Japan's area of responsibility, making 2004 the year with the most typhoons. However, government records in the last 100 years did not show increasing nor decreasing trends in typhoons. Fig. 3 shows the number of

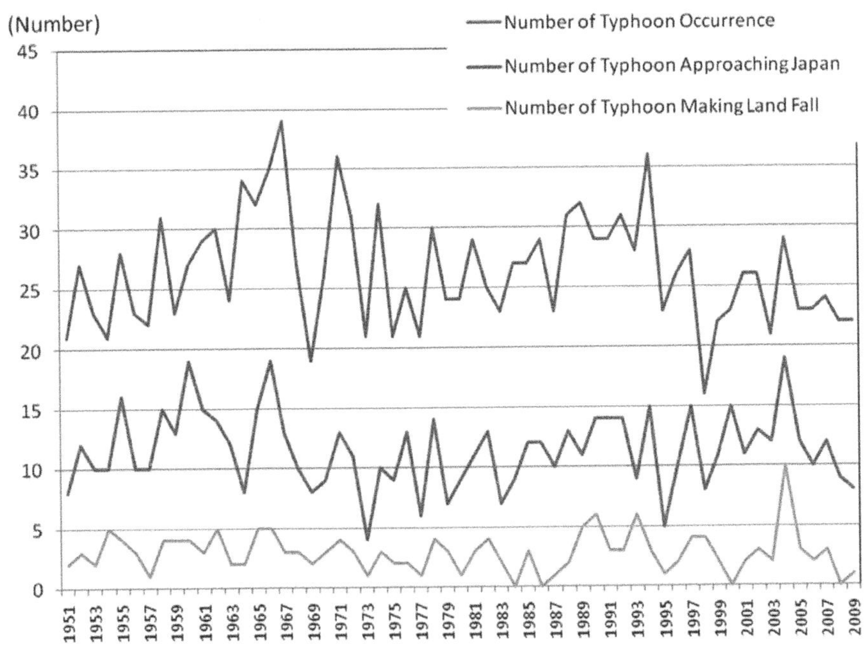

Fig. 3. Variation in Typhoon Occurrences in Japan (1951–2009). *Source:* Data from Japan Meteorological Agency.

typhoons that approached Japan, compared to those that made landfall. Through computer simulation and analysis, results indicate that the number of typhoon occurrences will decrease due to global warming. It is predicted, however, that the number of stronger typhoons with a maximum wind velocity of over 44 m/s will increase.

Sea Level Rise

In the last 100 years, the coastal areas of Japan did not show any tendency for abrupt sea level rise. From 1906 to 2008, sea level rose by 0.12 m on average, about two-thirds of the world average. It is predicted that the average global sea level rise will increase due to the effect of carbon dioxide emissions. However, from the analysis of data over a period of 100 years, there is no clear tendency of an increase in sea level although uncertainty was not considered.

CLIMATE CHANGE IMPACTS ON JAPANESE SOCIETY

Climate change will impact the social structure of Japan. The following discusses the implications of climate change in some sectors of Japanese society.

Water Environment and Resource

The change in the frequency and quantity of rainfall will have serious impacts on the water environment and resource. On one hand, the increase in heavy rainfall directly affects the level of flood risk. On the other hand, less rainfall directly affects the level of drought risk. The topography of Japan includes many mountains, steep rivers, and small plain areas. The distance between the mountain and the sea is very short, compared with other countries. Mountains have the capacity to hold water, but heavy rainfall often results in the release of stored water. With the frequent release of water, water shortage becomes a serious problem, especially after the rainy season.

Disasters

The impacts of climate change on natural disasters can be considered at two levels. The first pertains to the relationship between heavy rainfall and typhoons with flooding, landslides, and debris flow, among other things. The second concerns the direct impact of climate change on the rise in seawater level and the increase of strong typhoons affecting coastal areas. Many floods, landslides, debris flow, and other water-related disasters occur due to tidal waves or heavy rainfall. Current trends indicate that, in the future, floods will affect more areas and the rise in sea level will lead to lost sand dunes. As already predicted, the increase in the number of occurrences of floods, landslides, debris flow, and other disasters is directly proportional to the increase in the number of heavy rainfall and typhoon events.

Ecosystem

At present, the impacts of increased rainfall, climate change, and sea level rise on plant and animal life have been observed. Alpine plants have decreased in number, whitening of coral reefs has occurred, budding and wilting of flowers have hastened, and changing of leaf color has come late. It is expected that, in the future, these phenomena will increase.

Food Security

One positive impact of global warming on food security is that the productivity of farm products increases in direct proportion to the rise of the atmospheric carbon dioxide concentration. However, a sharp decrease in the quality of agricultural products has been reported. It is predicted that heavy rainfall and drought brought about by climate change will lead to a decrease in the quality and quantity of agricultural products. Moreover, the rise in seawater temperature will affect marine-related live stocks.

Health

Reports of an increase in heat-related illness and infectious disease–carrying insects have also been mentioned. Future predictions indicate an increase in the number of deaths caused by heat stress, especially among older age groups. In 2007, several cities in Japan recorded a number of patients suffering from heat injury. With regard to infectious diseases caused by insects, one prominent issue in 1950 was dengue, which affected the Tochigi prefecture and spread to the Tohoku area in 2000 (Kobayashi, Komagata, & Nihei, 2008).

Culture

Japan has four seasons in which people enjoy different activities and foods. This countrywide feature has led to a large sightseeing industry. It is important to consider agriculture and water resource in the varying seasons of Japan. Agriculture needs to be sustained to grow the varying and seasonal agricultural products and for water resources to be maintained for various water activities in the different seasons. Sports and leisure industries will be affected particularly by climate change. There are serious implications, especially to the winter sports industry, due to a decrease in snowfall. All these will have significant negative impacts on the important sightseeing industry in Japan.

CLIMATE CHANGE ADAPTATION APPROACHES IN JAPAN

It is important that the vulnerability of local areas be evaluated. Japan has sufficient scientific technology and social infrastructure for high adjustability. However, vulnerability occurs due to the impacts of floods and landslides brought about by typhoons, tidal waves, earthquakes, food insecurity, aging, and other natural disasters. This vulnerability increases as the impacts of climate change on social stability and security are experienced. Fig. 4 shows the adaptation framework of Japan's Ministry of Education, Culture, Sports, Science and Technology. An integrated evaluation of vulnerability needs to come from the viewpoint of the people. First, it is important to understand the present situation and develop effective adaptation measures. There is a need to prepare efficient policies and plans for adaptation to be effective. The policies and plans should consider different aspects of agriculture, environment, and society. It is important to include variables of uncertainty when developing these policies and plans.

Several forms of adaptation have been implemented in Japan. Adaptation measures on the water environment and water resource are done through preparation of emergency water supply systems and efforts to decrease the risk of water shortage. Adaptation measures for disaster events are developed through collection of meteorological data to study climate change impacts. The construction of new dikes and water-related infrastructure is undertaken to minimize flood disasters. Changes and plans for efficient land

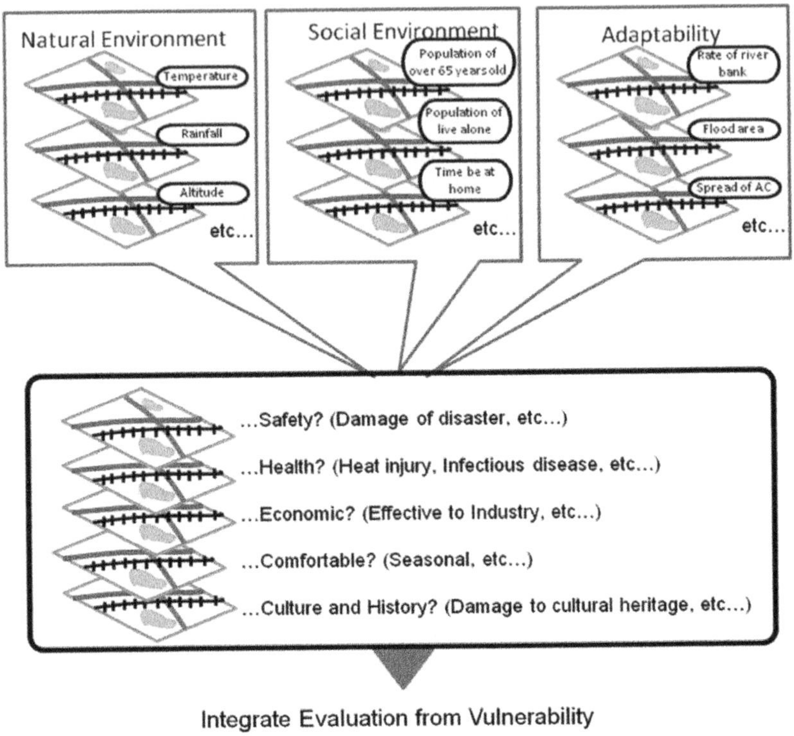

Fig. 4. Framework of Vulnerability Evaluation on Region. *Source:* MEXT et al., 2009.

use are being developed. Disaster risk management measures, such as early warning systems and recovery strategies, among others, are being implemented. Drainage systems are being developed as well as disaster risk management in the coastal area to address sea level rise. Adaptation measures relating to ecosystem management are done through monitoring of biodiversity and careful observation of instances of insect infestation. Adaptation measures on food security such as selective breeding are being done to make livestock and other farm products more resilient to high temperatures and strong sunlight. Residents' health is monitored to prevent heat injury and the spread of infectious disease. Increased plant life and regreening are being done, sometimes on building surfaces to increase green areas.

CASES FROM RECENT CLIMATE-RELATED DISASTERS

In the following section, a few cases will be described on the major recent climate-related disasters in Japan. Fig. 5 shows the locations of these recent disasters, which are distributed widely on the main island of Hokkaido and also on the Shikoku and Kyushu Islands. Table 1 shows the summary of seven recent climate-related disasters in Japan. Data shows that the main causes of damages were either typhoon or heavy rainfall, which is supported by the rainfall data (total or per day). Severe damages were observed in the housing sectors. Based on the analysis of the experiences of these seven climate-related disasters, we highlight the following issues.

Focus on Vulnerable Groups

The aging society is a big issue in Japan. The rate of aging society is 20.2%, based on 2005 data. The national government expects this to rise to 31.8% by 2030 (Ministry of Health, Labor and Welfare, 2009). This number is

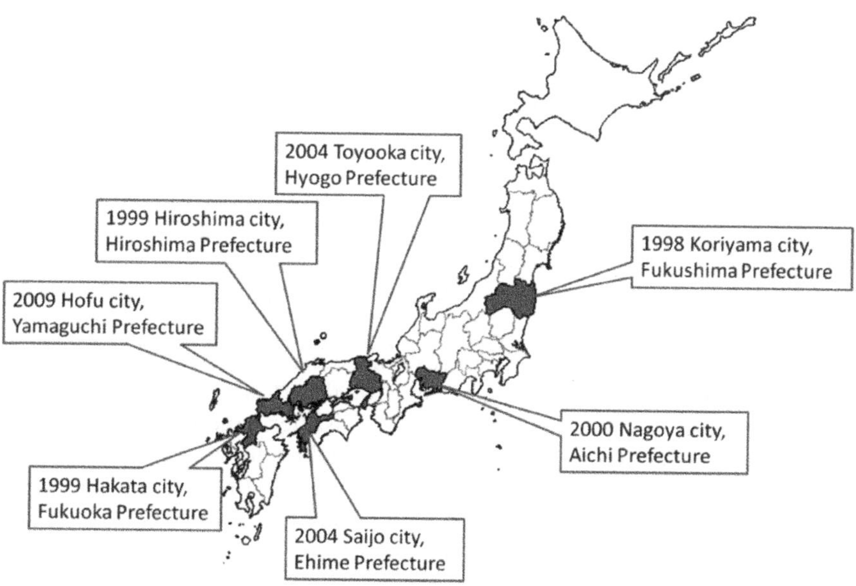

Fig. 5. Location Map of Case Study of Japan.

Table 1. Summary of Seven Recent Climate-Related Disasters in Japan.

Date	Place (Prefecture)	Causes	Total Amount of Rainfall/ Hourly Rainfall	Number of Dead and Missing People	Number of Houses Damage
Aug. 27–31, 1998	Fukushima Tochigi	Heavy rain fall	600 mm/ 24 hours	18	6,833
June 23–July 3, 1999	Fukuoka Hiroshima (+27 prefectures)	Heavy rain fall	1999 mm (Hiroshima)	40	19,315
Sep. 8–12, 2000	Aichi (+31 prefectures)	Heavy rain fall	589 mm/2 days	12	68,725
Sep. 25–30, 2004	Ehime (+31 prefectures)	Typhoon	412 mm	27	21,969
Oct. 20–21, 2004	Hyogo (+40 prefectures)	Typhoon	309 mm/day	92	73,603
July 19–26, 2009	Yamaguchi (+20 prefectures)	Heavy rain fall	549 mm	30	9,809

Source: Cabinet Office Disaster Databases (2010).

critical for disaster management for two reasons: One is that a person over 65 years old is not resilient during disasters, and another is that homes for the elderly increase every year (Ministry of Health, Labor and Welfare, 2010). The Japanese government has declared that disasters easily affect people who are handicapped and those who are less than six years old and over 65 years old. The growing aging society leads to low resilience of society in times of disaster. In the matter of housing, homes that care for the aged need wide spaces and large lands. However, the price of land is high in urban areas. Hence, facilities for the aged are often located in the mountains where land is cheap. In particular, homes for the aged are built on hills because they are relatively flatter. However, hilly lands are at high risk for impact by debris flow.

At the end of August 1998, Fukushima prefecture experienced serious heavy rainfall disaster (Inokuchi, 2001). In this disaster, five people who lived in a home for the aged died from a disaster caused by debris flow. In the same way, at the end of July 2009, debris flow disasters occurred in the Yamaguchi prefecture. Seven people died in this disaster (Ministry of Land, Infrastructure, Transport and Tourism, 2009). After the 2009 disaster, the Ministry of Land, Infrastructure, Transport and Tourism studied the location of care houses or homes for the aged (Ministry of Land, Infrastructure, Transport and Tourism, 2010). The study showed that

13,730 facilities are located in high-risk areas, with less than 30% equipped with some countermeasures against debris flows.

Because of climate change, the increase in disasters is predicted. For this reason, these facilities need to consider disaster risk management. Older people find it difficult to evacuate on their own. These facilities should make evacuation plans from improved disaster risk management.

Development of Residential Areas

Japan experienced rising economic growth from 1955 to 1975. The urban area spread under the influence of economic growth, with a corresponding population increase. Thus, residential areas expanded in the urban area. These locations were former fields and near the mountains or rivers. The area's residents do not have a deep relationship with other residents and do not know local history and indigenous knowledge. Some have their first experience of living near the mountains or rivers. Often, many do not understand the disaster situation. Hiroshima city is one such large city in Japan. This city experienced serious rainfall disaster in 1999.

Hiroshima city is a large city but, like other coastal cities in Japan, it lacks sufficient flat land. Between the 1970s and the mid-1990s, there has been a steady population growth, with people from different places settling in the city. As a result of this growth, the city has spread out over the mountain area. Therefore, Hiroshima city has the highest sediment disaster risk area in Japan. From past experiences and memory, older residents know where the risky parts are and when the disaster may happen, but new residents do not have that knowledge. In the new residents' area, the average age is 15 years. Among the older residents, the average age is 60 years (Takeuchi, 2004). The other issue is the geology of the area, which makes it vulnerable to sediment disasters.

On June 29, 1999, heavy rainfall (over 150 mm/3 hours) due to an atmospheric depression occurred in Hiroshima. Prior to 1999 rainfall event, similar heavy rainfall occurred twice in the same region (June 23–24 and June 26–27). Disaster damage included debris flow in 139 sites, the collapse of 186 buildings, and a resultant death toll of 31, with one person missing, and 154 houses damaged (Fujiwara, 2000). After the disaster, Hiroshima prefecture and Hiroshima city performed various soft and hard countermeasures. In Hiroshima city, 32 people died in the 1999 disaster: 16 males and 16 females. However, older men and child victims accounted for 71.8%. This number, including women victims, accounted for 84.3% of the casualties. The place of death was different for men and women; men died outside and traveling between house and work place; women, older people, and babies died inside houses.

The story of a 29-year-old woman who died inside her house is illustrative of the situation. The place where she died is a high-risk debris flow area. In the 1970s, this area was a paddy field. Furthermore, the location of the 29-year-old woman's house had previously been a pond supplying water to the paddy field. After the 1970s, the land use changed to residential and a new transport system was developed in the 1990s. It grew into a large residential area and, by the end of the 1990s, the pond was reclaimed and 10 houses were built on it. Additionally, it is located at the entrance of the valley, and prone to debris flow. Many new residents moved to this area from outside.

The family of the 29-year-old woman started living in this area three years before the 1999 disaster. Her family was comprised of her husband and two children (1 and 3 years old). On the day of the disaster, her husband went to work while she stayed at home with the children. This family did not have many contacts in the community and was not included in the community network. The community network is a social network consisting of members of the local voluntary disaster management organization and of the residents' association, which exists in each community. On the day of the disaster, there was an official announcement at 10 o'clock to inform residents through TV, radio, and the community network of the dangers of heavy rainfall (see Fig. 6) but this family was not part of the community network. Around 12 o'clock, residents received further information: about the changing color of river water (clear to brown), rumbling from the mountain, and a smell of burning, among other things.

For this woman, it was the first experience of living in a mountain area. She did not understand that the information was a prelude to debris flow, and many flows eventually occurred between 1 and 3 pm. In this area, communities have local voluntary disaster management organizations, and every year, they hold evacuation training programs. However, the training participants are male in the majority of cases. Her husband learned some risk information and other local information in this way but did not share it with other family members, and he left the community during daytime. She did not know about the evacuation shelter procedure in the community and stayed in the house. When the debris flow happened, she stayed on the first floor; her two children stayed on the second floor. One debris flow directly attacked her house, which was flooded with water, the height eventually rising to 35 m. Because the family was not included in the community

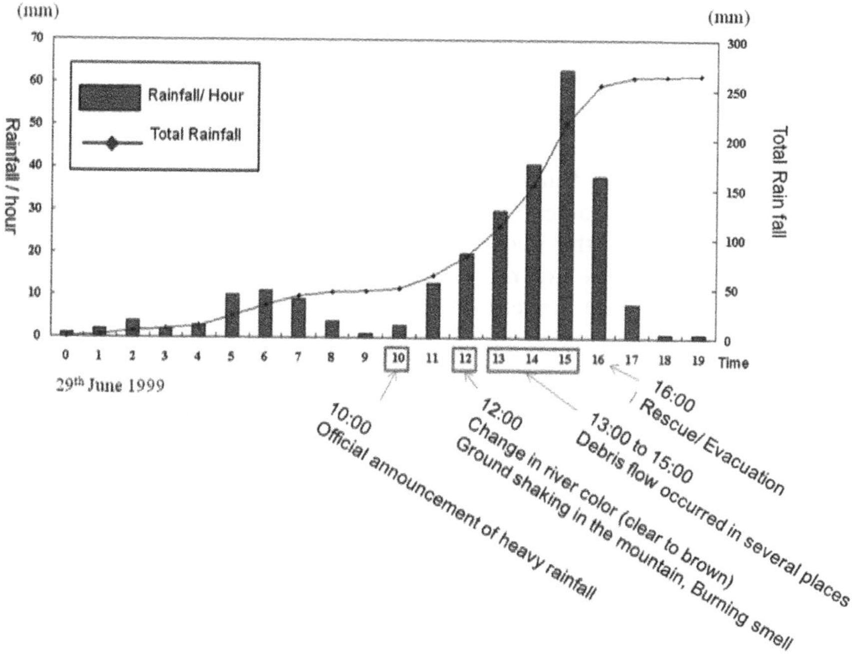

Fig. 6. Rainfall and Disaster Situation: June 29, 1999. *Source:* Takeuchi, 2005.

network, other community members did not know who lived in the house. Only the husband (who was out at work) knew this information, but he did not understand the disaster risk. The children were rescued from the second floor, but the 29-year-old woman was pulled from the debris after six hours, already dead.

By climate change, the increase of the disaster is predicted. For this reason, the new residential areas need disaster risk management more than ever.

Urban Flooding

With rising economic growth, the urban area spreads perpendicular to highly stratified buildings and an underground not spread on a level space. In addition, the government developed riverbanks and drainage systems in urban areas. Through these measures, the urban areas are protected from small-scale flood disasters. Therefore, some housing developed near the river area. However, big floods can overcome the riverbank and drainage system.

In this instance, serious damage occurs, especially near the river and underground areas.

Urban flooding is best illustrated by the experiences of Fukuoka and Nagoya. On June 29, 1999, the same day of the Hiroshima disaster, there was also heavy rainfall in the Fukuoka prefecture. Due to this rainfall, a flood disaster occurred in Hakata Station, the main station of Fukuoka prefecture. The flood came so fast that one lady had no time to run and evacuate and died at the underground channel. Similarly, in September 2000, Nagoya prefecture experienced heavy rainfall disaster. Both Fukuoka and Nagoya had built underground areas with shopping arcades and restaurants. Due to heavy and sudden rainfall, there was not enough time for proper precautions to evacuate people from the underground areas of these two cities. The general economic loss caused by this disaster, in which the metropolitan area of Nagoya was flooded, was the worst in 40 years.

Urban areas are often protected from small floods by riverbanks and drainage systems. However, the urban areas have a serious vulnerability to large-scale flood. This has major implications on large-scale flood disasters. Therefore, urban areas, especially the ones with very high development of underground structures, need to prepare disaster management measures considering climate change impacts.

Mountain Management and River Basin Linkages

Saijo city is located in the eastern part of Ehime prefecture. It has an area of 509.04 km^2, with a population of 116,059. The geography of the city is classified broadly into four parts: plain area along the coast, hilly area between Saijo City and Nihama City, hilly terrain ranging from the north side of the median tectonic line along the south side of the plain, and precipitous mountains in the southern side of the median tectonic line. There is Mt. Ishiduchi, the highest mountain in the western part of Japan. Two big rivers, Kamo River and Nakayama River, flow in the center of the city (Shaw, Takeuchi, & Badaoui, 2008).

In summer and autumn of 2004, six record-high typhoons (nos. 4, 6, 10, 11, 21, and 23) hit the Shikoku area. In this area, there was a maximum of 2 typhoons per year (in 1995), the average is 1.5 numbers of typhoon. In the Seto Inland Sea area where they have little heavy rain damage typically, there were many concentrated downpours, landslides, and high tide, and they damaged many precious possessions, like 61 people's valuable lives and houses. Six typhoons (nos. 15, 16, 18, 21, 22, and 23) caused damages to

Saijo City, and nos. 21 and 23 caused more damages. On September 29, 2004, typhoon no. 21 moved across the Shikoku area. Because of this, in Saijo City, they had record concentrated heavy rains, 75.5–150 mm of rainfall per hour. An avalanche of rocks, earth, and driftwood surged, which seemed to have occurred due to slope destruction of intermediate and mountainous areas, and destroyed a natural dam. A lot of driftwood got stuck at a bridge pier, and water was held back and overflowed. As the water level rose suddenly, surrounding houses were flooded. In the flat part, each area was flooded above and below floor level. In the mountainous areas, landslide disaster occurred frequently, roads were severed, many villages were isolated, and house destruction and human suffering were caused. The number of casualties in Ehime prefecture by typhoon no. 21 was 14, and this was the worst record in human suffering caused by a typhoon.

Frequent small slope failure by the concentrated heavy rains of typhoon no. 21 added to the damage. Although "deep-seated landslides," in which the entire ground slides, are not related to the form of forest, "shallow landslides," in which surface soil slides directly, result from the forest type and extent of its maintenance. In addition, in artificial forests that are not thinned for a long time, sunlight does not reach the ground and weeds and young trees do not grow well. When it rains there, surface soil is hit directly by raindrops and clogs, and rainwater that cannot soak through the ground runs on the surface. The "water road" caused by the erosion forms a valley and finally creates mudslides involving surface soil and uprooted trees. Abandoned thinned wood is also a problem. They are flushed into the river by the heavy rain, where they get stuck along bridge piers and cause water to overflow downstream.

Some designated shelters turned out to be dangerous. For example, in the mountainous area of Ofuki, some people evacuated to the community center that was a designated shelter. But one person noticed that the river near the center suddenly turned muddy, so they escaped to a different building. One minute after they evacuated, the center was buried in the mud. In the mid-mountainous area of Funakata, the designated shelter was at the head of a hill and was difficult to get to. So residents evacuated to a nearby meetinghouse. Therefore, the designated shelters built by the municipality should be reexamined, and residents should learn about the area.

Plain areas are rather urban with many young people. So, it is necessary to make a "disaster prevention network" between the plain areas and the mountainous areas, so as to help elderly people in the mountainous areas in case of a disaster. Just as driftwood can jam against a bridge pier and cause a plain area to flood, disaster in the mountainous areas can have bearings similar to that in the plain areas. Both plan and mountain residents need to know about the various circumstances. For these reasons, mountain watching is proposed to be implemented in Saijo City. Mountain watching is just like town watching but is conducted in the mountainous areas. The main participants are children, residents in the mountains, teachers, municipal officials, and forest workers. Participants watch the site damaged by the typhoon in 2004 and hear the story from victims.

A similar experience was observed in Toyooka city in Hyogo prefecture, with the same typhoon no. 23 of 2004 (October 20, 2004). The key cause of flooding during that typhoon was the runoff from nearby forested mountains surrounding Toyooka city, which blocked rivers and inlets to streams, and consequently resulted in a breach of the river embankments (Srinivas, Shaw, & Nakagawa, 2009). This understanding and the postdisaster assessments have provided the department with valuable lessons, as outlined previously, to better mitigate and be prepared for disasters in the future. Improving access to nearby forests and streamlining ownership of pockets of forested lands are cornerstones of the policies from which other cities can learn. For Toyooka, such an approach has brought the local community closer to the forests, and hence closer to monitoring potential degradation and maintenance problems. It also includes a more systematic integration of the city's image and vision into features derived from the forests, mountains, and agricultural areas that surround the city. A key challenge that the department continues to face, as will other cities adopting such an approach, will be the lack of awareness of the local communities and hence their active participation and involvement in the department's activities to reduce/mitigate the disaster risks and hazards that Toyooka faces.

CONCLUSION

The climate change is posing additional threats to the already vulnerable communities in Japan. The changing nature of the climate is affecting the typhoon frequency and intensity, as well as rainfall amount and intensity. Although the government has a specific legal and institutional framework at a national and local level to cope with natural disasters, sudden occurrences of disasters can make people, communities, and local governments unprepared and, in some cases, helpless. The recent disasters mentioned highlight certain aspects of climate change adaptation, which are related to the disaster risk reduction measures in Japan.

On the planning aspect, there needs to be proper planning and urban development, especially in the mountain towns and cities. Due to a lack of land area, in many cases, cities have grown into the mountain slope. New development areas are often found to be highly vulnerable due to their physical risk, as well as a lack of knowledge on mountain-related disasters, which, in fact, are related to social risk. Location of vital infrastructures like the schools, homes for the aging, and school cum residences for people with physical and mental challenges need to be properly located in safe areas, which have a strong link to the land use plan and development plans. In the major urban areas, with a highly developed environment with underground structures, proper evacuation plans are required to avoid major disasters during sudden and heavy rainfall.

On the natural/ecosystem management aspect, appropriate forest management is the key for reducing the damages due to typhoon-related mountain disasters. Proper forest management (with strong participation of the town people and communities near the forest) helps in enhancing the capacity for adapting to climate change in the long run. Most Japanese forests are monoculture forests, which lack a proper balanced ecosystem. Revitalizing the forest ecosystem will be one of the key adaptation actions for the mountains in Japan. The other areas are the coastal bio-shields that protect against coastal hazards and storm surges. Most of Japan is protected by seawalls and other hard measures, while the ecosystem-based approaches are still missing. To cope with the climate change and related coastal hazards, ecosystem-based coastal management will be a key adaptation action. The same principle applies for riverbank and related ecosystems, where natural buffers will be useful in some cases.

On the social aspects, due to a higher percentage of aged population, Japan needs to have additional coping mechanisms with specific focus on remote mountain villages. Communication and networking becomes one of the key issues, and needs to be incorporated into the adaptation strategy. As is evident from the case study of Hiroshima, risk communication to women is an absolute requirement for preparedness. Males generally dominate community meetings, and the discussion often is not passed to female members of the family. Therefore, risk information is very much limited to one gender group in the community (Takeuchi & Shaw, 2009). There are different ways to disseminate information to the female group (for instance, by encouraging more female participation in the community meetings, and by participatory education among school children). Recently, town watching or neighborhood watching is found to be a very useful tool for disaster education, and through interactions with children, and the Parent-Teacher

Association (PTA), it is possible to encourage wider participation among the mothers. Therefore, proactive school-based community education can play a significant role in climate change adaptation and disaster risk reduction.

ACKNOWLEDGMENTS

The authors acknowledge the support of the Global COE Program, "Sustainability/Survivability Science for a Resilient Society Adaptable to Extreme Weather Conditions (ARS)."

REFERENCES

Cabinet Office Disaster Databases. (2010). *Disaster statistics of Japan*. Available at www.bousai.go.jp. Accessed on July 12, 2010 (in Japanese).

Cabinet Office Government of Japan. (2002). *Disaster management in Japan*. Director General of Disaster Management, Government of Japan, Tokyo, p. 37.

Fujiwara, K. (2000). Analysis of sediment disaster at Hiroshima city on June 1999. *Hiroshima University of Economics*, *22*(4), 3–37.

Inokuchi, T. (2001). Report on the landslide disaster in the upper part of the Abukuma River, Fukushima prefecture, due to heavy rainfall in August 1998. *Natural Disaster Research Report of the National Research Institute for Earth Science and Disaster Prevention*, *37*, 107–136.

Japan Water Forum (JWF). (2006). *Flood fighting in Japan*, p. 15.

Kobayashi, M., Komagata, O., & Nihei, N. (2008). Global warming and vector-borne infections diseases. *Journal of Disaster Research*, *3*, 105–112.

Ministry of Education, Culture, Sports, Science and Technology (MEXT), Japan Meteorological Agency (JMA), & Ministry of the Environment (MOE). (2009). Integrate report of observation, prospect and effect of global warming – Effect of global warming of Japan, p. 65.

Ministry of Health, Labor and Welfare. (2009). *Progress of aging society*. White paper of Ministry of Health, Labor and Welfare, p. 38.

Ministry of Health, Labor and Welfare. (2010). *Number of the homes for elderly*. Available at http://www.mhlw.go.jp/toukei/saikin/hw/gyousei/07/kekka5.html

Ministry of Land, Infrastructure, Transport and Tourism. (2009). *Report of heavy rainfall disaster at Chugoku and Kyushu area on July 2009*. Available at http://www.mlit.go.jp/common/000051190.pdf

Ministry of Land, Infrastructure, Transport and Tourism. (2010). *Report of research result about location of care house or homes for aged*. Available at http://www.mlit.go.jp/report/press/river03_hh_000240.html

Nakano, T., Kadomura, H., Mizutani, T., Okuda, M., & Sekiguchi, T. (1974). Natural hazards: Report from Japan. In: G. White & I. Burton (Eds), *Natural hazards: Local, national, global* (pp. 231–245). USA: Oxford University Press.

Shaw, R., Takeuchi, Y., & Badaoui, R. (2008). Education, capacity building, public awareness for disaster reduction. In: K. Sassa and P. Canuti (Eds.), *Landslides disaster risk reduction* (pp. 499–515). Berlin, Germany: Springer.

Srinivas, H., Shaw, R., & Nakagawa, Y. (2009). Recovery from typhoon damages. In: R. Shaw, H. Srinivas & A. Sharma (Eds), *Urban risk: An Asian perspective* (pp. 355–373). UK: Emerald Publication.

Takeuchi, Y. (2004). Hazard map: Respondents' perception and requests in the case of Hiroshima, Japan. *Japan Society for Natural Disaster Science*, *23*(3), 349–361.

Takeuchi, Y. (2005). *An integrated community-level mitigation system against sediment disaster: A proposal based on the analysis of geomorphic process and residents' perception and behavior*. Doctoral thesis, Rissho University, Japan.

Takeuchi, Y., & Shaw, R. (2009). Gender dimension in risk communication: A perspective from sediment disaster in Hiroshima, Japan, Regional Development Dialogue (RDD). *United Nations Centre for Regional Development (UNCRD)*, *30*(1), 63–75.

CHAPTER 7

FOREST MANAGEMENT AS AN ADAPTATION OPTION IN MOUNTAIN AREAS OF JAPAN

Kumiko Fujita and Rajib Shaw

INTRODUCTION

According to a report by Food and Agriculture Organization, total forest area in the world as of 2005 is estimated at 3,952 million hectares (9,766 acres) or 30 percent of total land area (FAO, 2005a). The area of forest is unevenly distributed (FAO, 2005b). At the regional level, South America is the region with the highest percentage of forest cover, followed by Europe and North and Central America. Asia is the region with the lowest percentage of forest cover (FAO, 2005c). The forest environments are changing globally. Climate change is one of the reasons, and it affects world forest management. More attention has been devoted to timber as the major commercial aspect of forests rather than to other noncommercial resource values (Barnard et al., 1985). The purpose of forest management was once as simple as taking timbers, but it is now more complicated. When the forest environments change, the key trends of world forest management must also change. Now, generally speaking, the extent of forest resources refers to an overall maintaining of adequate forest cover and stocking of various forest types and characteristics, and the ultimate aim of monitoring the extent and characteristics of forest resources is to reduce unplanned deforestation,

Climate Change Adaptation and Disaster Risk Reduction: An Asian Perspective
Community, Environment and Disaster Risk Management, Volume 5, 127–145

ISSN: 2040-7262/doi:10.1108/S2040-7262(2010)0000005013

restore and rehabilitate degraded forest landscapes, manage forests sustainably, and evaluate the important function of carbon sequestration by forests, other wooded land, and trees outside forests, thereby contributing to moderating the global climate (FAO, 2005d).

This chapter aims to review the following questions: (1) What are the impacts of climate change on forests? (2) What are the measures for the impact on climate change? (3) How can Japan manage to adapt the climate change in forest sectors? and (4) What are the current forest management practices?

This chapter introduces a case study of Japan's current forest management, based on literature review and secondary data. The second part of the chapter introduces the impacts of climate change on forests in Japan, and this is followed by an introduction to climate change measures in world forestry. Then, we introduce Japan's efforts at adapting to climate change. Finally, the last part of the chapter addresses the comprehensive forest management plan for climate change adaptation and disaster risk reduction.

CLIMATE CHANGE AND FOREST IN JAPAN

Climate change is one of the factors expected to increase the frequency and intensity of weather-related natural disasters (Ibarrará, 2009). The mountains of Japan have suffered from various natural disasters such as floods, landslides, typhoons, volcanic eruptions, and earthquakes. Structural and nonstructural measures have been developed for preventing and mitigating each disaster. Recently, global warming and abnormal weather have affected the scale and type of disasters. Occurrence of localized torrential rain is increasing, and big-scale local floods and droughts have occurred (Meteorological Agency, 2008).

The average annual precipitation in Japan is about 1,800 mm (71 inches), and most of the precipitation is brought during rainy season, from early June to mid-July, and during the typhoon season, from July to October, as torrential rain and as snow in the winter season. Because increasing abnormal weather caused by the climate change of global warming is a hot issue, the Meteorological Agency provides analyzed observed climate data of more than 100 years, called abnormal weather risk map. In the map, data of daily precipitation is analyzed from 51 points in Japan from 1901 to 2006. The data shows the long-term trend of increasing days with daily precipitation more than 100 mm (almost 4 inches) (Meteorological Agency, 2008). In addition, it compares annual precipitation over 30 years, from

1931 to 2006, to show standard deviation. The national average annual precipitation is decreasing over the long term, but the average annual standard deviation is increasing every year. In addition, the gap of standard deviation is increasing in most of the 51 points (Meteorological Agency, 2008). Therefore, the difference between a high precipitation year and low precipitation year is growing. Thus, the trend of occurrence of localized torrential rain is increasing, and local areas tend to suffer enormous damage. In addition, the gap between high precipitation and low precipitation years tends to be wide, with big-scale local floods and droughts tending to occur.

Moreover, owing to global warming, temperatures in Japan have increased from the late 1980s, with the highest recorded temperature in the last 100 years occurring in 1990. The Meteorological Agency observed that the temperatures increased 1.07 °C in the last 100 years (Meteorological Agency, 2007). In addition, it is estimated that the temperatures every season in Japan increase after 100 years. It is predicted that the annual average temperature increase of about 2° to 3°C in 100 years and the number of torrential rain events will increase in most areas (Ministry of Agriculture, Forest and Fisheries, 2006a).

Japan has also suffered from frequent erosion and sediment movement, such as mountain collapses, debris flows, and pyroclastic flows, because of both primary causes such as steep mountains and vulnerable geology, and from provoking causes such as torrential rains, earthquakes, and volcanic eruptions. In addition to these "natural factors," "social factors," such as developed country in which many people highly utilize narrow land and gentle slopes, make Japan a "sediment disaster archipelago"(Ohta, 2006a). Therefore, Japan has a long history of coping with these disasters. There are mainly two measures against mountain disasters that are structural measures such as physical construction to reduce or avoid possible impacts of hazard, and nonstructural measures such as policies, awareness, and knowledge development.

Though comprehensive disaster reduction is achieved by implementing "disaster preparedness/risk reduction," "damage reduction/disaster response," and "recovery/reconstruction" in a balanced manner, the most important strategy is "disaster preparedness/risk reduction" (Meguro, 2004). Torrential rains in Hiroshima that occurred in 1999 were the trigger to enact the Act on Promotion of Sediment Disaster Countermeasures for Sediment Disaster-Prone Areas. The act aims to promote only nonstructural measures such as developing early evacuation systems. The damaged area, Hiroshima city, is also a typical area with sediment-related disasters like those in Kobe city, where housing land development has also been promoted in mountain areas; therefore, major disasters occurred in the developed mountain areas around Hiroshima city and Kure city (Watanabe & Fukuoka, 2003). The primary reasons for the disaster were developed housing land in dangerous areas, a lack of participation in disaster risk reduction by local residents, a lack of disaster risk management because of having no disasters for a long time, and a lack of forest management (Murata, Shibuya, & Nakai, 2001).

An early warning system is a well-developed nonstructural measure and risk reduction strategy for avoiding mountain disasters in Japan. Protecting the function of forests is also effective in disaster risk reduction. Maintaining intact forests prevents most of the surface soil erosion and decreases the surface collapse (Ohta, 2006b). In Japan, developed forest has no possibility of increasing the risk of soil collapse, but it is effective for preventing mountain disasters (Ono, 2008). Because forests in Japan fulfill functions of mountain disaster prevention, conserving and managing forests as they are prevents surface collapse (Ohta, 2006b). In addition, the forests in Japan are still growing and function as carbon sinks for a while (Ohta, 2006c).

Although the environment changes, the expectations of forests in Japan are also changing. An opinion poll on forests and life, conducted by the Cabinet Office, shows the changing public expectations toward forests (see Fig. 1).

The functions of forests, which contribute to preventing global warming by absorbing carbon dioxide (CO_2), have attracted attention since around 1999, and the forest functions for preventing global warming are most expected function toward forest in 2007. The expectation for disaster

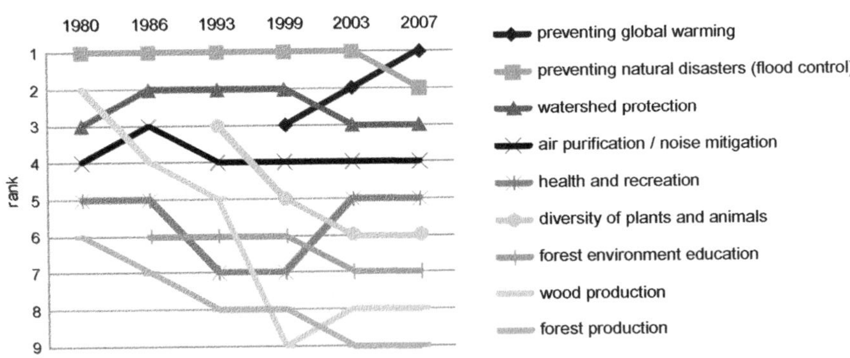

Fig. 1. Changes in Expectations of the Public toward Forest Policies. *Source*: Opinion Poll on Forest and Life.

prevention has been high. The expectation of the forests to function as an mechanism for flood control is raised even more nationwide, because the movement toward removing dams for conserving and revitalizing river ecosystems has been promoted all over the country (Nakamura & Kakizawa, 2009).

Forest resources serve various protective functions including influencing climate, protecting from wind erosion and avalanches, offering coastal protection, acting as air pollution filters, and protecting water resources.

A vigorous forest acts as a sink, or a fixer of carbon, whereas deteriorating forests give up carbon, or become sources of atmospheric carbon dioxide as they decay (Canada Forest Service, 1999). Because forests are able to absorb considerable carbon dioxide, reducing forest area accelerates global warming (Kaneko, 2008). In addition, forests reflect less heat back into the atmosphere than other types of land use and have functions of reducing wind velocity, moderating soil temperature, and increasing relative humidity. The leaves, branches, and trunks of forest trees are effective in preventing wind. Another important protective role of forests is coastal protection. Coastal vegetation has been widely recognized as a natural method for reducing the energy of tsunami waves (Tanaka, 2009) and tidal waves. Forests also function to mitigate and prevent landslides and surface soil erosion. The tree canopies block the rain and decrease the amount of rain when it reaches the ground; the fallen leaves and underbrush prevent surface erosion (Ohta, 2004a). In addition, the ground roughness and roots increase the erosion resistance; soil structure and developing roots function to keep soil penetrable; and forests function to stabilize micro-weather and to prevent occurrence of freezing and thawing inside forests (Gomi, 2006). Moreover, forests act as filters for air pollution by absorbing gas contaminants as well as liquid and solid particulates such as dust, ash, and pollen. Then, because forests covered by fallen leaves and underbrush are able to infiltrate rain into the soil and discharge water slowly, floods are mitigated, and water is purified and released after rains (Ohta, 2004b). This increases riverbank protection and erosion control functions.

In this section, we introduce mitigation in forestry because of the trend that is shifting from mitigation to adaptation. We also address what adaptation is and why it is important. The major reason why climate change mitigation is stressed climate change mitigation is stressed more, because it is able to simultaneously reduce impacts of climate changes in all climate change vulnerable areas. According to the "Climate Change 2007: Mitigation of Climate Change" of the Intergovernmental Panel on Climate Change (IPCC), forest management is highly appreciated for mitigating climate change. It says, in the long term, that a sustainable forest management strategy aimed at maintaining or increasing forest carbon stocks, while producing an annual yield of timber, fiber, or energy from the forest, will generate the largest sustained mitigation benefit (IPCC, 2007a). In addition, the extent of forest resources is also highly appreciated as one of the mitigation options in "Climate Change 2007: Synthesis Report" of IPCC. Forests are selected as an example of the seven key sectoral mitigation technologies, policies and measures, constrains, and opportunities, as Table 1 shows.

As Table 1 also shows the mitigation focus on plantations, but each country has different conditions in forest management and some countries have difficulty planting forest continuously. They may lack of investment capital. In addition, private forest owners may have different ideas for using forest resources. In addition, it is impossible to stop global warming and it is difficult to expect the mitigation effects to function in the short term, though

Table 1. A Selected Example of Seven Key Sectoral Mitigation Technologies, Policies and Measures, Constraints and Opportunities.

Sector	Forestry/Forests
Key mitigation technologies and practices currently commercially available	Afforestation; reforestation; forest management; reduced deforestation; harvested wood product management; use of forestry products for bioenergy, to replace fossil fuel use; *tree species improvement to increase biomass productivity and carbon sequestration; improved remote sensing technologies for analysis of vegetation/soil carbon sequestration potential and mapping land-use change*[a]
Policies, measures, and instruments shown to be environmentally effective	Financial incentives (national and international) to increase forest area, to reduce deforestation, and to maintain and manage forests; land-use regulation and enforcement
Key constraints or opportunities	Constraints include lack of investment capital and land tenure issues; *can help poverty alleviation*[b]

Source: IPCC (2007b).

[a]Key mitigation technologies and practices projected to be commercialized before 2030 are shown in italics.

[b]Normal font, constraints; italics, opportunities.

mitigation is implemented in all countries. Until then, we need to adapt our lives to the impacts of global warming.

Because current forest utilization and preservation is based on how forests developed under past climatic conditions (Spittlehouse & Stewart, 2003), global warming cannot be fully mitigated. Therefore, recognition of the importance of adaptation is needed to be increased.

However, adaptation alone is not expected to address all the projected effects of climate change, especially not over the long run, as impacts may increase in magnitude (IPCC, 2007c). Both strategies are needed to confront global warming. Adaptation is implemented for specific areas, in limited regions basically, and the effects will contribute to mitigate global warming. The projected magnitude of climate change, especially the anticipated rate of change because of human influences on the climate system, is beyond our current level of knowledge and adaptation mindset, and this "outrage" imposed by humanity on the climate system will require a significant paradigm shift and new levels of knowledge for forest managers (Maciver & Wheaton, 2005).

FOREST MANAGEMENT IN JAPAN

Because adaptation is a measure implemented in specific areas and places, this section introduces Japan's forest management against climate change. Forests are considered carbon sinks and areas of biomass production. Therefore, different approaches/policies are needed in implementing adaptation measures for forests.

FOREST AS CARBON SINK

Expected Forest Role as Carbon Sink

The Kyoto Protocol went into force on February 16, 2005, and Japan held up a promise to reduce 6 percent of GHG between 2008 and 2012. Then, because of its high forest rate, Japan focused its attention on the forests as a source for absorbing carbon dioxide (CO_2). Carbon dioxide, one of the GHGs, accounts for the highest percentage, 90.7 percent, among GHGs in 1990. Article 3, paragraph 3, and Article 4 of the Kyoto Protocol accepted that the "forest sinks" absorb this amount and is included in the GHG emission reduction target in a limited way. For the purposes of

Article 3, paragraph 3, eligible activities include human-induced afforestation, reforestation, and/or deforestation activities that started on or after January 1, 1990 and before December 31, 2012 – the last year of the commitment period. (Afforestation, reforestation, and forest management, are taken into account in the cap for forest management, as defined in the Marrakech Accords.)

Technical Issues

There is general agreement that terrestrial systems in the northern hemisphere provide a significant sink for atmospheric CO_2; however, estimates of the magnitude and distribution of this sink vary greatly (Goodale, 2002). Though we need to manage the forests properly as sinks and emission absorbers, a large amount of labor is needed to address them, especially in widespread areas such as regions and countries (Hirata, 2009a). New observation technology is needed, but some methods now available are effectively used in combination (Hirata, 2009b). Because forest management with long and medium-term vision and strategy is needed to be discussed based on future forecast by simulation, simulation models for forest carbon sink circulation need to be developed as soon as possible (Matsumoto & Mitsuta, 2009).

Issues of Ownership

Forest area in Japan is 25,150,000 ha (62,147,003 acres), and the percentage of forest area in total national land, which is called forest rate, is 67 percent. Most of the forests are distributed in mountain areas. In mountain villages, a small number of villagers has taken care of the vast area by conserving upper river basins, protecting watersheds, and conserving the natural environment. There is a big difference in ownership of the world's forest and Japan's forest. Though private ownership is on the rise, 16 percent of the world's forests were privately owned (FAO, 2005e) in 2005. In Japan, 58 percent was private forest (Ministry of Agriculture, Forest and Fisheries, 2007) in 2007, as Fig. 2 shows.

Forest management for the carbon sink needs to address income for forest owners, because private forest owners need to gain income from forest resources. In addition, Japan highly relies on the forest function of CO_2 absorption more than other measures against global warming in the

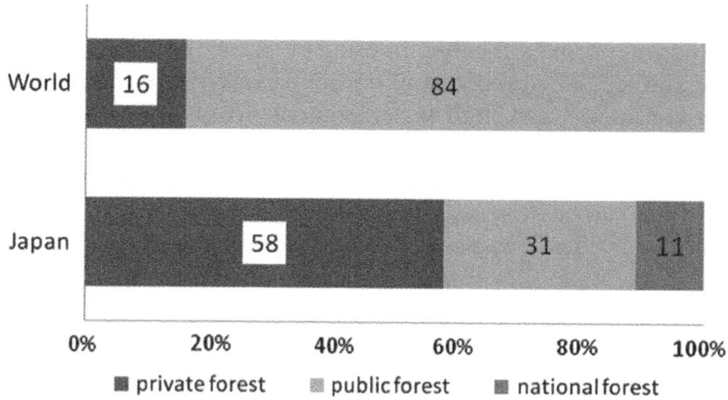

Fig. 2. Ownership of the Forest in the World and in Japan.

industrial and commercial fields, and relies more on the forest sink than other developed countries rely on it (Kobayashi, 2008). As a result, policies or rules such as domestic and international emissions trading (ET) systems are discussed.

Society's response to climate change is determined through the political process. From this point of view, certain policies, such as conversion of primary forests to rapidly growing plantations in an attempt to sequester as much carbon from the atmosphere as possible, will do more harm than good (Noss, 2001).

Biomass

Biomass Town

Japan has addressed using biomass energy and establishing a recycling-based society as a countermeasure against global warming. According to the definition by the Ministry of Agriculture, Forestry and Fisheries, "biomass town" is a region in which a network of all stakeholders, including those from other areas, establish a comprehensive utilization system of producing biomass and using the biomass effectively, and use or expect to use stable and proper biomass in the region. In other words, if a region becomes a biomass town, it promotes region-based global warming countermeasures, cyclic use of regional resources, new industry, and activating regional society. Creating a biomass town is considered to be progress.

Issues of Using Biomass

Because there are some difficulties in using biomass, the actual use of them is still being studied. The problems of collecting widely distributed resources, low calorie content, and high water content are not yet solved. Usually biomass is widely distributed in rural areas, and it costs a lot to collect and transport. Unused biomass such as nonfood parts of agricultural crops and thinned timber can be dried and used directly, but it is harder to use than oil and natural gas. The problems of unused biomass include its volume, difficulty in treating, low calorie content, unstable supply, and so on (Ministry of Agriculture, Forest and Fisheries, 2006b). When biomass with high water content is used, a problem emerges between the relation of energy for drying and produced energy.

Issues in Changing Expectation toward Forest Products

Before the period of high economic growth, from 1955 to 1973, the Japanese used to make charcoal from wood and used it as domestic fuel, but the fuel revolution started during the postwar reconstruction period, from 1945 to 1955. Charcoal and fuelwood had been replaced by electricity, gas, and oil as domestic fuel, and charcoal and fuelwood came to be considered as not suitable as an energy resource. Around the same time, the demand for timber as a building material increased enormously for restoring the devastated land after World War II. Because the supply could not keep up with the demand sufficiently, the wood prices rose, and the government implemented an expansive afforestation policy for rapid forestation. The afforestation boom spread in both national and private forests in Japan. About 4,000,000 ha (about 10,000,000 acres) in the present total artificial land, 10,000,000 ha (almost 25,000,000 acres) became artificial land in just 15 to 20 years (www.shinrin-ringyou.com/ringyou/) (Ouchi, 1987). Now, afforestation forest accounts for 46% in the private forest (Ministry of Agriculture, Forest and Fisheries, 2007). The total forest area from 1966 to 2007 has not increased nor decreased, but the artificial forest area increased by about 30 percent from 1966 to 2007.

It takes at least 40 years from when the trees are planted to when they are able to be use as a building material; therefore, while promoting afforestation, the import of lumber had been promoted gradually from about 1955 until 1964. Then relatively cheap and mass-produced timber was imported constantly and has been used a lot. As a result, the price of domestic timber has dipped after peaking around 1980 (www.shinrin-ringyou.com/ringyou/). Though the self-sufficiency rate in timber was more than 90 percent in 1955, the present rate is less than 20 percent, with the declining price of domestic timber. The expansive afforestation policy has not been reviewed or implemented since 1996.

In the 1990s, the world economy had been changed greatly by democratization and entry into the free market system by former socialistic countries. The so-called globalization era started. Timber products and trade were also globalized in the 1990s. As a result of the increased demand for imported timber, enormous artificial forests and debt are left (www.shinrin-ringyou.com/data/mokuzai_kyoukyu.php.). Fig. 3 shows the changing forest environment in Japan.

Species that are suitable for biomass may be considered as the next plantation crop. Again, planting species that are suitable only for biomass will do more harm than good, just like planting limited species that are suitable only for carbon sink. Comprehensive analysis is needed, including calculating the necessary amount of volume for carbon sink and biomass.

Forest Environment Tax

Properly maintained forests contribute to the mitigation of global warming as carbon sinks, prevent and mitigate mountain disasters, heal people, and so on. The current timber price has decreased because of the importation of cheap timber and the use of alternate materials, and the forest resources are less used and less treated. Because of this, the number of foresters decreased by one-fourth over about three decades (Kochi prefecture, 2010). Also as a result, there are many forests that need to be cut as soon as possible.

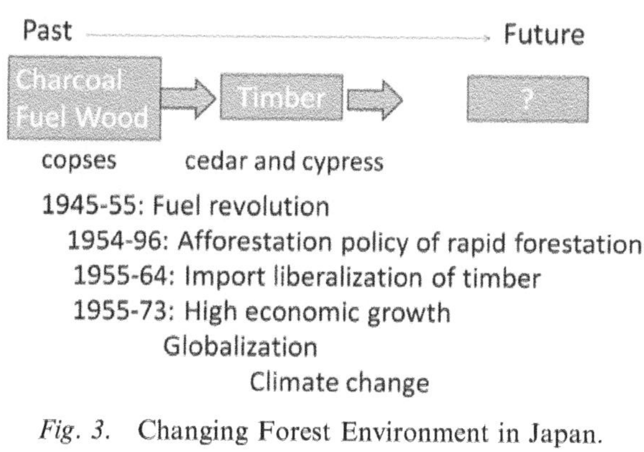

Fig. 3. Changing Forest Environment in Japan.

Forest Environmental Tax in Kochi Prefecture

The forest environmental tax is a prefectural tax, which is imposed thinly and broadly on prefectural residents and is used for developing forest environments in degraded/degrading forests and for maintaining public interest in forests for water source protection (Akiyama, 2005). The Kochi prefecture has a high forest rate at 84 percent, and it is the highest in Japan. Kochi is the first prefecture in Japan to introduce a forest environmental tax. The purpose of the tax is forest conservation by the participation of the prefectural inhabitants (Kochi prefecture, 2002a). The tax revenue is used to prevent the depression of forest functions: prevention of mountain disasters, including water source recharge, mitigation of disasters caused by weather phenomena, and the securing of biodiversity with the permission of local residents who receive the forest benefits. The tax is spent on forest maintenance projects such as thinning and for public relations activities (Tamaki et al., 2005). Each individual and corporation has been charged 500 yen on annually since 2003; after five years, the taxation term is evaluated (Kochi prefecture, 2002b).

The First Phase and Outcomes

There are two projects that receive these tax revenues, which are "the project for promoting forest management by the people's participation" and "the project for urgent forest environment conservation." The former project aims to educate the inhabitants about how the forest is important, to enlighten forest owners, and to manage model forests. The latter project aims to make mixed forests by thinning (Kochi prefecture, 2002c). Through these projects, the number of residents participating in the activities to protect mountain environments is increasing as are the number of volunteer organizations. In addition, the first forest environmental tax Kochi introduced affects areas all over the country, by establishing a "Shikoku mountain day" in four Shikoku prefectures and introducing similar taxes in other prefectures (Kochi prefecture, 2008).

As it was promised when the tax was introduced, the Kochi prefecture sent out questionnaires in the 2006 fiscal year, collecting opinions from prefectural residents about the forest environmental tax. As a result, more than 80 percent of the residents agreed to continue the tax after the 2008 fiscal year.

The Second Phase

The second phase started in the 2008 fiscal year, and is scheduled for five years, on the basis of the outcomes of the first phase, opinions from the residents, the prefectural forest environment, and the current national policy. In the second phase, it focuses on trimming young forests (from 11 to 35 years old) of cedar and cypress. Because the young cedar and cypress grow fast and are effective, trimming them when they are young is effective for preventing devastation, for preventing mountain disasters, and for improving water recharge function (Kochi prefecture, 2008). In addition, the second phase also focuses on supporting activities of forest volunteers and activities of elementary and junior high school students that promote awareness and a relationship to the forest environment through forest environment studies (Kochi prefecture, 2008). The project budget is estimated to be 160,000,000 yen in the 2008 fiscal year.

Collaborative Forest Restoration with Environmentally Progressive Companies

Japan held up a promise to reduce 6 percent of GHGs and focus attention on the forests as a source of CO_2 absorption. As a result, establishing an emissions trading (ET) system has been discussed in Japan. Because the Kochi prefecture expects Japan to establish the ET system, "Collaborative Forest Restoration with Environmentally Progressive Companies" was started in the 2005 fiscal year to promote forest restoration and exchange among the Kochi prefecture, cities, towns, villages, forestry cooperatives, and companies (Kochi prefecture, 2010). In this system, environmentally advanced companies, local people, and forest owners cooperate to revitalize untreated forests, with the purpose of "reforestation" and "developing a people-to-people link."

Because prefectural and regional policies and/or rules are made based on the national government policies, government policies such as ET are needed to be decided as soon as possible. After prefectural policy is determined, private owners can decide how to manage their forests.

FOREST MANAGEMENT AS AN ADAPTATION OPTION

Needs of Integrated Forest Management

Although environmental measures and disaster-prevention measures have been separately studied in different fields previously, the importance of collaborating environmental and disaster-prevention measures has been recognized, along with the physical environmental changes such as climate change (Sadohara, 2008). Generally, environment and disaster studies cover a wide range of phenomena, and there are many differences in each area – time scale, history, economics, social and cultural backgrounds – therefore, they are correlated strongly and the issues become extremely complicated and varied (Enomoto, 2008).

Japan's current forest management becomes more complicated compared to the past several decades, because forest environments change dramatically. Thus, forest management is needed now not only for disaster risk reduction, but also for climate change adaptation and mitigation. The climate change and disaster risk communities are paying increasing attention to differential vulnerability among demographic groups, particularly children and the elderly. The context in which people are aging will also influence future vulnerability to climate change (O'Brien et al., 2008). The average occurrence of sediment-related disasters, which are debris flow, landslides, and collapse, is 1,000 per year, based on data from 2002 to 2006, and the number of deaths is 60 per year. So-called disaster-vulnerable people, such as the elderly and children, account for about half of the victims of sediment-related disasters, encompassing 20 percent of the total population (Practical guide, 2000). In the mountain areas particularly, decreasing population and aging are big issues. According to a report on the population trend in Japan, the national population in 2030 is estimated to decrease by 7 percent from that of 2000, and the population in mountain villages needing restoration is estimated to decrease by 36 percent compare to the population of 2000 (National Institute of Population and Social Research, 2003). Moreover, in the restoration-necessity villages, the percentage of the population of people between 15 and 64 years old is estimated to be less than 50 percent, but the population of those 65 years and older is estimated to be more than 40 percent. As a result of this trend, it is predicted that the function of villages will decline, and absentee forest owners will increase more and more (Ministry of Agriculture, Forest and Fisheries, 2006b). Thus, social changes surrounding forests in Japan will also affect the forest management.

Necessity of Involving Local Residents in Industry-Academic-Government Cooperation

Because private forest accounts for 58 percent, the way that forest owners think is also important as they actually decide how to manage their forests.

Fig. 4 offers a comparison of the expected forest role in Japan and in the Reihoku area of Kochi. The Reihoku area is a typical mountain area with an 87.9 percent forest rate. A survey was conducted in the Reihoku area, and 347 respondents were randomly selected from a registered phone book list of 5,267. Of 155 who replied, 71 percent of them are forest owners.

Though the expectation toward wood production has been lowered since its peak in 1980, as Fig. 1 shows, the survey of the Reihoku residents shows a different result. The rank of "wood production" and "forest production" is quite high in Reihoku compared to the rest of Japan. It is considered so because residents in Reihoku have gained income from wood and forest products and still expect to gain income from them, though it seems impossible to sell as timber. In addition, the Reihoku respondents considered the forest's role as prevention against global warming less compared to other Japanese surveyed. If this gap is reduced by spreading awareness of the world trend and Japan's policy, forest owners may use forest resources differently and manage forest differently. Then a linkage system is needed between local residents and the industry-academic-government cooperation. The industry-academic-government cooperation has been encouraged and promoted, and each share knowledge and information among them. This knowledge and information should be shared more with the local residents. Including indigenous knowledge in disaster management, risk reduction, climate change adaptation, and sustainable development approaches of government will pave the way for its widespread acknowledgment and application (Shaw et al., 2009). Indigenous knowledge exists relevant to climate change adaptation and

Japan

What are your expected forest roles?
An opinion poll on forest and life, conducted by the Cabinet Office in 2007
(top 6)

1. Preventing global warming
2. Preventing natural disasters
3. Watershed protection
4. Air purification / noise mitigation
5. Health and recreation
6. Diversity of plants and animals

Edited an opinion poll on forest and life of figure 1

Reihoku, Kochi

What are your expected forest roles for the forest in Reihoku?
Questionnaire survey at Reihoku in 2009
(top 6)

1. Watershed protection
2. Preventing natural disasters
3. Wood production
4. Preventing global warming
5. Forest production
6. Air purification / noise mitigation &
6. Diversity of plants and animals

Questionnaire survey in 2009

Fig. 4. Difference of Expected Forest Roles in Japan and Reihoku.

disaster risk reduction, including from the natural resources/forestry sector (Shaw et al., 2009). Because the mountain villagers, especially forest owners, are considered to have knowledge and experience of disaster risk reduction in forests, it is important to work with them to plan integrated forest management for disaster risk reduction, climate change adaptation, and local issues.

It is essential for the administration to make an effort to provide information to local residents and for local residents to make an effort to know the government policy and trends. Forest owners especially will have more possibility to use forest resources other than forest and wood products.

Role of Industry-Academic-Government Cooperation

It is an issue that the conscience of Reihoku residents toward the forests as a prevention against global warming is low. Because they manage and live with forest, Reihoku residents need a significant paradigm shift and new levels of knowledge. Then, they need new direction for forest management. When the afforestation policy was issued, it was clear for forest owners to plant trees for use as timber, but government forest policy is not so simple now. Although we introduced the climate change adaptation regarding forests as well as biomass energy and the use of forest resources as a carbon sink earlier in this chapter, the necessary amount of forest and effective species for carbon sink or biomass are not introduced simply. Forest owners have difficulty making forest management plans. In addition, forest owners need to gain cash income from their forest resources. If the majority of forest owners choose single species or a few species for biomass, the same issues will occur as are evidenced by timber wood plantations. Planting different species is necessary for risk management in conserving forest resources and forest function. If the industry, academic, and government sectors cooperate in calculating the necessary amount, deciding the policy of ET and providing different ideas for the use of forest resources other than wood and forest products, forest owners will be able to decide their forest management plans.

Forest Management as a Part of World Forestry

In these days, the forests and surrounding environments change rapidly. Timber supplies had been a dominated international forestry issue for decades after the Second World War, but the world expectations regarding forests are distinctly different since the globalization of the 1990s. In addition, the expectations regarding forests are different in each country and region. The forest management issues are global challenges now. Each country needs to deal with forest management by considering the world forestry trends and other environmental issues and by analyzing typical issues in each country for solving world issues. Therefore, international industry-academic-government cooperation is also needed.

REFERENCES

Akiyama, T. (2005). *Forest environmental tax and its meaning in forest environment and Forestry* (in Japanese). Norinkinyu 2 (Norinchukin Research Institute): p. 33–101.

Barnard, J., et al. (1985). Survey for monitoring changes and trends in renewable resources: Forests and marine fisheries. *The American Statistician*, *39*(4), 2 (American Statistical Association): p. 363.

Canada Forest Service. (1999). Forest Health, Context for the Canadian Forest Service's Science Program. Ottawa, Canada: Canadian Forest Service, Natural Resources Canada, p. 8.

Enomoto, T. (2008). *Risk management for environment and disaster*. Architectural Institute of Japan (no. 6): p. 23.

FAO. (2005a). Global forest resources assessment 2005. *Food and Agriculture Organization*, p. 15. Rome, Italy.

FAO. (2005b). Global forest resources assessment 2005. Chapter 2: Extent of forest resources. *Food and Agriculture Organization*, p. 15. Rome, Italy.

FAO. (2005c). Global forest resources assessment 2005. Chapter 2: Extent of forest resources. *Food and Agriculture Organization*, p. 17.

FAO. (2005d). Global forest resources assessment 2005. Chapter 2: Extent of forest resources, Food and Agriculture Organization, p. 11.

FAO. (2005e). *Global forest resources assessment 2005–15 key findings*. Food and Agriculture Organization.

Gomi, T. (2006). Chapter II: Soil erosion and forest. *Journal of Forest Research*, *47*, 11.

Goodale, C. L., et al. (2002). Forest carbon sinks in the northern hemisphere. *Ecological Applications*, *12*(3), 891.

Hirata, Y. (2009a). Observing forest absorption and emission from the sky (in Japanese). Public lecture of Forestry and Forest Products Research Institute, p. 28.

Hirata, Y. (2009b). Observing forest absorption and emission from the sky (in Japanese). Public lecture of Forestry and Forest Products Research Institute, p. 23.

http://www.shinrin-ringyou.com/data/mokuzai_kyoukyu.php (accessed June 1, 2009).

http://www.shinrin-ringyou.com/ringyou/ (accessed June 5, 2009).

Ibarrará, M. E., et al. (2009). Climate change and natural disasters: Macroeconomic performance and distributional impacts environment. *Development and Sustainability*, 11(3) (Springer Netherlands): p. 549.

IPCC. (2007a). *Climate change 2007: Mitigation of climate change, WG III report for the IPCC.* Summary for Policymaker and Technical Summary, p. 69.

IPCC. (2007b). *Climate change 2007: Synthesis report, in the fourth assessment report.* IPCC, p. 60.

IPCC. (2007c). *Climate change 2007: Synthesis report, in the fourth assessment report*, p. 56.

Kaneko, S. (2008). Forestry and global environment: Theoretical and historical study for commons. *Josai University Economic Bulletin* 4: p. 70.

Kobayashi, N. (2008). Global warming and forest (in Japanese). Japan Forestry Investigation Committee. ISBN: 978-4-88965-180-5. p. 57.

Kochi prefecture. (2002a) New tax (Forest Environmental Law) (in Japanese). Kochi prefecture, Forest Bureau, p. 1.

Kochi prefecture. (2002b). New tax (Forest Environmental Law) (in Japanese). Kochi prefecture, Forest Bureau, p. 2.

Kochi prefecture. (2002b). New tax (Forest Environmental Law) (in Japanese). Kochi prefecture, Forest Bureau, p. 1–2.

Kochi prefecture. (2008). Forest environmental tax for conserving mountain and forest, 2008. Available at http://www.pref.kochi.jp/~seisaku/kinobun2/hp_1/02zei/pdf/2008.4-panhu.pdf (accessed April 2008).

Kochi prefecture. (2010). Collaborative forest restoration with environmentally progressive companies (in Japanese). Available at http://www.pref.kochi.jp/^junkan/kyoudouno_mori/haikei/kyoudounomori_hello.html

Maciver, D. C., & Wheaton, E. (2005). Tomorrow's forests: Adapting to a changing climate. *Climate Change*, *70*(Springer 2005), 274. The Marrakech Accords, Decision 11/CP.7.

Matsumoto, M., & Mitsuta, Y. (2009). Role of the forest as carbon sink (in Japanese). Public lecture of Forestry and Forest Products Research Institute, p. 11.

Meguro, K. (2004). Disaster resilience social system (in Japanese). Architectural Institute of Japan (no. 2): p. 59.

Meteorological Agency. (2007). 3-2. Annual precipitation in abnormal weather risk map. Available at http://www.data.kishou.go.jp/climate/riskmap/index.html

Meteorological Agency. (2008). (in Japanese). Japan Meteorological Agency. Available at http://www.data.kishou.go.jp/climate/riskmap/index.html

Ministry of Agriculture, Forest and Fisheries. (2006a). White paper on forest and forestry (in Japanese), p. 67.

Ministry of Agriculture, Forestry and Fisheries. (2006b). White paper on forest and forestry (in Japanese), p. 116.

Ministry of Agriculture, Forestry and Fisheries. (2007). White paper on forest and forestry. Chapter 1: Challenge of new forestry (in Japanese).

Murata, S., Shibuya, H. & Nakai, M. 2001. *Questionnaire survey of human consciousness on the heavy rainfall of Hiroshima-city in June 1999*. Sojyo University Research Report 26(1): 121–131.

Nakamura, F., & Kakizawa, H. (2009). Evaluating forest functions (in Japanese). Hokkaido University Press.

National Institute of Population and Social Research. (2003). Demography of cities, towns and villages in Japan estimated in 2003 (in Japanese). Available at http://www.ipss.go.jp/p-info/e/psj2003/PSJ2003.html

Noss, R. F. (2001). Beyond Kyoto: Forest management in a time of rapid climate change. *Conservation Biology*, *15*(3), 586.

O'Brien, K., et al. (2008). *Disaster risk reduction, climate change adaptation and human security*. In Global Environmental Change and Human Security Report (3): p. 15.
Ohta, T. (2004). Japan's forest and mountain management in the 21st century. *Journal of Geography, 113*(2)203–211: p. 207.
Ohta, T. (2006a). Sediment-related disasters and new forest management. *Journal of Forest Research, 47*, 34.
Ohta, T. (2006b). Sediment-related disasters and new forest management. *Journal of Forest Research, 47*, 35.
Ohta, T. (2006c). Sediment-related disasters and new forest management. *Journal of Forest Research, 47*, 33.
Ono, K. (2008). "Environment" "Disaster" and Forest Vegetation (in Japanese). Architectural Institute of Japan (no. 6): p. 26.
Opinion Poll on Forest and Life. (in Japanese), the Cabinet Office. Available at http://www.rinya.maff.go.jp/j/press/kikaku/pdf/071120-03.pdf.
Ouchi, Y. (1987). Historical process of afforestation policy (in Japanese). *The Japanese Forest Economic Society, 111*, 3.
Practical Guide for Sediment-related Disaster Prevention Law (in Japanese). (2000). Taisei Shuppan, p. 14.
Sadohara, S. (2008). Systematic approach and challenge for integration of "Global Environment" and "Disaster Risk Reduction" (in Japanese). Architectural Institute of Japan (no. 6): p. 5.
Shaw, R., et al. (2009). Indigenous knowledge for disaster risk reduction, Policy Note, p. 14.
Spittlehouse, D. L., & Stewart, R. B. (2003). Adaptation to climate change in forest management. *BC Journal of Ecosystems and Management, 4*(1) (Forest Research Extension Partnership): p. 4.
Tanaka, N. (2009). Vegetation bioshields for tsunami mitigation: Review of effectiveness, limitations, construction, and sustainable management. *Landscape Ecological Engineering, 5*(71–79), 71.
Watanabe, A., & Fukuoka, S. (2003). Characteristics of the heavy rain in Hiroshima disaster 1999 and the significance of information of radar rain gage. *Journal of Japan Society for Natural Disaster Science, 22*(2), 166–185.

CHAPTER 8

MAINSTREAMING CLIMATE CHANGE ADAPTATION AND DISASTER RISK REDUCTION: A MALAYSIAN APPROACH

Joy Jacqueline Pereira, Tan Ching Tiong and Ibrahim Komoo

INTRODUCTION

Since independence, Malaysia has generally registered continuous economic growth and this development has brought about numerous benefits including improved social amenities and a trend toward greater urbanization of the population. Economic development in Malaysia has contributed to environmental degradation and uncontrolled physical development, especially in the urban areas. Protection of the environment has become a necessity rather than a luxury in order to maintain public health and well-being as well as to sustain the economic growth. As in most developing countries, there are many challenges facing the country, especially so in urban areas, where the human, physicochemical and biological environments are interlinked (Pereira & Komoo, 2004). One major challenge is the increasing occurrence of geological and flood-related disasters, causing property damage and high cost of maintenance as well as loss of lives, in extreme cases. In part, this is a manifestation of poor planning, and many of

Climate Change Adaptation and Disaster Risk Reduction: An Asian Perspective
Community, Environment and Disaster Risk Management, Volume 5, 147–167

ISSN: 2040-7262/doi:10.1108/S2040-7262(2010)0000005014

the problems related to hazards in urban areas are often exacerbated by human activities.

Natural disasters have been observed with an increasing trend since the last century, and climate change is expected to aggravate the situation (Cruz et al., 2007). A holistic and integrated approach is crucial in managing the risks of climate change and disasters. This chapter commences with a brief discussion on how climate change is expected to influence hazards and disasters in Malaysia. The subsequent sections focus on efforts to mainstream climate change and disaster risk reduction into national development in the country through policies, institutional arrangements, and planning responses; initiatives to bridge the science–governance interface; platforms to link researchers, practitioners, and other stakeholders; as well as efforts to enhance education and awareness. Issues and challenges to integrate climate change adaptation and disaster risk reduction are implicitly discussed in these sections.

THE PROGNOSIS

The process of urbanization has resulted in the present global scenario, where 3% of the land surface is occupied by cities and 50% of the world's population lives in cities (UNEP, 2006). Asia hosts 60% of the world's population, with 48% of its population being urban and a majority living in the coastal zones. Urban centers are drivers of economic development, requiring human capital and consuming natural resources while exerting pressure on the environment. The advent of climate change is expected to increase the current pressures on natural resources and the environment, especially so in Asia.

In Asia, populated coastal areas in South, East, and Southeast Asia are expected to be vulnerable due to increased flooding (Cruz et al., 2007). Changes in the hydrological cycle are projected to increase flooding from the rivers in some megadeltas. Records indicate that over the past decade the occurrence of hydrometeorological disasters is about five times higher compared with geological disasters in Asia (Emergency Event Database [EM-DAT], 2009). The average number of people affected per million inhabitants is about a 100 times higher for hydrometeorological disasters compared with geological disasters. By the 2050s, climate change will most likely contribute to the already high levels of hydrometeorological disasters and the number of people affected in the region, particularly with respect to flooding.

In Malaysia, the most common hydrometeorological disaster is flooding. Windstorms have also been reported since the 1970s, but they have affected a very small number of people. Records indicate that flooding is the major disaster in the country, affecting the greatest number of people over the last century (Liew, 2009). Flooding generally occurs during the high rainfall seasons, affecting major rivers, particularly in the east coast of Peninsular Malaysia. Flash-floods are also a common occurrence, particularly in urban areas. For example, between 1990 and 1996, over 100 incidents were reported in the Klang Valley alone, where Kuala Lumpur is located (Pereira & Komoo, 2004). Flash-floods usually occur with little or no warning after heavy rain and can reach the peak level in a short time. A number of reasons contribute to flash-floods and these include increased impermeability due to an increase in the built-up areas; increased accelerated erosion from exposed surfaces resulting in sedimentation of rivers and streams, which subsequently cause flash-flood along flat low-lying river channels; and poor maintenance of drainage facilities in built-up areas (Jamaluddin, 1986).

Geological disasters have been reported in Malaysia since the early 1900s. The early reports of disasters related to rock-falls and subsidence in limestone terrain, coinciding with urban expansion into limestone terrain. Since the onset of economic growth and expansion of urban areas into hilly terrains, landslides and slope failures have become more prevalent, affecting a significant number of the population. For example, between 1990 and 1996, about 20 incidents were reported around Kuala Lumpur alone, with total fatalities amounting to 48 people (Pereira & Komoo, 2004).

In Malaysia, localized climate projection indicates a substantial increase in monthly rainfall over the North East Coastal region, and a decrease in monthly rainfall in the West Coast of Peninsular Malaysia may be expected by 2050 (National Hydraulic Research Institute Malaysia [NAHRIM], 2006). By end of the century, a more significant change in the annual rainfall may be expected in the western regions of Sabah and Sarawak (Wan Azli, Mohan, & Kumarenthiran, 2008). Simulations of future river flows in several watersheds in the east coast of Peninsular Malaysia indicate increases in hydrologic extremes, i.e., higher high flows and lower low flows when compared with historical levels (NAHRIM, 2006).

With the onset of climate change, the number of disasters and people affected is anticipated to increase. Climate change is a process that influences all hydrometeorological hazards (Table 1). Generally, geological hazards are not influenced by climate change (Schmidt-Thome, 2006). However, landslides and subsidence are to a certain extent influenced by the water table level that is sensitive to rainfall, which is, in turn, susceptible to

Table 1. Hazards Acknowledged in the Hyogo Framework for Action and their Responses to Climate Change.

Geological hazards	Not influenced by climate change
Earthquakes	
Tsunamis	
Volcanic eruptions	
Landslides	Influenced by climate change
Subsidence	
Hydrometeorological hazards	
Avalanche	
Extreme temperature	
Drought	
Forest fire	
Flood	
Storms and storm surges	
Biological hazards	May be influenced by climate change
Technological hazards	
Environmental degradation	

Source: Modified after Schmidt-Thome (2006).

climate change. Increases or decreases in water levels of rivers may also have consequences for water quality. Higher and extreme runoffs may result in increased risk of flooding, which, in turn, heightens the risk of landslide disasters in urban areas. Increased flooding, particularly in areas previously not exposed to the hazards, could lead to dispersal of contaminants and toxins into rivers where wastewater treatment plants are overwhelmed. Such occurrences are already suspected in recent times and are currently under investigation. The possibility of circulation of environmentally hazardous substances in surface water, where industrial sites and landfills are affected, cannot be ignored. The consequences could be disastrous if such risks are not addressed.

Flood mitigation projects have been given priority to address the flash-flood situation, particularly around Kuala Lumpur. Such projects have served to lessen flash-flood disasters in limited parts of the city. In addition, much effort has been made to help planners and decision makers to understand and appreciate risk assessment and disaster reduction due to geological hazards. This includes creating effective linkages between geologists, planners, and decision makers to increase the awareness of the essential contribution of geological understanding to sound planning. Notwithstanding this, much more needs to be done, especially to address the new challenges that are anticipated in the onset of climate change.

In order to achieve sustainable human development, it is necessary to ensure that the planning process includes socioeconomic considerations, management of resources, suitability of land taking into account its potential geohazards, and environmental impacts. In addition, the planning process also has to take into account the changing climate and its potential impacts so as to reduce vulnerability and ensure resilience of any proposed socioeconomic development agenda. It would be more cost-effective to take adaptation measures early on, especially for critical infrastructure with long economic life. Current planning of critical infrastructure should take into account adaptation to the impacts of climate change to reduce the risk of disasters.

The issue of climate change and its interplay with the environment is complex, connected, and highly variable in spatial and temporal scale. There is also a high level of uncertainty involved. Decision making in the context of climate change demands an appreciation of the "big picture." The "big picture" draws on physicochemical knowledge systems as well as policy and planning instruments, including institutional arrangements, as they are currently practiced.

POLICY AND INSTITUTIONAL ARRANGEMENTS

The Climate Change Platform

Malaysia is a Party to the United Nations Framework Convention on Climate Change (UNFCCC) since 1994. The Ministry of Natural Resources and Environment is currently the National Focal Point to the UNFCCC. A National Steering Committee on Climate Change (NSCCC) was established at the Ministry to guide national responses on climate change. Under the stewardship of the NSCCC, Malaysia submitted to the UNFCCC in 2000 its Initial National Communication that, among others, contains impacts assessment and relevant adaptation measures (Ministry of Science, Technology and Environment [MOSTE], 2000). With growing concerns on the impacts of climate change, a Cabinet Committee on Climate Change was formed in early 2008 (Suboh, 2008). The Ninth Malaysia Plan (2006–2010) and several existing national policies encompass programs that may directly address or indirectly contribute to managing issues of climate change adaptation based on sectoral-specific context and needs.

To ensure concerted holistic responses to address climate change, a National Policy on Climate Change was formulated based on stakeholder consultation. The Policy calls for mainstreaming of measures to address climate change challenges for sustainable development; integration of balanced adaptation and mitigation responses into national policies, plans, and programs to strengthen the resilience of development; and strengthening of institutional and implementation capacity through coherent coordination of policy responses (Pereira, 2008). The National Policy seeks to mainstream climate resilient development into national, state, and local levels of government. Climate resilient development in the context of the Policy refers to development that takes into account measures to address climate change and extreme weather in line with national priorities. Aspects related to promotion of low-carbon economy and mitigation of climate change are equally balanced with efforts to climate-proof development by intensifying climate change adaptation. Thus, disaster risk reduction is given prominence in the Policy. The Policy has since been approved by the Malaysian Cabinet in the late 2009. The Policy is expected to be implemented in 2010.

The Disaster Risk Reduction Platform

Disaster risk reduction has always been a concern of the Malaysian Government in view of the severity of its impact on lives and livelihood as well as the development of the nation. There are several existing laws that support disaster mitigation and management in Malaysia. The National Security Council of the Prime Minister's Department issued Directive No. 20 on "Policy and Mechanism on National Disaster and Relief Management" on 11 May 1997 to guide relevant management activities according to the level and complexity of a disaster (Komoo, Aziz, & Lim, 2009a). The Council is responsible for coordinating activities that are implemented by the Disaster Management and Relief Committee at federal, state, and local levels that comprises various agencies. Government agencies are responsible for different aspects of disaster risk reduction while maintaining their core responsibilities. For example, the Malaysian Meteorology Department provides information and early warning on adverse weather, sea, and seismicity conditions as well as for tsunamis. The Department of Drainage and Irrigation oversees aspects related to flood mitigation, including monitoring of river flow and local rainfall. Relief and rehabilitation come under the purview of the Social and Welfare Department, while search and rescue is conducted by the Special Malaysia Disaster Assistance and Rescue Team (SMART). In addition, there are several nongovernment and civil society organizations that work closely with the National Security Council focusing on disaster relief including search and rescue operations.

Malaysia adopted the Hyogo Framework for Action (HFA) in 2005. The National Security Council is the National Focal Point to the HFA and coordinates its implementation within the country. In line with its function as the National Focal Point, the National Security Council and the United Nations International Strategy for Disaster Reduction (UNISDR) Asia Pacific hosted the Third Asian Ministerial Conference on Disaster Risk Reduction in Kuala Lumpur on December 2–4, 2008. The Ministerial Conference acknowledged the evolving nature of disaster risk amidst the dynamic climate scenario and called for the promotion of climate change adaptation wherever appropriate. This call was a follow-up to the Second Asian Ministerial Conference on Disaster Risk Reduction in 2007, which called for effective measures on integrating disaster risk reduction and climate change adaptation under the principle of common but differentiated responsibilities (Pereira, Aziz, Alatas, & Er, 2009). The Organisation of the Ministerial Conference in Kuala Lumpur served to heighten awareness on climate change among disaster risk reduction practitioners in Malaysia. However, there is still much to be done in this respect.

Existing laws and policies are sectoral in nature and the need for comprehensive legislation and law enforcement to minimize disaster impacts and encourage preventive measures has been acknowledged by the government (Che Moin, 2007). Emphasis should be increased on addressing gaps in governance, risk reduction and early warning, knowledge management, and education. For example, there is need to integrate disaster mitigation planning into the overall national development plans and projects for sustainable development. Capacity to develop and undertake risk assessment in reducing disaster vulnerability for translation into sectoral policies and measures needs to be enhanced. In addition, an integrated database management system needs to be developed. Regular educational and community awareness programs are also required on disaster prevention measures, specifically for disaster risk mitigation strategies. For this purpose, Malaysia should formally institute a National Platform on Disaster Risk Reduction. The establishment of a National Platform would facilitate stakeholder engagement to enhance implementation of disaster risk reduction activities in the country. The benefits of a National Platform are manifold and include advocating disaster risk reduction at different levels; increasing levels of knowledge and skills related to disaster risk reduction; acting as a catalyst for national consultations and consensus building; and coordinating, analyzing, and advising on areas that require concerted action, among others (Pereira et al., 2009). Furthermore, the establishment of a National Platform would facilitate in mainstreaming of disaster risk reduction into policies, planning, and development programs at various levels.

Linking and Mainstreaming

In the context of climate change, the term "mainstreaming" is used to refer to integration of climate change vulnerabilities or adaptation into a single sector, such as water management, disaster preparedness, and emergency planning or land-use planning (Klein et al., 2007). Mainstreaming also refers to the involvement of administrative actors and other stakeholders, whose main tasks are not connected to climate change adaptation or disaster risk reduction, to initiate their own actions to promote these goals (Pereira, 2009). Examples of such actions include formulation and implementation of broad development strategies, macropolicies, sector-specific policies, changes in institutional arrangements or organizational structures, vulnerability or hazard assessments as well as development project design and implementation. Horizontal mainstreaming takes into account promotion of such goals throughout all ministries and agencies at the same level of administration, i.e., federal, state, or local. Vertical mainstreaming refers to the integration of climate change and disaster risk reduction into a single sector through federal, state, and local levels. In order to effect mainstreaming of climate change adaptation and disaster risk reduction in the country, the custodian of both these issues has to be formally linked first to interact on a routine basis. There should also be coordination and clarification of roles and responsibilities.

The interaction between the national focal points of climate change and disaster management in Malaysia has been rather limited. It is noteworthy that the Malaysian Meteorological Department (MMD) has played a bridging role (Tan, Pereira, Shaw, & Komoo, 2009). Officers of the MMD occasionally participate in national delegations to meetings of both the UNFCCC and HFA. In addition, the MMD, with support from the Institute for Environment and Development (LESTARI-UKM) and other parties, also took the lead to organize high-profile regional and national events since 2006, linking climate change adaptation and disaster risk reduction. These events provided a platform for policy makers from both focal points and their respective cadre of stakeholders to interact. The increasing recognition of the relationship between climate-related hazards and disaster risk reduction has served to improve ties between the two groups of practitioners at the national level albeit on an ad hoc basis. Institutional arrangements for routine communication and exchange of information between the two national focal points have yet to be formalized. It is anticipated that implementation of the National Policy on Climate Change, which will see the engagement of a broad range of stakeholders as well as the establishment of a formal National Platform on Disaster Risk Reduction for stakeholder engagement, will serve to improve the situation.

There are many ways to mainstream the climate change agenda (Klein et al., 2007). One way to achieve mainstreaming at the country level is through policy formulation. Mainstreaming of climate change adaptation and disaster risk reduction through policy formulation can facilitate economic planning and project implementation. The issue of climate change is expected to become an agenda in the mainstream through the implementation of the National Policy on Climate Change. Disaster reduction can be linked and incorporated to climate change adaptation goals and indicators when implementing the National Policy, giving both issues more prominence at national, state, and local levels. In addition, economic planning including budgetary allocations for both the issues can be given higher priority at all levels. Decision makers can also request that all future projects take into account vulnerabilities to climate change adaptation and disaster risk reduction. In addition, allocations can be made for all existing development projects to consider climate risks or address vulnerability to climate variability and change for disaster prevention (Fig. 1). Notwithstanding this, there are challenges to mainstreaming the agenda. In Malaysia, these include provision of relevant information for decision makers, uncertainty associated with climate projections, status of federal–state relations, and compartmentalization of government agencies at all levels, among others.

SPATIAL PLANNING RESPONSES AND DEVELOPMENT CONTROL

Legislative Context

The planning process is critical to achieve sustainable human development, which encompasses socioeconomic stability security, wise management of resources, and environmental conservation. Development plans assist in the

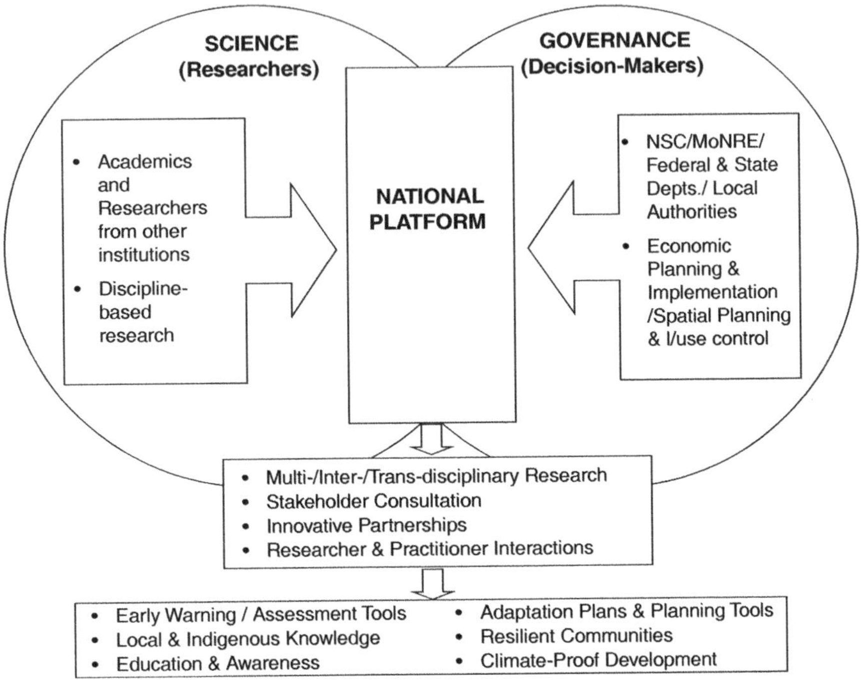

Fig. 1. A Conceptual Diagram Illustrating the Domains of Science, Governance, and the National Platform, to Link CCA and DRR.

translation of socioeconomic objectives into spatial and physical forms. The Town and Country Planning Act 1976 (Act 172) provides the legal basis for the formulation of development plans in Peninsular Malaysia. Such development plans include the National Physical Plan, State Structure Plan, Local Plan, and Special Area Plan. In terms of hierarchy, the National Physical Plan is the most macro, while the Special Area Plan is relatively micro. In the case of Malaysia, ignorance of geological processes and their omission from spatial planning considerations in the past have resulted in unsustainable development (Pereira & Komoo, 1998). It has been an expensive lesson in human and economic terms (for instance, the Highland Towers tragedy, Taman Sri Muda floods). The Town and Country Planning Department took the lead in responding to the urgent need for geohazard risk identification in the planning process to ensure a safe environment for the well-being of present and future generations.

The Town and Country Planning Act 1976 (Act 172) was amended to the Town and Country Planning (Amendment) Act 1995 (Act A933), with the objectives of strengthening the administrative mechanism of planning at the state and local council levels in order to improve efficiency and meet the changing needs for planning, i.e., to balance environmental conservation with physical development and reduce damage to property and loss of life due to uncontrolled development. *Section 21A* of the revised Act requires the submission of a development proposal report, when applying for planning permission under Section 21(1). The development proposal report should contain description of the land including its physical environment, topography, landscape, geology, contours, drainage, water bodies and catchment as well as natural features of the land, among other requirements. The applicant has to show to what extent the development proposed would affect the above elements and adjacent areas. The applicant is also required to suggest measures to conserve and improve the physical environment. The requirement of a development proposal report increases the power of the local authorities and enables the evaluation of suitability of the proposed project in terms of planning, for conservation and physical development, by incorporating elements of safety, health, and well-being (Ismail, 1996). The Town and Country Planning (Amendment) Act 1995 (Act A933) is the first line of defense in terms of environmental protection, and it can be also be used as a planning response to climate change and disaster risk reduction. In addition, the Act requires the participation of the public in reviewing the plans that are formulated, making it suitable as a tool to enhance stakeholder engagement with respect to climate change and disaster risk reduction.

Planning Process and Tools

The Town and Country Planning (Amendment) Act 1995 (Act A933) exerts primary control over the planning process in Malaysia. There are essentially two aspects and these are land-use planning and development control. Land-use planning at the local authority level results in structure plans and local plans, which takes into account economic needs, population density, existing land use, and other factors associated with the development process as well as environmental aspects, such as conservation of wetlands and catchment areas. Development control involves the regulation and control of planning for projects. This takes into account consultation with and input from several technical departments, prior to the concession of land conversion or planning permission. The local authority consults several technical departments before permission is granted.

The provision of appropriate and easily understood information regarding vulnerable areas in the future will aid local authorities in making a decision regarding an application for planning permission. Planning tools such as the Guidelines on Integrated Environmental Sensitive Areas (ESAs) could be expanded in scope to incorporate elements of climate change adaptation and disaster risk reduction (Pereira, Halimaton, & Komoo, 2006; Halimaton, Pereira, & Komoo, 2007). Integrated ESAs is a very useful tool to ensure land use that supports the aspirations of sustainable development, particularly when used in conjunction with existing land-use planning mechanisms at the state or local levels, as they will contribute greatly to the development of an effective National Adaptation Plan (Fig. 1).

However, the implementation of Integrated ESAs is facing many barriers (Pereira et al., 2006). A major challenge is to convince decision makers that it is useful for making decisions on applications for land-use conversion as well as for applications for planning permission, which are submitted routinely to state and local authorities. As an interim measure, vulnerable areas could be identified and ranked separately, to be integrated into the process of formulating development plans so that these aspects are taken into account. This will contribute greatly to the development and review of structure and local plans at the state and local authority levels. Strategic Environmental Assessment and Sustainability Assessment are useful tools for assessing the sustainability of development plans at the macro level. Such tools could also be used to assess the effectiveness in managing the vulnerable areas with respect to sustainable development. Risk identification should be taken into account and integrated into the overall planning and development process, and the capacity of all planning and executing agencies at the state and local levels should be strengthened in this respect.

RESEARCH TO BRIDGE THE SCIENCE–GOVERNMENT INTERFACE

The current contribution of science in the implementation of policies to facilitate good governance is generally not very encouraging (Pereira & Komoo, 2003). The awareness of the relevance and significance of scientific information, particularly among policy and decision makers is very low. The contribution of science in seeking solutions to global sustainability problems is not commonly highlighted. Scientific information should be presented in a form that can be widely understood and fits into the governance structure, systems, and processes that are undertaken (Komoo, Aziz, Mokhtar, Jamaluddin, & Lim, 2009b). In addition, information on costs and benefits in applying science would facilitate its uptake in the governance arena. There is much knowledge to be generated to bridge the science–governance interface with respect to the pressing issues of sustainability in Malaysia (Pereira & Chen, 2008; Chen & Pereira, 2008; Komoo et al., 2009b).

Knowledge generation to bridge the science–governance interface requires integrative scholarship involving multidisciplinary, interdisciplinary, and transdisciplinary approaches (Nordin, 2004). Multidisciplinary approaches are goal-oriented and involve the study of an issue from the perspective of several disciplines, with minor changes in the methodology of individual disciplines. Interdisciplinary approaches require the transfer of methodology from one discipline to another; where at advanced levels, this leads to changes in epistemology. Transdisciplinary approaches are simultaneous integration of knowledge from various disciplines, perspectives, involving various stakeholders, researchers, and practitioners. This approach requires a common framework for assimilation. Climate change adaptation and disaster risk reduction requires such integrative scholarship to generate and apply science for informed decision making in governance (Fig. 1). There should be explicit focus on real-world actions rather than theoretical assessments of potential responses to hypothetical climate scenarios.

Currently in Malaysia, there are pressing issues regarding climate change adaptation and disaster risk reduction that should be the focus research questions. For example, what is the status of critical infrastructure in Malaysia? How vulnerable are roads, railways, dams, ports, airports, water supplies, and wastewater treatment, which have been designed to stay in service for a long period but have not been designed to take into account the impacts of changes in climate, particularly to floods or drought? How will changes in precipitation alter groundwater levels and pore-pressure conditions, which in turn have an impact on landslides and mudslides? Can stability be expected to deteriorate in sloping weak rock terrain, and where would the deterioration be substantial? Where are the areas that are regarded as stable today that will need to be the subject of stability improvement measures should climatic conditions change? Such areas need to be identified particularly if they are host to human settlements. Many developed countries are investing in research activities for identifying areas that are susceptible to such risk in the event of flooding due to increased precipitation. The findings of such work will form the basis for their national adaptation action plan to prepare for the impacts of climate change. Malaysia should embark on a similar path.

In addition, systems must be developed to identify, monitor, provide early warning, and take measures to reduce risk of disasters upon the onset of climate change. In order to have these systems in place, gaps in scientific knowledge need to be addressed jointly by both researchers and practitioners. Currently, the government has initiated research on adaptation measures in several sectors. Unfortunately, a study to link climatic hazards and climate change adaptation has not commenced. Such a study is critical as any impact on human settlements and infrastructure may not only cost loss of lives, but also have the potential to trigger a national crisis if economic and civil activities are disrupted. The goal of such a study should be to enhance the resilience of vulnerable communities in Malaysia to present and future hazards. This requires an integrated approach to hazards and risk management under different climatic regimes and establishment of a vulnerability assessment methodology and development of early warning systems taking into account social, cultural, ethical, political, economic, and legal considerations. The availability of various technologies for adaptation and disaster risk reduction to promote human well-being and security should also be looked into.

INNOVATIVE PARTNERSHIPS AND STAKEHOLDER ENGAGEMENT

Efforts to address climate change include a broad range of issues, such as emissions targets, strategies to promote low-carbon economy, expansion of financing instruments, adoption of common policies, and implementation of development-oriented actions at sector-specific as well as federal, state, and local levels. However, there is general acknowledgement that a certain level of climate change can no longer be avoided, prompting urgent consideration on adaptation to prevent adverse impacts. Disaster risk reduction will be a major component of such measures as climate change may alter exposure and drivers of vulnerability to hazards and exacerbate disaster risks. If climate change adaptation measures are to be efficient and effective, they must build on and expand existing disaster risk reduction efforts. Similarly, sustainability of disaster risk reduction approaches will be affected without considering the impacts of climate change (Venton & La Trobe, 2008). The disaster management community has a long history of dealing with disaster events, and therefore, such past experience on disaster risk reduction should be capitalized on for climate change adaptation (O'Brein, O'Keefe, Rose, & Wisner (2006); Venton & La Trobe, 2008). Measures typically associated with climate change adaptation will support disaster risk reduction through reducing long-term vulnerability and influencing development potential.

A critical step is to institutionalize a platform to promote communication and collaboration between researchers and practitioners in the climate change adaptation and disaster risk reduction communities. Such a platform can be initially created through existing research activities for exchange of information, which aim to encourage continuous systematic dialogue and subsequently lead to promotion of joint development of climate change adaptation–disaster risk reduction strategies and actions. To support the collaboration mechanism, assessment and mapping of past disasters and climate change projection should be undertaken in an integrated manner and localized to the extent possible. Analysis should encompass identifying drivers of past and current vulnerabilities and assessing success of current adaptation to present-day climate risks, with a view to enhance capacity building and resilience. The experience thus far demonstrates that participatory initiatives with a wide range of stakeholders can produce a high level of acceptance, involvement, and knowledge sharing (Pereira et al., 2009). This is a key factor for successful initiatives aiming to implement disaster risk reduction practices in some risk-prone parts of the world. Based on this experience, climate change adaptation and disaster risk reduction research initiatives in the country have been streamlined and consolidated, to promote as much as possible the broadest engagement of stakeholders (Fig. 1). In addition, linkages from regional to national platforms have also been given emphasis to facilitate exchange of experience.

The Southeast Asia Disaster Prevention Research Institute (SEADPRI-UKM) was established by the Government of Malaysia at the Universiti Kebangsaan Malaysia (UKM) in 2008 to provide research inputs to various issues related to science and governance in order to reduce underlying risk factors to disasters at all levels of planning (Mohd. Najib, 2009; Southeast Asia Disaster Prevention Research Institute [SEADPRI-UKM], 2009). SEADPRI-UKM supported the National Security Council Malaysia in organizing and documenting the proceedings and outcomes of the Third Asian Ministerial Conference on Disaster Risk Reduction in December 2008 (Pereira et al., 2009). Researchers from SEADPRI-UKM also worked in close collaboration with the Malaysian Network for Climate, Environment and Development (MyCLIMATE), one of the nine core research groups of LESTARI-UKM, in developing the national policy on climate change for

Malaysia and promoting the mainstreaming of climate change adaptation into selected sectoral policies. MyCLIMATE, which is supported by the Ministry of Natural Resources and Environment, serves as the platform to bring together and promote interaction between policy and decision makers, researchers, and practitioners from government, universities, business, and industry organizations as well as nongovernment and community-based organizations. The activities of SEADPRI-UKM and MyCLIMATE of LESTARI-UKM are expected to enhance the platforms and enable further synergy between stakeholders from the policy, operational, and research levels in integrating climate change adaptation and disaster risk reduction.

At the regional level, SEADPRI-UKM is a member of the Asian University Network of Environment and Disaster Management (AUEDM), which is a regional education and research platform that promotes sharing and synergies of educational product, experiences, and actual implementation of disaster risk reduction efforts (Shaw, 2008). The network is based in the Graduate School of Global Environmental Studies, Kyoto University, with the research objective on exploring the integration of climate change adaptation and disaster risk reduction. The network is currently documenting the Asian experience on various issues on climate change adaptation and disaster risk reduction, which will become a knowledge base for students and researchers as well as scientific inputs to practitioners and policy makers. Efforts are also underway for AUEDM to work closely and build synergies with the upcoming Adaptation Research and Policy Network for Asia and the Pacific (ARPNAP), which will be based at LESTARI-UKM and aims to build capacity of researchers in the region to provide adaptation policy-relevant information. This link is expected to enhance synergies between stakeholders involved in both climate change adaptation and disaster risk reduction.

Researchers from SEADPRI-UKM serve as members of the Malaysian Delegation to political conferences on disaster risk reduction, i.e., the Global Platform and Asian Ministerial Conferences under the auspices of the United Nations International Strategy for Disaster Reduction (UNISDR). MyCLIMATE is actively providing research support to the Malaysian Focal Point to the United Nations Framework Convention on Climate Change (UNFCCC). Together, both SEADPRI-UKM and MyCLIMATE are continuously engaged with policy and decision makers at the national level to enable better grasp of both the climate change adaptation and disaster risk reduction issues. The engagement also provides opportunity to identify country and regional needs, influence actions, and advance implementation. Researchers from both entities are also involved in the forthcoming Special Report of the Intergovernmental Panel on Climate Change (IPCC) on "Managing the Risks of Extreme Events and Disasters to Advance Climate Change Adaptation." The Special Report, to be completed by 2011, will contribute to the preparation of the IPCC's Fifth Assessment Report and UNFCCC negotiations as well as implementation of the Hyogo Framework for Action. There will also be active participation from Malaysia at the upcoming preparation of the Fifth Assessment Report the IPCC.

EDUCATION AND AWARENESS

Education has an important role to play in climate change and disaster risk reduction, but it must not just address response but educate on matters pertaining to risks, to empower communities to be prepared and address current and future disasters. Education must take into account all knowledge and put them into practice. Indigenous, local, and traditional practices are important and should be integrated to strengthen awareness, community responses, and preparedness. An educated community can make informed decisions and take part in the decision making process to better implement programs on climate change and disaster risk reduction.

An assessment of the status of education with respect to climate change and disaster risk reduction should be given high priority in Malaysia. There is a need to mainstream these issues into the formal education system through curriculum development at primary, secondary, and tertiary levels. Nonformal education to promote public awareness and participation, retrain practitioners, enhance technical knowledge, conduct state-of-the-art and participatory action research, bridge gaps in the science–governance interface, and strengthen institutional arrangements to targeted stakeholders should also be a priority (Fig. 1). Communication strategies should be put into place and the role of the media in this regard should be strengthened.

Currently, MyCLIMATE has embarked on an initiative with the United Nations University to undertake an assessment of higher education needs for climate and ecosystems change and adaptation in Malaysia. This will serve as the basis for initiating an assessment of higher education needs for climate change and disaster risk reduction. In addition, SEADPRI-UKM is also collaborating with MERCY MALAYSIA to develop awareness training programs on climate change and disaster risk reduction, targeting various stakeholders including local communities. The implementation of the training will be led by MERCY MALAYSIA to capitalize on the existing networks and mechanisms that have already been developed.

CONCLUDING REMARKS

Climate change is very likely during the 21st century. The projected change and increase of hydrometeorological hazards is also likely to have consequences on certain geological hazards. Unless there is serious intervention to reduce vulnerability and build resilience in a changing climate, increases in sociopolitical and economic costs are inevitable. Thus, it is critical to ensure that climate change adaptation and disaster risk reduction is a priority at national, state, and local authority levels with a strong institutional basis for implementation. The use of knowledge, innovation, and education should be promoted to build a culture of safety and resilience, reduce the underlying risk factors, and strengthen disaster preparedness for effective response at all levels. Efforts to identify, assess, and monitor risks as well as enhance early warning in a changing climate must be strengthened.

Mainstreaming of climate change and disaster risk reduction into national development in the country can be effected through appropriate policies, institutional arrangements and planning responses, research initiatives to bridge the science–governance interface, platforms to link and create synergies between researchers, practitioners, and other stakeholders, as well as concerted efforts to enhance education and awareness. Two institutions within Universiti Kebangsaan Malaysia, SEADPRI-UKM and LESTARI through MyCLIMATE, are promoting and influencing initiatives to mainstream climate change and disaster risk reduction through research, and making a significant difference to the country in the context of sustainable development.

ACKNOWLEDGMENTS

This paper is part of a research project entitled "Climate Resilient Development – Linking Science and Governance for Sustainability" funded by Universiti Kebangsaan Malaysia (Arus Perdana) and led by Prof. Dr. Joy Jacqueline Pereira. The authors would like to acknowledge the research team on climate change at the Institute for Environment and Development (LESTARI), Universiti Kebangsaan Malaysia, for their support and contribution of ideas.

REFERENCES

Che Moin, U. (2007). Disaster mitigation support and management in Malaysia. In: J. J. Pereira & M. Subramaniam (Eds), *Rapporteurs report for the national seminar on socio-economic impacts of extreme weather and climate change*, June 21–22, 2006, Putrajaya, organised by Ministry of Science, Technology and Innovation, Malaysia (p. 62).

Chen, S. P., & Pereira, J. J. (2008). Role of geoscience in governance for sustainable development and human security – issues and challenges. In: J. J. Pereira & S. P. Chen (Eds), *Role of geoscience in governance for sustainable development* (pp. 63–79). Bangi: LESTARI Publishers.

Cruz, R. V., Harasawa, H., Lal, M., Wu, S., Anokhin, Y., Punsalmaa, B., Honda, Y., Jafari, M., Li, C., & Huu Ninh, N. (2007). 2007: Asia. In: M. L. Parry, O. F. Canziani, J. P. Palutikof, P. J. van der Linden & C. E. Hanson (Eds), *Climate change 2007: Impacts, adaptation and vulnerability. Contribution of working group II to the fourth assessment report of the intergovernmental panel on climate change* (pp. 469–506). Cambridge, UK: Cambridge University Press.

Emergency Event Database (EM-DAT). The OFDA/CRED international disaster database. Available at www.emdat.be. Retrieved on August 25, 2009.

Halimaton, S. H., Pereira, J. J., & Komoo, I. (2007). Sustainable city planning – emphasis on management of environmentally sensitive areas. *Journal of the Malaysian Institute of Planners*, *5*, 61–76.

Ismail, I. (1996). Pindaan Akta Perancangan Bandar dan Desa dan Laporan Cadangan Pemajuan [Revision to the town and country planning act and development proposal plan]. *11th Senior Officials Meeting of the Town and Country Planning Department, Kuching, Sarawak*, October 2–5, 1996. Organisers: Town and Country Planning Department of Peninsular Malaysia and the Government of Sarawak.

Jamaluddin, M. J. (1986). Human perception and responses to the flash-flood hazards in parts of Kuala Lumpur. In: S. Sani (Ed.), *A study of the urban ecosystem of the Kelang valley region, Malaysia, Vol. 1 – The Kelang valley region: Some selected issues* (pp. 84–104). Bangi: Universiti Kebangsaan Malaysia.

Klein, R. J. T., Huq, S., Denton, F., Downing, T. E., Richels, R. G., Robinson, J. B., & Toth, F. L. (2007). Inter-relationships between adaptation and mitigation. In: M. L. Parry, O. F. Canziani, J. P. Palutikof, P. J. van der Linden & C. E. Hanson (Eds), *Climate change 2007: Impacts, adaptation and vulnerability. Contribution of working group II to the fourth assessment report of the intergovernmental panel on climate change* (pp. 745–777). Cambridge, UK: Cambridge University Press.

Komoo, I., Aziz, S., & Lim, C. S. (2009a). Incorporating the hyogo framework for action into landslide disaster risk reduction in Malaysia. In: *Proceedings of the third Japan-Malaysia symposium on geohazards and geoenvironmental engineering, organised by JSPS/VCC Core University Program "Environmental Science,"* Kyoto University, Kyoto (pp. 1–6).

Komoo, I., Aziz, S., Mokhtar, M., Jamaluddin, T. A., & Lim, C. S. (2009b). *Sustainability science for sustainability governance – A study on science governance symbiosis. Science Council of Asia Joint Project Final Report*. LESTARI Publishers, Bangi.

Liew, V. K. (2009). Malaysia's official statement. In: *Second session of global platform for disaster risk reduction*, June 16–19, 2009, Geneva.

Ministry of Science, Technology and Environment (MOSTE). (2000). *Malaysia initial national communication*, p. 131.

Mohd. Najib, A. R. (2009). Inaugural message. In: J. J. Pereira, S. Aziz, S. M. S. H. Alatas, & A. C. Er (Eds.), *Multi-stakeholder partnership for disaster risk reduction – From national to local. Conference Report of the Third Asian Ministerial Conference on Disaster Risk Reduction*, LESTARI Publishers, Bangi (pp. xv–xviii).

National Hydraulic Research Institute Malaysia (NAHRIM). (2006). *Study of the impact of climate change on the hydrologic regime and water resources of Peninsular Malaysia – Final report*. Malaysia: Ministry of Natural Resources and Environment (p. 184).

Nordin, M. (2004). *Basis and structures of integrative scholarship – towards the construction of a holistic and comprehensive approach in environment and development* (32p.). [In Malay]. Bangi: LESTARI Publishers.

O'Brein, G., O'Keefe, P., Rose, J., & Wisner, B. (2006). Climate change and disaster management. *Disasters*, *30*(1), 64–80.

Pereira, J. J. (2008). National policy on climate change (Draft 1 – September 10, 2008) in the *Consultation Workshop on the Draft National Policy on Climate Change*, September 24–25, 2008, Putrajaya.

Pereira, J. J. (2009). Mainstreaming climate change: Development of the national policy on climate change. *Presentation at workshop on mainstreaming climate change in agricultural and water sectors*, August 10, 2009, Kuala Lumpur, organized by LESTARI-UKM & IGES-Japan.

Pereira, J. J., Aziz, S., Alatas, S. M. S. H., & Er, A. C. (Eds), (2009). Multi-stakeholder partnership for disaster risk reduction – From national to local. *Conference report of the third Asian ministerial conference on disaster risk reduction*, LESTARI Publishers, Bangi (p. 183).

Pereira, J. J., & Chen, S. P. (2008). Geoscience governance – issues and challenges. In: J. J. Pereira & S. P. Chen (Eds), *Role of geoscience in governance for sustainable development* (pp. 19–26). Bangi: LESTARI Publishers.

Pereira, J. J., Halimaton, S. H., & Komoo, I. (2006). Management of environmentally sensitive areas. In: S. H. Halimaton & H. Norliza (Eds), *Planning a developed state-development planning for sustainability* (pp. 13.1–13.13). Shah Alam: Selangor State Government.

Pereira, J. J., & Komoo, I. (1998). Using geological information as planning tools for urban centers-the case of Klang valley, Malaysia. In: D. P. Moore & O. Hungr (Eds), *Engineering geology: A global view from the pacific rim – 8th Congress of the IAEG* (Vol. 4, pp. 2573–2578). Rotterdam: Balkema.

Pereira, J. J., & Komoo, I. (2003). Mainstreaming geoscience inputs for sustainable development: Prospects for the Malaysian scenario. *Warta Geologi*, *29*(6), 233–238.

Pereira, J. J., & Komoo, I. (2004). Sustainable management of the physical component of urban ecosystems: The Malaysian perspective. In: D. Miller & G. de Roo (Eds), *Integrating city planning and environmental improvement-practicable strategies for sustainable urban development* (2nd ed., pp. 115–126). Aldershot: Ashgate Publishing Ltd.

Schmidt-Thome, P. (2006). *Integration of natural hazards, risks and climate change into spatial planning practices.* Espoo: Geological Survey of Finland.

Shaw, R. (2008). Purpose and background: Asian University Network of Environment and Disaster Risk Management (AUEDM). In: *First workshop of AUEDM: Focusing on higher education and disaster reduction and climate change adaptation*, (pp. 14–15). Japan: Kyoto University Graduate School of Global Environmental Studies.

Southeast Asia Disaster Prevention Research Institute (SEADPRI-UKM). (2009). *SEADPRI Newsletter* (1), 1–2.

Suboh, M. Y. (2008). Opening speech at the *Forum and Exhibition on Climate Change in the Asia Pacific Region*, Kuala Lumpur, April 28, 2008. Available at http://www.nre.gov.my/ucapan/archive/2008/04/28/majlis-perasmian-forum-dan-pameran-quote-climate-change-in-the-asia-pacific-quote.aspx. Retrieved on December 1, 2008.

Tan, C. T., Pereira J. J., Shaw, R., & Komoo, I. (2009). Climate change adaptation and disaster risk reduction: Gaps and opportunities in Malaysia. *Proceedings of the third Japan-Malaysia symposium on geohazards and geoenvironmental engineering*. JSPS/VCC Core University Program "Environmental Science", Kyoto University, Kyoto (pp. 23–26).

UNEP. (2006). One planet, many people: Atlas of our changing environment, UNEP/GRID, USGS/EROS data center, Sioux falls, USA. Available at http://www.na.unep.net/OnePlanetManyPeople/powerpoints.html.

Venton, P., & La Trobe, S. (2008). *Linking climate change adaptation and disaster risk reduction* (16pp.). UK: Tearfund.

Wan Azli, W. H., Mohan, S., & Kumarenthiran, K. S. (2008). Climate change scenario climate change scenario and the impact of global warming on the winter monsoon. In: *Second national conference on extreme weather and climate change: Understanding science and risk reduction*, October 14–15, 2008, Putrajaya, Malaysia.

CHAPTER 9

CLIMATE CHANGE ADAPTATION IN NEPAL: ISSUES AND STRATEGIES

Jishnu Subedi

INTRODUCTION

Three years ago, Naina Shahi's husband left their small village in rural Nepal to seek work in neighboring India, leaving her to bring up their three children alone. The dry winters and unpredictable monsoons Nepal has experienced in recent years had hit crop production on the couple's land plot in the foothills of the Himalayas, forcing them to look for other ways to feed their family (A report in September 4, 2009 issue of Republica; Cozens (2009)).

The majority of the population in Nepal lives below the poverty line and their livelihoods depend on rain-fed agriculture and forest resources that are highly climate-sensitive. The livelihoods of the people will be severely affected by climate change. Some changes are already being witnessed, and perhaps many unknown changes are yet to be observed. Moreover, the recent events show that extreme weather events – floods, drought, and storms – have become more frequent and more severe. "Consistent rises in annual mean temperature, less frequent but more intensive rainfall events, increasing frequency and intensity of floods, changes in monsoon on – and offset, growing threat from Glacial Lake Outburst Floods (GLOF), longer

Climate Change Adaptation and Disaster Risk Reduction: An Asian Perspective
Community, Environment and Disaster Risk Management, Volume 5, 169–196

ISSN: 2040-7262/doi:10.1108/S2040-7262(2010)0000005015

dry spells and drought events, and increasingly stronger storms have already been experienced in Nepal in the past decade" (GoN/UNDP, 2008).

Nepal is a least developed country (LDC) party to the United Nations Framework Convention on Climate Change (UNFCCC). Nepal, considering its diverse topography, geo-spatial location, and regional differences, is party to UNFCCC under a special category as specified in UNFCCC Article 4.8. Article 4.8 of the UNFCCC states that:

> In the implementation of the commitments (under Article 4.0), the Parties shall give full consideration to what actions are necessary under the Convention, including actions related to funding, insurance and the transfer of technology, to meet the specific needs and concerns of developing country Parties arising from the adverse effects of climate change and/or the impact of the implementation of response measures, especially on countries with areas (i) prone to natural disasters; (ii) fragile ecosystems, including mountainous ecosystems; and (iii) land-locked countries.

Nepal ratified the UNFCCC in May 1994, and taking active part in the Conference of the Parties (COP) meetings thereafter. In September 2005, Nepal had access to the Kyoto Protocol and had entered into force since December 2005. The Initial National Communication (INC) (MoPE/UNEP, 2004) of Nepal made the case that Nepal is highly vulnerable to the potential negative impacts of climate change. Climate change is not only decreasing agriculture productivity and affecting livelihoods of the people but is also causing damage and loss of human lives and property due to increased frequency and severity of hydrometeorological disasters. These issues are affecting people already under harsh condition of deprivation and poverty.

The growing urban population in Nepal is already in short supply of many of basic urban facilities including water supply, sewage system, and electricity. Waterborne diseases are on the rise as the floodwater stagnates in and around homes in Kathmandu (see Fig. 1). "One of the key factors which affects all these cities is the lack of sufficient drainage, which leaves people vulnerable to flooding after heavy rains. Twenty-five percent of households in Kathmandu have been flooded due to inadequate drainage as storm water no longer drains away as it used to and the technical systems put in place are not sufficiently flexible to deal with a changing climate. Furthermore, the sewage network in Kathmandu is only able to operate at 40 percent of its capacity due to sludge and other debris blocking the system" (IIED (International Institute for Environment and Development) [IIED], 2009).

Nepal needs to take immediate and stringent measures for mitigation of and adaptation and coping to adverse impacts of climate change. The entry

Fig. 1. Passersby Walk through a Road in Kathmandu Where Water Is Stagnant due to Blockage of Drainage from Solid Waste.

point for this can be expanding upon and replicating successful experiences of community-owned initiatives (COIs), such as community forestry and micro-hydro projects. The achievements of these initiatives can be witnessed in terms of better livelihoods, social mobilization, and income generation for local people and institution building at a grassroots level. This chapter gives background of climate change issues in the context of Nepal, initiatives in Nepal to address climate change, achievements in COIs and the strategic approach for mitigation, and adaptation to climate changes.

GENERAL BACKGROUND OF NEPAL

Nepal, a landlocked country, is around 850 km long (east-west) and around 200 km wide (north-south). The country experiences a wide range of climatic variations, from subtropical in the south to alpine in the north, as the

Table 1. Summary of Climatic Conditions of Nepal.

Ecological Belt	Climate	Average Annual Precipitation (mm)	Mean Annual Temperature (°C)
Mountain	Arctic/alpine	Snow/150–200	<10
Hill	Cool/warm temperate	275–2,300	10–20
Terai	Subtropical	1,100–3,000	20–25

Source: GoN/UNDP (2008).

elevation changes from 64 to 8,850 m above sea level within a span of about 200 km (see Table 1). It experiences heavy rains from June to September due to the southeast monsoon, which account for 80% of the total rainfall. The winter rains from November to January and pre-monsoon rains from April to May account for the rest of the rainfall. Rain concentration in a short period triggers landslide problems in higher mountainous terrain and floods in lowlands. In average, landslides and floods kill about 400 people each year, and thousands of people are affected adversely.

Nepal has more than 6,000 rivers and tributaries, most of them running from north to south. The Himalayas are the source of the rivers and they flow south in the steep topographic conditions. The rivers carry about 225 billion cubic meters of water, offering huge potential for electricity, irrigation, and domestic water use. However, only a small part of the water has currently been utilized, and a major part of the resources go untapped. The country is covered by around 39.6% forestland, out of which communities own and manage 40%.

Nepal has a population of 280 million and a gross domestic product (GDP) of about US$270. The country ranks 142 among a total of 177 countries in terms of the Human Development Index. Nepal's economy is basically agriculture-based, with more than 80% of the population still engaged in agricultural activities. The share of agricultue in terms of the GDP is 40%. About 40% (31% according to MoF/GoN, 2009) of the people still live below the poverty line (UNDP, 2007). The hydropower potential of the country is hugely underutilized (less than 1%) and the major source of energy is still biomass covering around 87%. Around 40% of the population has access to electricity (urban access is 87% and rural access is 27%) (NPC/UNDP, 2005). Share of energy consumption according to source type is shown in Fig. 2.

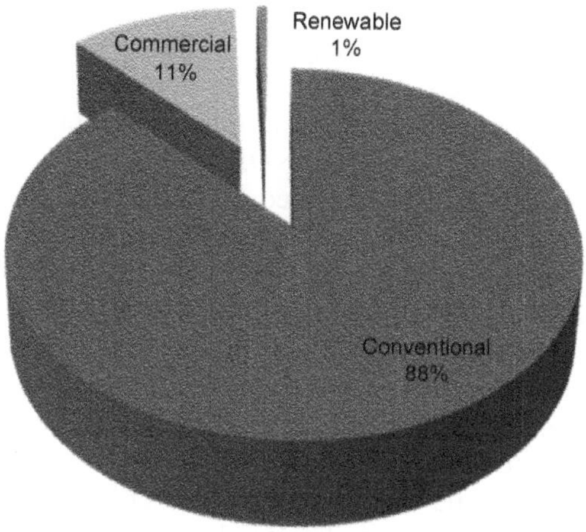

Share of different sources in Energy Consumption

Sources	%
Conventional	*88*
Fuel wood	89.2
Agricultural residue	4.1
Livestock residue	6.6
Commercial	*11*
Petroleum products	64.2
Coal	17.5
Electricity	18.6
Renewable energy	*0.7*

Fig. 2. Share of Energy Consumption According to Sources. *Source:* MoF/GoN (2009).

IMPACT OF CLIMATE CHANGE IN NEPAL

The fourth assessment report of the Intergovernmental Panel on Climate Change (IPCC) concluded that the warming of the earth's climatic system is unequivocal; during the 21st century, the average temperature could rise by 1.1°C to 6.4°C and sea level could rise between 18 and 59 cm. The change in the earth's climatic system, caused most likely by anthropogenic greenhouse gas (GHG) emissions, would lead to uncertainty in weather conditions and increase in frequency and severity of disasters such as floods, landslides, and droughts.

Nepal contributes only 0.025% of the global production of GHGs; however, Nepal ranks as one of the most vulnerable countries in the world in terms of impact due to climate change. High dependency on a climate-sensitive economy such as a rain-fed agriculture system and forest resources, a fragile geomorphology, and steep slopes within a short span render the country highly vulnerable to climate change. As has been witnessed in the past several years, Nepal is already experiencing the adverse impacts of climate change that range from unusual weather behavior and decline in crop production to severe water-induced disasters. "The country has witnessed the visible effects of climate change such as accelerated snow and glacial melting processes, increasing size of glacial lakes up to critical geostatic thresholds, formation of new glacier lakes, rainfall variation, impacts on forest and biodiversity, and unexpected droughts which have direct bearing on human health and food security. . . . These unusual phenomena have potential adverse impacts on biodiversity and ecosystem health and affect Nepal's achievement of the Millennium Development Goals in general and MDG7 in particular" (GoN/UNDP, 2008).

Impact of climate change in Nepal can be observed as impacts on the agricultural sector, impacts on livelihoods, impacts on the health sector, impacts on biodiversity, and impacts in water resources such as retreat in glacial lakes and changes in hydrological regime. In agriculture, the studies have shown mixed results with marginal increases in production of rice and a decline in the production of wheat and maize. The results, however, point that there are many uncertainties in the impact of climate change to the agriculture sector and more intensive and contextualized research is necessary. A growing risk of malaria, kala-azar, and Japanese encephalitis outbreaks is identified as a potential impact of climate change on health. Subtropical and warm temperate regions of Nepal would be more vulnerable to malaria and kala-azar. Increase in temperatures would make the subtropical region of Nepal more vulnerable to Japanese encephalitis as well (MoEST/UNDP, 2008).

The climate change is expected to have significant adverse impacts also on biodiversity and the Himalayan ecosystem. The rich biodiversity may be endangered as effects such as upward shifting of zonal vegetations occur because of warming effects. The INC to the United Nations Framework Convention on Climate Change (UNFCCC) prepared in 2004 underscored the fact that the warming effect will be significant in the subalpine and alpine regions. The INC has predicted that there will be possible extension of tropical and subtropical moist forest, while cool temperate forests might change to warm temperate moist forests as an effect of a 2°C temperature increase and 20% increase in precipitation (MoPE/UNEP, 2004). "Rising temperatures will mean increased photosynthesis, resulting increases in biomass production and productivity, changes in biological phenomena and soil-carbon stock. Mountains are a storehouse of biodiversity. Biodiversity hotspots with endemic genetic resources and species habit will be threatened by climate change" (Sharma, 2009).

The Himalayas, which are the source of fresh water for nine major river basins in the region, are very sensitive to changing climatic patterns. The Himalayan ecosystem is unique, complex, and fragile. The rise in temperature and precipitation pattern will reduce snow cover in the Himalayas, and glaciers will melt. Nepal has over 3,250 glaciers covering a total area of 5,310 km^2. Glacier melt in the Himalayas is projected to

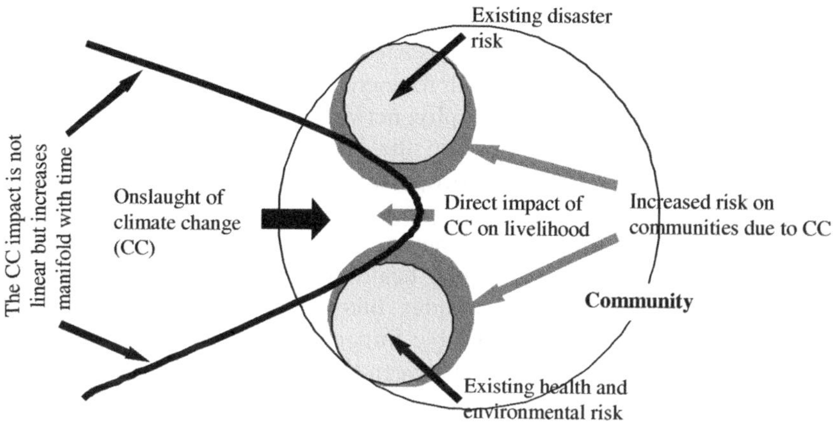

Fig. 3. Impact of Climate Change on Communities.

increase flooding and rock avalanches, and affect water resources within the next two to three decades. "The global warming has resulted in the retreat of glaciers in the Nepal Himalaya, where over 20 glacial lakes have become consistently more hazard-prone. Scientists have formulated a strong demand to monitor glaciers and install effective weather warning systems in order to reduce climate-induced disaster losses" (GoN/UNDP, 2008). Receding glaciers have increased the risk of sudden flooding following glacial lake outbursts.

In summary, the impact of climate change in Nepal can be classified into three broad categories (see Fig. 3):

1. Impact of climate change to livelihood of people: This is largely because of change in climatic pattern such as precipitation concentration, temperature increase, and prolonged drought. The change in climatic pattern causes decline in agriculture products and decline in yield of livestock. Land degradation, drought, and desertification also result from climate change. If unabated, the impact of climate change increases exponentially. The impact on livelihood is more pronounced in Nepal, where rain-fed agriculture still holds the major share in the economy and nearly 40% of the population still lives below the poverty line.
2. Increased disaster risk: The increased severity of hydrometeorological disasters such as floods, droughts, and landslides has already been

witnessed in many parts of the world. Nepal witnesses nearly 400 deaths each year due to water-induced disasters. The frequency and severity of these disasters can be expected to increase due to of climate change impact because of steep topography, networks of snow-fed rivers running north-south and originating from the Himalayas and galloping in the Ganges, and poverty-stricken people earning livelihoods from disaster-prone areas.

3. Increased health and environmental risks: Climate change–induced factors exacerbate already existing health anomalies. The climate system is an integral part of the complex life-supporting processes and the exposed climate always has visible impact on human health and well-being. "There is a close linkage between the environment and livelihoods. The majority of the people still depend on forests for firewood as well as for timber, medicinal plants, and other forestry products. They also use forests to graze livestock and to collect fodder to feed the large number of livestock raised to supply the manure essential for agricultural crops" (MoEST/UNDP/GEF, 2008). The impact in health can be from direct impact due to weather extremes, from various processes of environmental change and ecological disruption that occur in response to climate change, and indirect health impact originating from other climate-induced extremes (Kien & Shaw, 2009; McMichael, Woodruff, & Hales, 2006). The climate change-induced variation – often disruption – in the local ecosystem and local weather conditions also affect the communities, directly and indirectly (see Fig. 4).

Fig. 4. Drying Up Rivers in the City Areas Are Turning into Sewage Areas, Causing an Environmental Nuisance. Drying up traditional water spouts are also manifesting in urban drought.

CLIMATE CHANGE AND PEOPLE'S PERCEPTION

> ...[U]npredictability is perhaps the most immediate challenge that climate change brings – and already people are struggling to cope with weather patterns they no longer recognize. "It's cold when it should be warm and warm when it should be cold," said a fisherman in Tonga (IFRC, 2009).

Climate change is neither a new phrase nor a new phenomenon for scientific communities. There already exist myriads of tested and proven scientific evidences that unequivocally point to global warming and climate change. The phrase *climate change,* however, often doesn't correspond with the understanding of the communities who are directly affected by its adverse impacts. The change in climatic pattern is evident to local people, as witnessed in decline in agricultural products, erratic livestock behavior, rising temperature, and change in monsoon patterns. However, connecting all these patterns not observed before climate change impact gives the communities an impression of its invincibility. Translating the language of climate change to their day-to-day concerns and addressing their concerns in their own way is essential for reducing adverse impacts on communities.

The local perception of climate change has been reported in a survey of people in the eastern Himalaya region, which "covers an area of 524,190 km^2, extending from Eastern Nepal to Yunnan in China" (Sharma, Tsering, Chetri, & Shrestha, 2009). Most of the people from the region who responded to the research "associated climate change with an increased risk of floods and landslides and an increase in temperature, land degradation, drying up of water sources, outbreaks of pests, and food shortages. Respondents recollected personal experiences of changing climate conditions and were of the opinion that farmers would be the hardest hit. Many others reported phenological changes in threes . . . to be good indicators of climate change" (Sharma & Tsering, 2009). The climatic trends and people's perception re given in Table 2.

LIVELIHOODS AND CLIMATE CHANGE ADAPTATION

Nepal already faces immense challenge to address the country's abject poverty and lack of alternative livelihood options. The onslaught of climate change and associated increased vulnerability of people have made the task further challenging and complicated. The majority of the population is dependent on rain-fed agriculture and forest-based livelihoods – both of

Table 2. Overview of Climate Change Trends, People's Perception, Impacts, Vulnerable Entities in the Eastern Himalayas.

Climatic Trends/Projections	People's Perception
• There is increased magnitude of warming with elevation, with areas at >4,000 m experiencing the highest warming • Observed warming is 0.01–0.06°C/yr • Annual mean temperature is expected to increase by 2.9°C by the middle of the century • The Hadley Centre regional climate model (HadRM2) projects milder winters with enhanced precipitation	• Perceived as a consequences of diverse human activities causing pressure on resources • Associated in context of warming weather condition, hazards, outbreak of pests, food and water shortage • Adaptation toward change in cropping pattern • Better farmland productivity at high altitude, less at lowlands • Phenological changes in trees (times of budding, flowering, leaf fall, and so forth)

Source: For complete table, see GoN/UNDP (2008).

which are sensitive to climate change. The growth of the industrial and technology sectors lags far behind the pace of population growth.

The urban sector faces scarcity of water, increasing air and surface pollution, and power shortages; the rural sector is marred by low agricultural productivity, lack of alternative livelihood options, and internal conflict. Although the government recognizes that poverty alleviation can only be sustainable through sound environmental management, the government fails to address already existing social, environmental, and developmental problems. The new challenges and uncertainties introduced by climate change not only make the task more difficult but also may unwind whatever progresses have been made in the past.

The diverse topography, varying traditions and cultures among the communities, rapid climatic variations, and different habitats within a small distance mean the impacts of climate change on livelihoods of communities is not easy to ascertain. The Thematic Assessment Report on Climate Change underscores this challenge (MoEST/UNDP/GEF, 2008) as follows:

> The knowledge on the impacts of climate change on livelihood in Nepal is very much limited as there are no analytical means to assess it. Nepal is basically an agricultural country. About 80 percent of the people are dependent on agriculture for their income and employment. Agriculture and livestock farming is the most common livelihood for the majority of people living in rural areas. Their livelihood may well be threatened due to impacts of climate change on crops production and livestock raising. Indeed, climate change will weaken the livelihoods of poor people by eroding their livelihood assets.

> Poor people are vulnerable to loss of physical capital (because of damage to shelter and infrastructure), human capital (because of malnutrition and diseases), social capital (because of displacement of communities), natural capital (because of loss of productivity in agriculture and fisheries) and financial capital (because of more disasters and lower income). Degradation of livelihoods by climate change will thus leave poor people with less of the assets they need to withstand shocks and stresses.

Few initiations have been made at the government and nongovernment levels for local-level adaptation and coping mechanisms against adverse impacts of climate change. The Department of Hydrology and Meteorology (DHM), with support from Advancing Capacity on Climate Change Adaptation (ACCA) and in collaboration with Himalayan Climate Centre and Nepal Red Cross Society, is implementing a community-based climate change adaptation program in Syangja district. Likewise, Practical Action Nepal has initiated documentation of adaptation measures in the Chitwan district in community-based management of hydrometeorological hazards and early warning systems from 2004. The initiative "links the impacts of climate change with the promotion of improved livelihood options such as infrastructures, conservation of water sources and forests. . . .The project has installed a meteorological station in the village to record weather data and to monitor the climatic conditions of the area. Through training, workshops, and exposure tours it also enhanced the capacity of community members in raising improved breeds of goats, sloping agricultural land technology (SALT), vegetable farming, and animal health. The community now has better understanding about community forest management, disaster preparedness, climate change and its adverse impacts, and alternative opportunities for income generation" (Practical Action Nepal, 2008).

These initiatives are only a small step but can serve as pilot projects to expand to wider regions. Because of large variation in climatic pattern throughout the country, localized and contextualized approach is necessary. The country needs to adopt a strategy that can both accommodate an integrated framework at the macro level and a flexible and contextualized adaptation and coping strategy at the micro level. On a local level, "adaptation measure to climate change could be approached by intensifying the conservation of drought-resistant crop varieties, by improving cropping practices to conserve water, and by promoting crop diversification. Several aspects of the vulnerability and impact analysis in agriculture sector can also be improved, for example, development of improved climate scenarios, development of more suitable crop models, and search for alternative analytical approaches" (MoEST/UNDP/GEF, 2008).

CLIMATE CHANGE: A REGIONAL CHALLENGE

Nepal stands in a strategic location geospatially in the region, with Hindu Kush Himalayas (HKH) in the north and alluvial plains in the south. "The HKH is a huge mountain system. It contains all or parts of eight countries, 10 immense river systems, and about 200 million people in the mountains alone, and impact on perhaps more than 1.3 billion people in the surrounding lowlands. . . .The HKH extends from west to east over a distance of about 3,500 km, the variability – from arid mountains with less than 0.4 mm of annual precipitation in the west to mountains with more than 10 m of rain in the east" (Messerli, 2009). The impact of climate change has regional dimension as well. The Himalayas are a source of water that supports the livelihoods of about 1.3 billion people. The impact of climate change on the mountain range means that not only the livelihoods of people dependent on this ecology will be adversely affected, but the frequency and severity of natural disasters such as floods, cyclones, droughts, and landslides will also be increasing throughout the region. Although the level of development varies among the countries in the region, the majority of the population in the region is largely dependent on climate-sensitive sectors, such as rain-fed agriculture, for their livelihoods; its ecosystem is fragile and the rapidly changing topography makes the region prone to hydrometeorological disasters.

The climate change witnessed in the region is evident in the temperature change, change in monsoon patterns, and glacial melting. The temperature change in the eastern part of the HKH region is given in Table 3. Sharma et al. (2009) state that: "In general, there is a southwest to northeast

Table 3. Regional Mean Temperature Trends for the Period 1977–2000 (Per °C Per Year).

Regions	Seasonal				Annual
	Winter (Dec–Feb)	Pre-monsoon (Mar–May)	Monsoon (June–Sep)	Post-monsoon (Oct–Nov)	
Trans-Himalayas	0.12	0.01	0.11	0.10	0.09
Himalayas	0.09	0.05	0.06	0.08	0.06
Middle mountains	0.06	0.05	0.06	0.08	0.08
Siwalik	0.02	0.01	0.02	0.08	0.04
Terai	0.01	0.00	0.01	0.07	0.04

Source: Shrestha and Devkota (2008) (cf. Sharma et al., 2009).

trending diagonal zone of relatively less or no warming for annual and seasonal trends. This zone encompasses the Yunnan province of China, part of the Kanchin state of Myanmar, northeastern states of India. Eastern Nepal and eastern Tibet record relatively higher warming trends than the lowland areas. The warming in winter is greater than the normal scenario and more widespread. Additionally, a significant positive trend with altitude has been observed throughout the region. High-altitude areas have been exposed to comparatively greater warming effects than those in the lowlands and adjacent plains."

Countries in South Asia are already witnessing the impacts from the changes induced in the Himalayas, along with the changes eminent in the local regions. A regional cooperation mechanism has already been created in SAARC (South Asia Association for Regional Cooperation), and countries in the region came together in Kathmandu to formulate joint vision for the region before the Copenhagen meeting in December 2009. To reduce the impact of climate change needs, people need to realize the urgency of the situation, reach out to the vulnerable communities, reduce the adverse impacts, and harvest the potential of the region, building upon the good practices already achieved. Climate change isn't confined to one country and it affects the people across the countries. The interwoven river systems and hugely interdependent weather system across the region in South Asia can be responded to only by cooperation. To have better understanding of the changing climate patterns, to predict the rainfall behavior, and for better temporal and spatial understanding of the occurrence of drought and rainfall requires a coordinated approach in the larger geographic space.

Regional coping mechanisms are inevitable for effective responses to reducing the adverse impacts on livelihood of people. On the one hand, the countries in South Asia are facing decline in crop yield (ADB, 2009a) due to climate change; on the other hand, the region is yet unable to harness the potential, such as hydropower, spread across the region (ADB, 2009b). So far, the threats in the region have been underestimated and the potentials have been undermined. The immediate task ahead for South Asia is to build a common understanding of the magnitude of the impending risks, to establish knowledge infrastructure, and to establish a regional vision for the future.

Realizing the need for a common platform to address the issues and build a common understanding before the Copenhagen meeting in December, the countries in the Himalayas and the South Asia region gathered in a conference in Kathmandu in August. The conference, named "From Kathmandu to Copenhagen: A Vision for Addressing Climate Change

Risks and Opportunities in the Himalaya Region," aimed to provide a forum for the countries of the South Asia Himalayas and other countries in the region to share knowledge and experience about common risks that climate change brings and the immense development opportunities that could be fostered, and to develop a common message to the global community regarding the climate change challenges the region faces. The gathering concluded with a common vision for the region, stressing the need to accelerate sustainable social and economic growth in accordance with the principles and provision of the UNFCCC (see Box 1).

One area where the region can take immediate action is in management of water resources in the region. The huge network of rivers across the region supports lives and livelihoods of people. "Management of rivers to improve irrigation and reduce the impacts of floods and droughts will help large numbers of people adapt to their changing environment. Sustainable management of watersheds, forests, and huge untapped hydropower resources will not only provide safety nets but also reduce carbon levels" (Kathmandu to Copenhagen (2009)).

CLIMATE CHANGE: MITIGATION AND ADAPTATION/COPING

The strategies adopted toward addressing issues of climate change can be classified into two categories: mitigation of climate change and coping against or adaptation to impacts of climate change. The mitigation strategy (see Fig. 5(a)) includes retarding the onslaught of climate change and reducing the progressive increase of climate change impact (as explained in Fig. 3). The first issue needs a global approach of reducing the emissions of greenhouse gases, whereas the latter demands localized action in urban planning, in improved forest management, and in clean environment.

The adaptation and coping strategy (see Fig. 5(b and d)) can be divided into three broad categories: (1) reduction of the impacts on communities by contextualized adaptation and coping strategies such as a robust agriculture system; (2) disaster risk reduction strategies to reduce impacts of disasters; and (3) environmental management such as protection of biodiversity, clean environment, and reduction of negative health impacts. These mitigation and adaptation strategies will be effective only when they are an integrated part of the development efforts.

Box 1. Common Vision from the South Asia Regional Climate Change Conference, August 31 to September 1, 2009, in Kathmandu.

1. South Asia, including the Hindu Kush Himalayan (HKH) region, is a climate change hot spot that influences the lives of half of the world's population. Climate change in this region will affect people and ecosystems from the mountains, to the coast, to the sea.
2. The South Asia region is highly vulnerable to the adverse impacts of climate change and is characterized by critical knowledge gaps, especially of mountain ecosystems within and across its constituent units.
3. The countries of South Asia need to accelerate sustainable social and economic growth in accordance with the principles and provisions of the UN Framework Convention on Climate Change.
4. The South Asian countries must therefore come together to enhance their climate change responses including through the generation of required data and adaptation at all levels with incremental adaptive steps meshing with agreed regional and global efforts to address the impacts of climate change. The enhanced climate change responses require additional financial and technical resources.
5. Water resources are particularly impacted by climate change and immediate and urgent actions are required to address these challenges through management practices, including basin-wide approaches at appropriate levels, and provisions of additional financial and technical resources in accordance with the provisions of the UNFCCC.
6. The inhabitants of the South Asian region are among those likely to be worst affected by climate change. Special strategies need to be evolved and additional resources provided to address the specific vulnerable communities of the region.
7. The countries of South Asia need to address the challenges of climate change through collaborative actions aimed at enhancing capacity building, including activities regarding data collection and sharing and research about climate change impact through SAARC and other institutions as agreed.
8. Financing mechanisms on adaptation and technologies should sufficiently meet the urgent and immediate needs of financial requirements of the South Asia region in a predictable, easy, and direct manner. Finance must come from Annex 1 Parties to the UNFCCC.

9. Recognition and payment for credits from the forestry sector must include a comprehensive approach to sustainable management and conservation of forests, and enhancement of forest carbon stocks.
10. Development and transfer of clean technologies including capacity building should take place with a view to ensure green development. Transfer of technologies should be provided by Annex 1 countries to non-Annex 1 countries and they should not be constrained by the high upfront costs of intellectual property rights. Appropriate indigenous technologies should also be promoted through channeling of funds to developing countries in accordance with the Convention's provisions. Both public and private sectors should be encouraged in addressing the impacts of climate change through provisions of clean technologies.

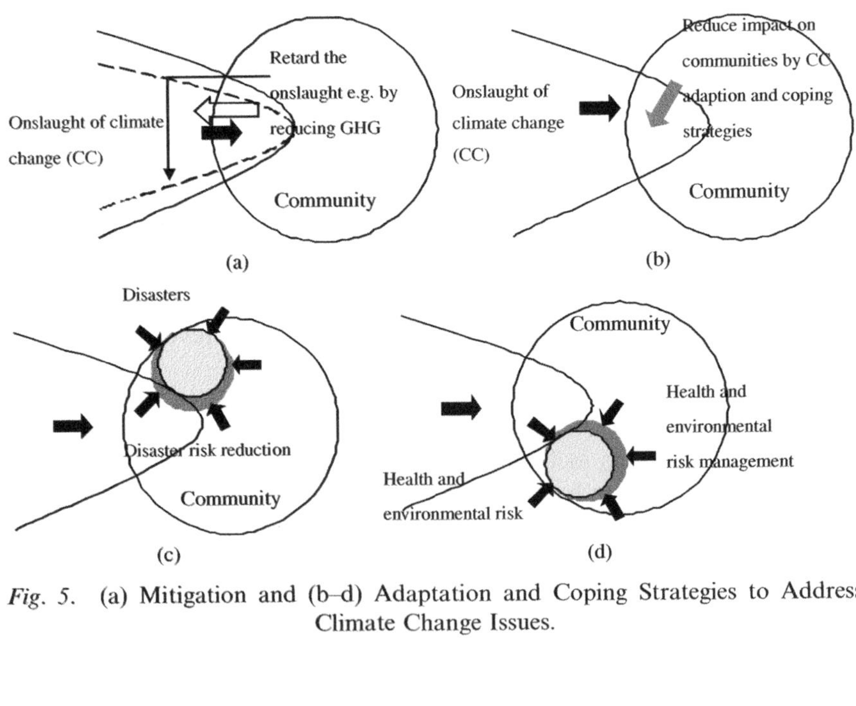

Fig. 5. (a) Mitigation and (b–d) Adaptation and Coping Strategies to Address Climate Change Issues.

INSTITUTIONAL STRUCTURES AND MECHANISMS IN NEPAL

There have been many initiatives, both from the government and nongovernment sectors, related to addressing climate change issues in Nepal. Additionally, there are many other initiatives, not directly addressing climate change issues, but contributing toward mitigation and adaptation to climate changes.

The Ministry of Environment, Science and Technology (MoEST) is the focal point to UNFCCC. The Alternative Energy Promotion Centre (AEPC) and the DHM are two key institutions under the umbrella of MoEST. MoEST has also constituted the 23-member Climate Change Network (CCN) to address information, coordination, and communication issues related to climate change in Nepal. The CCN has been constituted to (1) identify working areas on climate change among the government, NGOs, and private and donor organizations; (2) conduct policy/field-level research and studies and implement activities based on the capacity and expertise; (3) promote Clean Development Mechanism (CDM)-related activities, and launch public awareness and capacity building programs; (4) develop position papers for the parties meeting and enhance negotiation capacity; and (5) also develop the Climate Change Clearing House for easy information sharing (MoEST/UNDP/GEF, 2008). The CCN involves interministerial departments and organizations, interested bilateral and multilateral agencies, International Non-Governmental Organizations (I/NGOs), academia, and private sector. The CCN, however, "is presently grappling with the problems of voluntarism and irregular participation. There is no clear-cut mechanism of funding for joint initiatives or actions of the network, and the MoEST is presently not able to make a strong case for budgetary allocations with Treasury (due to a lack of visibility of climate change amidst other government priorities). . . . There is no common and sustained platform of information, knowledge, and learning on climate change matters that would enable the CCN to grow into a strong, agenda-based platform with appropriate visibility to inform and support policy decisions" (GoN/UNDP, 2008). Strengths and weaknesses of existing institutional structures and mechanisms to address climate change are compared in Table 4.

The government of Nepal recently formed a high-level committee to address climate change issues under the chairmanship of the prime minister. Under the leadership of the prime minister, the committee has an opportunity for raising the political awareness of key environmental issues, and potentially securing political commitment from the various line ministries to actively participate in policy coordination. Although having

Table 4. Strengths and Weaknesses of Existing Institutional Structures and Mechanisms to Address Climate Change.

Strengths	Weaknesses
• Climate Change Network was established, which fills a technical capacity gap of the government • Well-established community forestry that helps combat climate change • Data is being collected by Department of Hydrology and Meteorology • Alternative energy and clean development mechanism is being promoted to mitigate the impacts of climate change • Identification of potential energy development projects • Nepal Agricultural Research Council (NARC) conducts agricultural research for adaptation to climate change • Cleaner vehicles are being promoted • The National Health Research Council was established to conduct studies and raise public awareness on impacts of climate change on the health sector	• Policy is not established to address climate change • The departments lack capacity to conduct vulnerability assessments or to contribute to the analysis and monitoring of sectoral policies and their interventions to assess climate change impacts • Lack of scientific research and an insufficiency of data stations • Environmental compliance is not monitored adequately • The MoEST lacks human resources

Source: MoEST/UNDP (2008).

such a high-level participation in the committee demonstrates positive commitment of the government, there are very likely chances of the committee being nonfunctional as has been witnessed in the Environment Protection Council and National Coordination on Sustainable Development, both of which have not been convened from 1998 and 2002, respectively.

In the policy and legislative front also, a lot of initiations have been made including bringing Sustainable Development Agenda for Nepal (SDAN) in 2003, to guide the sustainable development path until 2017, and implement many sectoral policies, strategies, plans, and legislations to facilitate the implementation of SDAN targets (MoEST/UNDP/GEF, 2008). The government is in the process of formulating climate change policy by 2009. Although there is no specific policy, strategies, or laws relating to climate change so far, several important policies, strategies, and laws have been enacted that are linked with environment and climate change. The interim

Table 5. Key Legislations, Policies, and Strategies Linked with Climate Change.

1	National Conservation Strategy (1988)
2	Master Plan for Forestry Sector (1989)
3	Perspective Energy Plan (1991–2017)
4	Nepal Environment Policy and Action Plan (1993)
5	Agriculture Perspective Plan (1997)
6	Revised Forests Policy (2000)
7	Renewable Energy Perspective Plan (2000–2020)
8	Hydropower Development Policy (2001)
9	Water Resources Strategy (2002)
10	Nepal Biodiversity Strategy (2002)
11	National plan (Tenth Plan, 2002–07); Three-Year Interim Plan (2007–2010)
12	Irrigation Policy (2003)
13	Sustainable Development Agenda (2003)
14	Water-Induced Disaster Management Policy (2003)
15	National Water Plan (2005)
16	Rural Energy Policy (2006)

constitution (2006) recognizes the "right to clean environment" as one of the fundamental rights. The major national environmental policies include the National Conservation Strategy (1988), the Nepal Environment Policy and Action Plan (1993), the Sustainable Development Agenda (2003), and National Plans (Tenth Plan, 2002–07; Three-Year Interim Plan, 2007–2010) (see Table 5).

There are many successful initiatives toward greenhouse gas mitigation, which are also beneficial directly to increase living standards of the people. Although reducing emissions of GHGs in Nepal will not contribute significantly on a global level, this will have direct positive impact on the surface climate. One of the major sources of pollution for the microclimate in city areas is the smoke emitted from brick kilns. Replacement of the burnt bricks in construction by other alternatives will have a positive impact on reduction of GHGs emissions (see Fig. 6).

During the last 15 years, Nepal has given highest priority in the promotion of alternative and renewable energy technologies as a strong means to poverty reduction, particularly in rural Nepal. Besides this, the government is promoting compliance toward environmental standards, improving industrial efficiency and providing incentives to greener vehicles.

The key activities government of Nepal has carried out to address climate change and global warming are listed in Table 6.

Fig. 6. Brick Factories Are One of the Major Sources of Pollution in Kathmandu (left, photo source: Sonam Wangchuk). The prototype for a school building constructed replacing the burnt bricks with compressed stabilized earth blocks (on right) is an example of how emission of greenhouse gases can be lowered.

Table 6. Key Activities of Climate Change in Nepal.

Date	Event
October 1994	Inventory of GHGs for the energy sector, based on 1990 data
1999	Preparation of an implementation strategy on environment-related conventions
August 2000	Workshop on UNFCCC and institutional design of implementation mechanism of Kyoto Protocol
	Implementation of Promotion of Renewable Energy, Energy Efficiency and GHG Abatement (PREGA)
2004	Initial national communication report with the parties to UNFCCC
May 2004	Nepal ratified UNFCCC
September 2005	Accesses Kyoto Protocol and it has entered into force since December 2005
2006 onward	Organization of public awareness activities on capacity building and negotiation skills
2007	Proposal for preparation of National Adaptation Program of Action (NAPA)
November 2008	MoEST entered into agreement with UNDP Nepal to implement NAPA
2008	Implementation of Strengthening Climate Change and Environment Project
October 2008	Completion of National Capacity Needs Self-Assessment (NCSA)
2008	Climate change policy formulation initiation
2008	Formation of Climate Change Network (CCN)
2008	Second national communication report
2008	The government joined Pilot Program Climate Resilience (PPCR) and Cool Earth Programme
August 2009	Formation of high-level Climate Change Council
August 2009	Nepal hosted a Regional Climate Change Conference

GOOD PRACTICES: COMMUNITY-OWNED INITIATIVES (COIS)

A community-focused, localized approach and sustainable strategy is necessary for communities to build resilience to climate change and disaster risks. Nepal already has experience with successfully implementing a community-owned forest management system and locally owned micro-hydropower projects. The experience of involving communities in management of forests and its evolution as a robust institution can provide guidelines for any other community-owned initiatives and is also an entry point for adaptation and coping against the adverse impacts of climate change. Similarly, the success of micro-hydro has potential for replicating a localized business model and replacing conventional energy sources with renewable energy sources. These initiatives are more community-owned initiatives (COI) than community-based, and COI is the only effective strategy for adapting to and mitigating effects of climate change. A brief description of community-owned forest management and micro-hydro is given in the following section.

Community Forest

Nepal's successful community forest management program, has led to additional carbon sequestration and watershed protection, and created a buffer zone between protected areas and communal agricultural land (Webersik & Thapa, 2008). In Nepal, 35% of the populations are members of community forest users groups (CFUGs). The community forestry has helped not only to provide basic needs of local communities for fuelwood, fodder, and timber but has also contributed to a decrease in the rate of deforestation that "has substantially decreased (from 8,000 to 800 ha per year) due mainly to the implementation of community forestry" (Kanel, Poudyal, & Baral, 2005). Nepal's successful march toward handing the forests over to the local community materialized with the establishment of the Master Plan for the Forestry Sector (MPFS), which was approved in 1989. The plan still remains the main policy and planning document for the continuing development of the forestry sector. Based upon the Master Plan Forest Act, Forest Rules, and Forest Sector Policy were introduced in 1993, 1995, and 2000, respectively.

The rights and responsibilities of CFUGs are established in the legislation, and they are authorized by the law to operate as independent and self-governing institutions. "The focus of this legislation is on institutionalizing CFUG as an independent and self-governing entity, nationwide expansion of

Table 7. Forestry in Nepal at a Glance.

Forest area	4.27 million hectares (29%)
Bush and shrubs	1.56 million hectares (10.6%)
Community forestry	
Number of community forest user groups	14,559
Area of handed-over community forests	1.23 million hectares
Number of beneficiary households	1.66 million
Number of women-managed community forests	795
Leasehold forestry	
Number of leasehold forests user groups	4194
Area of handed-over leasehold forests	23,423 hectares
Beneficiary households of leasehold forests	36,478
Partnership forestry	
Demonstration program operating districts	Bara, Parsa, Rautahat, Mahottari
Number of partnership forests	4
Areas of partnership forests (hectares)	Bara, 2,058; Parsa, 3,138; Rautahat, 1,473; Mahottari, 2,006

Source: MoF/GoN (2009).

community forestry, providing utilization and management rights to the local community, and creating an accountability forum for community development. It has also limited the role of the district forest office to that of supporter, facilitator, monitor, and regulator of community forestry" (Kanel et al., 2005). As of 2009, the total share of forest (including shrubs and bushes) is 39.6% and out of which 40% is handed over to the communities (see Table 7). The involvement of community in forest management was initiated to reduce deforestation that has gradually evolved into forest management for both conservation and income generation. The achievements of community forestry can be seen in terms of better forest conditions, social mobilization, and income generation for rural development and institutional building at the grassroots level. This model of local peoples' participation in natural resource management has also been expanded to watershed management and buffer zone management, where conservation of watersheds and biological diversity is the prime objective (Kanel et al., 2005).

Alternative Energy Promotion and Micro-Hydro Development

Steep topography and snow-fed perennial networks of river provide Nepal with immense potential for development of hydropower. Out of feasible potential of 83,000 megawatts, about 42,000 megawatts is considered technically (and economically) viable (Asian Development Bank/ICIMOD, 2006). "There is enormous potential for uses of various renewable energy sources including biogas, micro-hydropower, solar energy (photovoltaic and thermal), improved water mills, improved cooking stoves and wind energy etc." The government of Nepal, since the year 2000, has been providing subsidy for the promotion and development of rural energy, for improving the living standard of low-income rural people through maximum utilization of this source of energy.

The ownership of micro-hydro in Nepal is either by small groups of individuals or by communities. In recent years, with micro-hydro being promoted by different agencies, the share of community-owned and -managed micro-hydro is increasing. The success of micro-hydro and interest of communities in it has two underlying implications to mitigation of and adaptation to climate change at the local level. The first implication is that micro-hydro, as a source of renewable energy, contributes to replacement of conventional sources of energy that ultimately lead to reduction in greenhouse gases. The second implication is that it provides an opportunity for better livelihoods and improved resiliency in rural areas. The success of micro-hydro also underscores the fact that local entrepreneurship, which is essential for building communities resilient to disaster and climate change, is a viable option (Fig. 7).

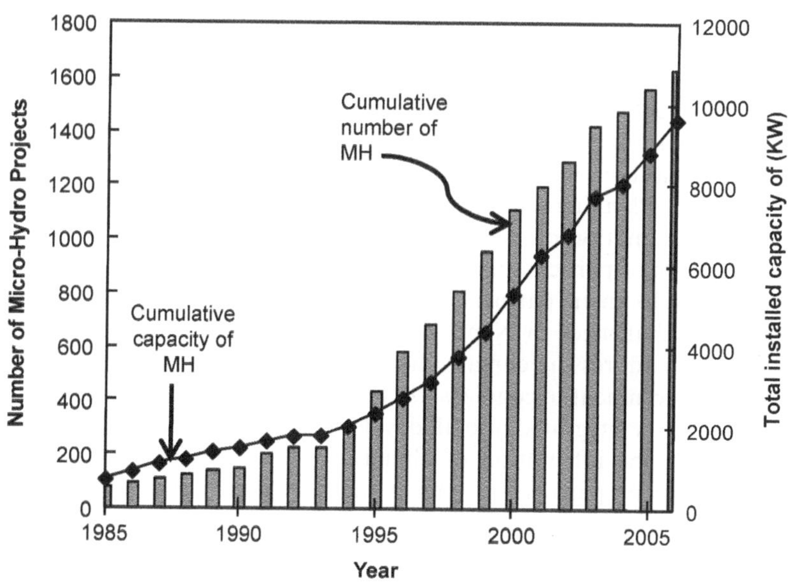

Fig. 7. Installed Plants of Micro-hydro and Their Cumulative Capacities. *Source:* Household energy network (2008).

Box 2. Compact on Climate Change in Nepal, Ministry of Environment, Government of Nepal

1. An understanding between the government of Nepal, the Ministry of Environment, and development partners on ways to address climate change challenges
2. Purpose
3. This compact commits the undersigned to a set of principles to guide how development partners will support the government of Nepal to implement a series of actions designed to (1) identify and assess climate risks, particularly to the most vulnerable people and sectors, (2) elaborate, test, and implement adaptive responses, and (3) establish the basis for a climate-resilient economy.
4. Background
5. The government of Nepal is managing concurrent initiatives on climate change that are supported by different developing partners. These include the National Adaptation Programme of Action (NAPA), Reducing Emissions from Deforestation in Developing Countries (REDD), the Pilot Programme for Climate Resilience (PPCR), the strengthening capacity for managing climate change and the environment, and the Second National Communication. The Ministry of Environment (MoE) is the focal point of the government on climate change, with the Ministry of Forests and Soil Conservation (MoFSC) programme. The Ministry of Finance (MoF) has the responsibility to coordinate foreign aid flows, and the National Planning Commission (NPC) leads in policy formulation and coordination on development planning.
6. Principles of participation
7. The development partners fully appreciate the urgency of climate change challenges in Nepal and express their willingness to assist Nepal in addressing climate change vulnerabilities.
8. The government recognizes that the need to ensure the climate resilience is integrated into all development work. Each line ministry will lead the integration of climate resilience in their sector. The development partners recognize the need for harmonization and coordination in these efforts.
9. The Ministry of Environment, with the support of the Ministry of Finance and National Planning Commission Secretariat and in close

consultation with other ministries of the government will lead, being the focal point of the United Nations Framework Convention on Climate Change. The development partners will participate in collaborative actions on climate change.

10. The development partners will organize themselves, with defined roles and responsibilities of each agency and facilitating partner. Communication will be maintained directly with the government of Nepal on the availability and allocation of technical and financial resources on climate change. Such communication will be through the normal channels and will also be given to the Ministry of Environment.
11. The development partners agree to harmonize and align actions on climate change toward assisting the government of Nepal in the provision of human resources and institution and capacity development; the expeditious delivery of urgent and immediate adaption needs including climate-related disaster risk reduction; a programmatic approach to climate adaptation prioritizing the most vulnerable people and sectors; with achievement of climate resilience through investments in low emission development pathways; and, the sustainable development of clean and renewable energy, water, forestry, and agricultural sectors.
12. Participation of the development partners does not indicate a specific future financial commitment.

CLIMATE CHANGE AND THE WAY AHEAD

Nepal Agriculture Research Council will be strengthened to conduct imitable and advanced research as per the demand of the farmers. To minimize the negative impact on the agricultural production due to change in climate, emphasis will be given to the research of the eco-friendly agricultural system (MoF/GoN, 2009).

Realizing the need to mainstream climate change issues in all development projects and to build resilient communities to climate change, the government of Nepal has issued a compact that necessitates commitments of development partners of Nepal for (1) identifying and assessing climate risks, particularly to the most vulnerable people and sectors, (2) elaborating, testing, and implementing adaptive responses, and (3) establishing the basis for a climate-resilient economy (see Box 2).

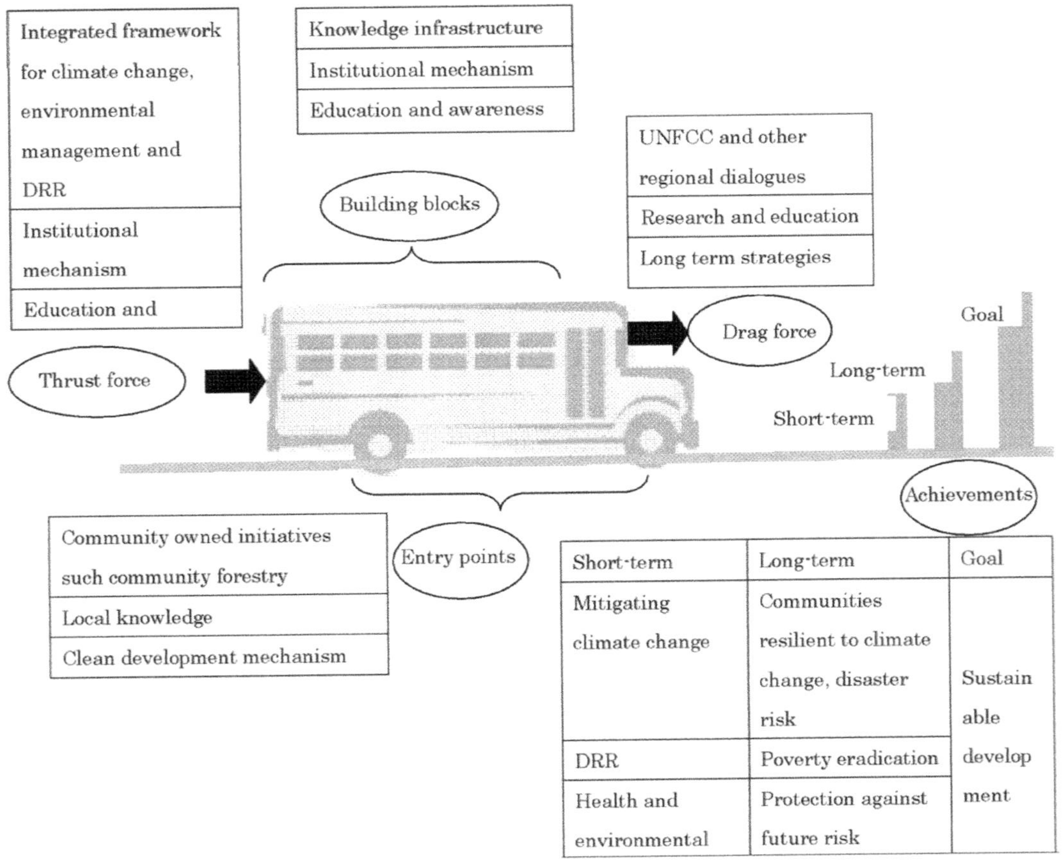

Fig. 8. Different Elements Adding to the Dynamic System to Address Issues of Climate Change.

The dynamic mechanism of strategic entry points, building blocks, driving forces, and achievements (short-term and long-term goals) is shown in Fig. 8. The strategic entry point for Nepal in addressing climate change issues can be already existing robust institutions at the community level, such as community forest users' groups (CGUGs) and community-owned initiatives such as micro-hydro and community forests. The existing policy, legislatives, and institutional frameworks provide excellent building blocks. To give the process enough momentum, political commitment and millennium development goals (MDGs) are thrust forces, whereas research initiatives and forums like UNFCCC and regional meetings act like a dragging force. The process produces short-term achievements in the form of better livelihoods, better-prepared communities against adverse impacts of climate change and disasters, and improved health and environmental management. The long-term achievements are communities resilient to climate changes and disasters, poverty eradication, and communities that are able to protect themselves against emerging new risks. The goal of the whole process is sustainable development.

REFERENCES

ADB. (2009a). *Building climate resilience in the agriculture sector of Asia and the Pacific*. Mandaluyong City, Philippines: Asian Development Bank, 2009.

ADB. (2009b). *Improving energy security and reducing carbon intensity in Asia and the Pacific*. Mandaluyong City, Philippines: Asian Development Bank, 2009.

Asian Development Bank/ICIMOD. (2006). Environment Assessment of Nepal: Emerging Issues and Challenges. Kathmandu, Nepal.

Cozens, C. (2009). *Nepal villagers on climate change frontline*. Republica (September 4): p. 7.

GoN/UNDP. (2008). National adaptation program to climate change. *UNDP project document*. Kathmandu: Government of Nepal/United Nations Development Program.

Household energy network. (2008). Micro-hydro development in Nepal: An overview and achievements. Available at http://www.hedon.info/Micro-hydroDevelopmentInNepalwww.hedon.info/Micro-hydroDevelopmentInNepal (accessed September 20, 2009). Kathmandu.

IFRC. (2009). World Disaster Report, p. 96. Geneva, Switzerland: International Federation of Red Cross and Red Crescent Societies (IFRC).

IIED (International Institute for Environment and Development). (2009). Climate change and the urban poor: Risk and resilience in 15 of the world's most vulnerable cities. London, UK: IIED.

Kanel, K. R., Poudyal, R. P., & Baral, J. P. (2005). Nepal community forestry in RECOFTC 2005. First Regional Community Forestry Forum – Regulatory Frameworks for Community Forestry in Asia – Proceedings of a Regional Forum held in Bangkok, Thailand, August 24–25 (pp. 69–84).

Kathmandu to Copenhagen. (2009). Information for From Kathmandu to Copenhagen: A vision for addressing climate change risks and vulnerabilities in the Himalayas. A regional climate change conference, August 31 to September 1, 2009. Kathmandu, Nepal. Available at http://www.kathmandutocopenhagen.org/ (accessed September 20, 2009).

Kien, T. M., & Shaw, R. (2009). Climate change adaptation and human health: Linkages of climate, disaster and health issues in Vietnam. In: R. Shaw & R. R. Krishnamurthy (Eds), *Disaster management: Global challenges and local solutions* (pp. 444–467). Hyderabad, India: Universities Press.

McMichael, A. J., Woodruff, R. E., & Hales, S. (2006). Climate change and human health: present and future risks. *Lancet*, *367*, 859–869.

Messerli, B. (2009). Biodiversity, environmental change and regional cooperation in the Hindu-Kush Himalayas. In: *Biodiversity and climate change in the Himalayas* (pp. 3–6). Kathmandu: ICIMOD.

MoEST/UNDP. (2008). Thematic assessment report on climate change. National Capacity Self-Assessment Project. Ministry of Environment, Science and Technology (MoEST) Government of Nepal, United Nations Development Programme (UNDP), Kathmandu, Nepal.

MoF/GoN. (2009). *Economic survey fiscal year 2008/09*. Kathmandu: Ministry of Finance (MoF), Government of Nepal (GoN).

MoPE/UNEP. (2004). *Initial national communication to the conference of the parties of the United Nations Framework Convention on Climate Change*. Kathmandu: Ministry of Population and Environment (MoPE)/UNEP.

NPC/UNDP. (2005). Nepal Millennium Development Goals: Progress Report 2005. National Planning Commission (NPC), His Majesty's Government of Nepal and United Nations Development Programme (UNDP), Kathmandu, Nepal.

Practical Action Nepal. (2008). Disaster risk reduction and climate change. Available at http://practicalaction.org/climatechange_adaptation?id = climatechange_nepaladaptation (accessed September 15, 2009).

Sharma, E., & Tsering, K. (2009). Climate change in the Himalayas: The vulnerability of biodiversity. In *Sustainable Mountain Development* No. 55 (Biodiversity and climate change in the Himalayas), pp. 10–12. ICIMOD, Kathmandu.

Sharma, E., Tsering, K., Chetri, N., & Shrestha, A. (2009). Biodiversity in the Himalayas: Trends, perceptions, and impacts of climate change. In *Proceedings of the international mountain biodiversity conference on biodiversity conservation and management for enhanced ecosystem services: Responding to the challenges of global change*, Kathmandu, November 16–18, 2008. ICIMOD, Kathmandu.

Sharma, P. (2009). *Climate change: Implications, planning and federalism in Nepal*. Republica (September 4): p. 8.

UNDP. (2007). *Human development report 2007/2008-Country fact sheets-Nepal*. New York: United Nations Development Programme (UNDP).

Webersik, C., & Thapa, M. (2008). Nepal Climate Change and Security Factsheet, United Nations University Institute of Advanced Studies (UNU-IAS). Available at http://www.ias.unu.edu/. Retrieved on September 15, 2009.

CHAPTER 10

CLIMATE CHANGE ADAPTATION AND DISASTER RISK REDUCTION IN PAKISTAN

Amir Nawaz Khan

INTRODUCTION

Pakistan lies between latitudes 24° 37′ N and longitudes 62° 75′ E, covering a total land area of 796,096 km^2. The country shares its borders with Iran to the west, India in the southeast, Afghanistan in the northwest, and China in the north. The Arabian Sea lies to its south (UNISDR, 2005; Khan, 2004d). The country has four provinces: the Punjab, the North West Frontier Province (NWFP), Sindh, Baluchistan, and two federally administrated territories – the Federally Administered Tribal Areas (FATA) and the Northern Areas. The Northern Areas have been recently given the status of Gilgit-Baltistan province. The structure of the provisional government is still in its formation stage. In addition, the territory of Azad Jammu and Kashmir (AJK) is under the administration of the government of Pakistan. Each province or territory is further divided into administrative units known as districts (GOP, 2003; Khan, 2004d).

The territory of Pakistan encompasses portions of the Himalaya, Hindu Kush, and Karakoram mountain ranges, making it home to some of the world's highest mountains peaks, including K2, which is the world's second highest peak at 8,611 m above sea level (Blood, 1995; Khan, 2004d).

Climate Change Adaptation and Disaster Risk Reduction: An Asian Perspective
Community, Environment and Disaster Risk Management, Volume 5, 197–215

ISSN: 2040-7262/doi:10.1108/S2040-7262(2010)0000005016

Physiographically, Pakistan can be divided broadly into two main regions, namely the plains of the Indus River and its tributaries and a huge complex of mountains and plateaus lying in the north and northwestern boundaries. The plains are more or less level, consisting mostly of irrigated agriculture and arid, semiarid deserts. The mountain complex consists of broad valleys, partly irrigated with high, steep, and rugged mountains and plateaus. About 60 percent area of Pakistan is covered by mountainous complex (Khan, 2004d). Most of Pakistan has a generally dry climate and receives less than 250 mm of rain per year, although the northern and southern areas have noticeable climatic differences. The average annual and seasonal temperature considerably varies from region to region in Pakistan. In northern mountainous areas, the temperature in the winter's coldest months is normally below 0°C. In the plain areas of Punjab, Sindh, and Baluchistan plateau, the temperature reaches up to 50°C in the warmest months. On the other hand, in the coastal areas, seasonal temperature variation is comparatively small and has pleasant weather conditions throughout the year. Mid-December to March is dry and cool; April to June is hot, with 25–50 percent relative humidity; July to September is the wet monsoon season; and October to November is the dry season (Blood, 1995; Khan, 2004d).

The Survey of Pakistan has classified the country into eight climatic zones (GOP, 2003). According to the Köppen Geiger classification of climatic zones, where zones are defined on the basis of monthly temperature and precipitation data, there are 11 distinct as well as overlapping climatic zones in the country (Khan, 1993c). These zones range from having mild, moist winters and hot, dry summers in the north to semiarid and arid zones in the west and parts of the south. The northeastern mountainous and submountainous areas receive more than 1,700 mm annual precipitation with a major share (over 1,000 mm) from the summer monsoon. On the other hand, the extremely arid plains of southwest Baluchistan receive only 30 mm of rainfall during the whole year. Thermal regimes exhibit extreme diurnal, seasonal, and annual variations: Temperatures can fall as low as −26°C over the northern mountains and go as high as 52°C over the central arid plains. In the semiarid plains, temperatures as high as 42°C are recorded at various stations in the months of May and June (GOP, 2003; GOP, 1987; Khan, 2004a). After giving a brief introduction, this chapter discusses the history of natural disasters in Pakistan. The third section looks at the glacial extension and climate change in the country. Climatic changes and variability in its impacts and vulnerabilities are discussed in fourth section. The fifth section deals with climate change and disaster management policies and strategies in Pakistan. Finally, conclusions and recommendations are given in the final section.

HISTORY OF NATURAL DISASTERS

Pakistan is one of the most hazard-prone countries in Asia. A series of natural and human-induced disasters threaten the sustained economic growth of the country by causing severe shocks, that is, the earthquake of October 2005 and the flash floods of 2008. Apart from earthquakes, most of these hazards are seasonal in nature. Experience, history, available data, hazard mapping, and assessment exercises conducted by government officials, field researchers, and other stakeholders have identified that Pakistan is vulnerable to disaster risks from a range of hazards including glacial surges, avalanches, cyclones/storms, droughts, earthquakes, epidemics, floods, glacial lake outbursts, landslides, pest attacks, river erosion, and tsunami. Human-induced hazards that threaten the country include transport, industrial oil spills, urban and forest fires, civil conflicts, terrorism, and internal displacements of communities due to multiple factors. High-priority hazards in terms of their frequency and scale of impact include flooding, earthquakes, droughts, windstorms, desertification, and landslides. These hazards have been causing widespread damages and losses in the country as shown in Table 1 (GOP, 2007a; Khan, 1992, 1993, 2004a, 2004b, 2005a, 2006; Khan & Atta-Ur-Rahaman, 2006, 2008).

Table 1. Top Ten Natural Disasters of Pakistan, from 1980 to 2008.

Name of Hazard	Year of Occurrence	Number of People Affected	Number of People Killed	Economic Damages (US$ × 1,000)
Flood	2005	7,000,450	520	30,000
Flood	1992	6,655,450	1,334	1,000,000
Flood	1992	6,184,418	NA	NA
Earthquake	2005	5,128,000	73,338	5,200,000
Drought	1999	2,200,000	NA	NA
Storm	2007	1,650,000	NA	1,620,000
Flood	1996	1,300,000	NA	NA
Flood	2003	1,266,223	NA	NA
Flood	1995	1,255,000	1,051	NA
Flood	1988	1,000,000	1,000	NA

NA, Data not available.
Source: EM-DAT (2008).

Floods have been recognized as a major natural disaster. The country has a long history of flooding from the Indus River and its tributaries. River and flash floods of 1928, 1929, 1955, 1957, 1959, 1973, 1976, 1988, 1992, 1995, 1996, 1997, 2001, and 2008 are the sad, yet memorable events entailing tremendous damages to life and property. Similarly, the recent flash floods in Peshawar vale (NWFP) and Baluchistan caused huge damages to life and property (GOP, 2007a; South Asian Flood [SAF], 2009; Khan, 2004c, 2005a). Four major earthquakes hit Pakistan in the 20th century, including the 1935 Quetta earthquake, the 1945 Makran coast earthquake, the 1976 Northern Areas earthquake, and the 2005 earthquake, which was the most devastating in the history of Pakistan. The 2005 earthquake affected an area of some 30,000 km^2 and caused over 73,000 deaths, left many more people seriously injured, destroyed homes, schools, hospitals, and infrastructure on a massive scale (GOP, 2007b; ICRC, 2003). Pakistan also has a history of tsunami disaster. Pakistan experienced a big wave on November 28, 1945, due to an earthquake of a magnitude 8.3 on the Richter scale, offshore the Makran Coast. The tsunami produced sea waves of 12–15 m tall and killed at least 4,000 people in Pasni and the adjoining areas (Georg, 2009). From 1971 to 2001, about 14 cyclones have been reported, of which the year 1991 was the severe one that caused damages to life and property in the Badin district particularly (ICRC, 2003; Regional Specialized Metrological Centre [RSMC], 2009).

Pakistan has limited surface and groundwater resources. Already, annual water availability per capita has been reduced from 5,600 m^3 at the time of independence (1947) to a meager 1,200 m^3 during 2001 (Kahlown, 2002). Historical climatic data indicate that the frequency of drought occurrence has been much higher in Baluchistan and Sindh provinces, which lie in hyper-arid climatic classification as compared to Punjab and NWFP. During the recent past (1998–2001), Pakistan faced one of the worst droughts of its history due to extremely low rainfall. Pakistanis, in severely affected areas, faced a famine-like situation; consequently, over 3.3 million families were affected. As a result, hundreds of thousands of families had to migrate to "safe areas," while hundreds of people lost their lives due to lack of food and thirst. Moreover, about 30 million livestock were also affected due to the scarcity of water and fodder (Pomee, Zaheer-ul-Ikram, Khan, & Ali, 2005).

Pakistan is predominantly an arid to semiarid country, with 68 million hectares of land lying in regions where the annual rainfall is less than

300 mm. One-fourth of the country's land area, which is suitable for intensive agriculture, is seriously subjected to wind and water erosion, salinity, waterlogging, flooding, and loss of organic matter. About 11 million hectares are affected by water erosion, and 3–5 million hectares by wind erosion. The amount of soil removed by wind is about 28 percent of total soil loss. Due to deforestation, forest cover is shrinking by 3.1 percent and woody biomass by 5 percent annually (7,000–9,000 ha taken away annually). Uncontrolled free grazing of livestock, aridity, and prolonged drought in arid lands have affected the biodiversity in various regions of the country (GOP, 1987; Khan, 2004d). About 15.5 million hectares are affected by waterlogging and 5.0 million hectares by salinity. Recent drought has affected 2.2 million people and 7.2 million livestock (GOP, 2002a). In addition to the region of Kashmir, the Northern Areas, the Galiat region, and parts of NWFP are particularly vulnerable to landslide hazards (Khan, 1992, 1994a, 1994b, 1995; GOP, 2007b; Yousaf, 2007; UNISDR, 2005). The current militancy situation in the NWFP and FATA, most particularly in the western mountainous areas, exacerbate the deforestation.

A number of factors are responsible for the vulnerabilities of Pakistani society to various hazards and disasters. The major dynamic pressures that increased vulnerabilities of Pakistani society to disasters are fragile natural environment, explosive population growth, lack of awareness and education, environmental degradation resulting from poorly managed urban and industrial development processes, deforestation, climate change and variability, poor construction practices, poor livestock and agricultural management, weak early warning systems, and poverty. The nonavailability of safer land for construction, scattered settlement patterns, and harsh climatic conditions further intensify vulnerabilities in the mountainous regions of the country. For both climate change adaptation and disaster risk reduction (DRR), the existent aim is efficient development planning and programming: Managing risks and uncertainties for all shocks and stresses is just fine business, particularly in the face of mounting evidence that disasters are hampering development and poverty alleviation. Realizing the importance of disaster risk reduction and climate change adaptation for sustainable social, economic, and environmental development, the government of Pakistan and the Hyogo Framework for Action (HFA) have embarked upon establishing suitable policy and legal and institutional arrangements, and implementing strategies and programs to minimize risks and vulnerabilities (HFA, 2005). A brief discussion on various strategies and policies is follows.

GLACIAL EXTENSION OF PAKISTAN AND CLIMATE CHANGE

The glaciers of Pakistan are mostly in the northern part of the country in the Karakoram, Himalayas, and Hindu Kush ranges, where a substantial proportion of the annual precipitation falls as snow, particularly at high altitudes above 3,000 m. In the higher reaches, snowfall builds up from year to year to form glaciers that provide long-term reservoirs of water stored as ice. Pakistan has more glaciers than any other land outside the North and South Poles. Pakistan's glacial area covers some 13,680 km^2, which represents an average of 13 percent of mountainous regions of the upper Indus Basin. In fact, in the laps of the Karakoram of Pakistan alone, there are glaciers whose total length would add up to more than 6,160 km^2. The 135 glaciers are both named and unnamed. The Siachen Glacier is 75 km. The Hispar (53 km) joins the Biafo at the Hispar La (5.2 km) to form an ice corridor that is 116.87 km long. The Batura Glacier is 58 km in length. The most outstanding of these dynamic rivers of ice is the 62-km Baltoro Glacier. This mighty glacier fed by some 30 tributaries constitutes a surface of 1,291.39 km^2 in the western mountains region (GOP, 1987; Khan, 2004d; Pakwatan, 2009).

Glaciers have two important aspects of climate change. They provide valuable information of past climate that is preserved in the core, particularly in trapped air bubbles. The studies of this scientific information enable scientists to analyze past climate change and predict future trends for the recent climate changes. The second important aspect of glacial studies is the analysis of the current situation for climate change, as glaciers are very sensitive to temperature, which is the basic element of weather condition and ultimately the climate. The study of glaciers and weather changes over a long period can provide a basis for current climate changes. At the global level, glaciers are receding and wasting, which is the sign of global warming and climate change. In some cases, it has also been reported that glaciers may not be receding or wasting but the snow melting and feeding capacity have increased with time, which is the root cause of associated hazards in the region (Handwerk, 2006; Hewitt, 2009).

It has been reported that the global mean temperature is expected to increase due to global warming, from 1.5°C to 6°C before the present century ends (GOP, 2007a; Pachauri, 2009). This increase will certainly change the climate dramatically, and the result will be most obvious at two opposite regions, that is, coastal areas and glaciers. It will have many implications on national, regional, and local levels. It has been observed that these changes are responsible for the conspicuous increase in storms, floods, and raging forest fires that have severe consequences for the existing infrastructure, including built systems, natural as well as human systems. The World Glacier Monitoring Service collects data annually on glacier retreat and glacier mass balance. From this data, a world glacier inventory has been compiled since the 1970s. Initially, mainly based on aerial photographs and maps, this compilation has resulted in a detailed inventory of more than 100,000 glaciers covering a total area of approximately 240,000 km^2. Glaciers worldwide have been found to be shrinking significantly, with strong glacial retreating in the 1940s, stablilizing or growing conditions during the 1920s and 1970s, and again retreating from the mid-1980s to the present. These observations are true for Glaciers of Pakistan also, which is part of great Himalayan mountain system (National Snow and Ice Data Center [NSIDC], 2009). According to the Intergovernmental Panel for Climate Change's (IPCC) Fourth Assessment Report on Climate Change and the Copenhagen Conference on Climate Change – 2009, glaciers have been the "most visible" indicators of climate change. Many of the Himalayan glaciers are retreating faster than the world average (Dyurgerov & Meier, 2005). This process is more likely to lead to reduction in capacity of natural snow storage. Although the period of time is certainly questionable, the impact of climate change and its consequences are generally agreed upon by most scientists (Samjwal, Mool, & Sharesta, 2007; GOP, 2007a; Institute for Social and Environmental Transition [ISET], 2008; NSIDC, 2009). Recent studies suggest that some glaciers of Pakistan are growing due to shortened periods of summer and winter seasons and rapid change in seasonal temperatures (Handwerk, 2006; Hewitt, 2009).

Glaciers of Pakistan are a major source of fresh water, and more than 50 million people directly depend upon this water. Changes in the glaciers will certainly affect this whole population (GOP, 1987, 2007a; Samjwal et al., 2007; Jianchu et al., 2007; Hewitt, 2009). Some important aspects of the glacier impacts are the bulk of melt water yield that coincides with the summer monsoon (90 percent of glacier melt comes in about two months of the year), which increases the risk of flood hazards; most of this water goes to the sea due to lack of storage capacity. Variability of summer weather includes huge ups and downs of water yield between the sunny (high flows) and stormy (low flows) weather, the effects of which are clearly visible in the shoulder season yields; this water yield is critical for agriculture, hydropower, dam storage, and city uses. The history of natural hazards affecting water supply and infrastructure shows that the effects have been worst with greater snowfall (avalanching, etc.), cold (less melting and water availability, shorter summers), and advancing glaciers (great ice dams, GLOFs, land loss). Associated hazards like glacial lake outburst floods (GLOFs), glacial surges, catastrophic rockslides, and flash floods are increasing with time. Lack of proper monitoring and forecasting capabilities for the glacial regions of the upper Indus Basin increases the risk of all associated hazards. Future important issues concerned with the climate change and glaciers of Pakistan are the study of avalanche-fed glaciers; on-ice dust, dirt, and debris; snow and ice temperatures; seasonal regimes and "shoulder seasons"; and glacial hazards (Mason, 1934; Hewitt, 2009).

CLIMATE CHANGE AND VARIABILITY IN PAKISTAN: IMPACTS, VULNERABILITY AND RISK

Climate change impacts are worldwide (Samjwal et al., 2007; Jianchu et al., 2007; Hewitt, 2009). For Pakistan, they are most critical because a large part of its population is chronically exposed and vulnerable to a variety of natural hazards and disasters. Global warming and climate change will continue for many decades to come, resulting in dangerous consequences for nations like Pakistan, unique in its vulnerability contexts. Impacts of climate change, variability, and extreme events will lead to severe stress on the overall development, environment, and human society for generations ahead.

It is now well established that emissions due to anthropogenic sources have, over the last 50–60 years, resulted in the excessive emission of greenhouse gases into the earth's atmosphere, increasing its temperature, causing global warming, and changing climate patterns. Although climate change affects lives and livelihoods across the world, its impact is greatest on the rural poor, especially those households living in areas frequently struck by natural disasters. These disasters can actually destroy the natural resource base of their areas. South Asia has seen an increase in the incidence of cyclones, floods, droughts, and storms. These have the potential to destroy crops, cause human and livestock fatalities and loss of infrastructure, and lead to large-scale human migration. The IPCC reports that the number of cyclones in South Asia has decreased since the 1970s, although their intensity has increased. This was seen in the case of immense damages caused in coastal Sindh in 1999 and 2007 (IPCC, 2007).

Millions of people living in South Asia's rural and coastal areas are under threat due to rising temperature and sea level, with Pakistan among the top 12 most threatened countries in the world (The NEWS, 2009). The temperature in the country's coastal areas has risen since the early 1900s by 0.6°C to 01°C (Khalid, 2009; Pachauri, 2009). The climate change impacts include sea level rise, variability of monsoon, increased frequency and intensity of extreme weather events, melting of glaciers and snow, and stress on agriculture and livestock. It has been observed that the last 12 years were the warmest since 1850, with 1998 being on the top, which recorded 0.6°C increases in the average global temperature during the last century (Khalid, 2009).

It has been suggested that the impacts of climate change will not be uniform in the Indus Basin. It means that the temperature increase in both summer and winter will be higher in northern Pakistan than in the southern part. Studies further revealed that temperature increases in the north and south will be higher in winter than in summer. Furthermore, the changes will increase variability of monsoons, adding to severe water-stressed conditions in arid and semiarid regions of the country. Among the possible effects of climate change are the likelihood of increased frequency, and severity of occurrence of extreme events such as floods and droughts, etc (ISET, 2008).

Because Pakistan is particularly susceptible to such events and has experienced large-scale destruction on these accounts in the recent past, the country needs to be particularly vigilant to the causes of changes and variability in climate (Haque, Islam-ul, Anum, & Ruhmma, 2009). That is the reason that Pakistan, in line with an agreement with Bali, is trying to make appropriate actions to reduce its carbon emission. It is despite the fact that Pakistan contributes only 0.43 percent of the world's total greenhouse gases and is 135th in the world's ranking for GHG emissions (GOP, 2007a; Pachauri, 2009). It has been suggested that there is a chance of decline in irrigated wheat yield in the semiarid areas of Pakistan, in the range of 9–30 percent for a temperature increase of 1–4°C. Precipitation has decreased 10–15 percent in the coastal belt and hyper-arid plains over the last 40 years while summer and winter rains have increased in northern Pakistan. About 80 percent of the nation's food and fiber needs are met by irrigated agriculture. The canal irrigation network of the country is mainly dependent on glaciers of the Hindu Kush and Himalayas, which are believed to be receding. Climate change is expected to result in changes in land and water resources that will subsequently affect agriculture productivity. Thus, agriculture and food security in Pakistan already under stress would be further threatened due to projected increased intensity and frequency of drought resulting from changes in climate (GOP, 2007a; Hewitt, 2009). Climate change could influence monsoon dynamics and cause summer precipitation levels to drop, as well as delays in the start of the monsoon season. There may also be longer breaks between rainy periods (Alam, 2009). Moreover, Oxfam conducted a study on "Climate Change, Poverty, and Environmental Crises in the Disaster-Prone Areas of Pakistan" in three districts, located in the Punjab, Sindh, and Baluchistan provinces. The result can be generalized for the rest of the country as well. The study identified increased incidence, frequency, and intensity of extreme climatic events: more intense and heavier rainfall in coastal areas, more intense cyclones, more intense flooding in flood-prone areas along the Indus, and more pronounced droughts in the arid areas of Khuzdar (Baluchistan). In most areas, rainfall patterns have become very erratic, making it difficult for communities to predict local rainfall patterns. It has also been pointed out that duration of the cropping period has shrunk perceptibly in Southern Punjab and Baluchistan, with a forward shift in sowing time and an earlier harvest. Summers have become hotter and winters much warmer across the areas studied. In some areas, communities have noticed some degree of cooling during the monsoon season over the last 30 years. Fish and prawn catches in coastal areas (freshwater fisheries) have shrunk due to seawater intrusion and the increase in sea surges and cyclones (which bring seawater into land depressions far inland). It implies that high-sea fishing remains the only solution, but few communities can afford the necessary equipment. It was also found that the groundwater table is falling very rapidly in the drought-prone areas of Khuzdar (Baluchistan). Traditional coping mechanisms used to deal with water shortages, declines in fish catch, and reduced agricultural produce are no longer enough to counter the immense impact of climate change. Seasonal out-migration has been observed in the areas studied, implying that incomes from traditional sources are no longer enough to support families.

CLIMATE CHANGE AND DISASTER MANAGEMENT POLICIES AND STRATEGIES IN PAKISTAN

Because Pakistan is severely suffering from natural and man-made disasters as well as climate change and variability, these problems have been tabulated in their various policies and strategies, both directly and indirectly.

Disaster Risk Reduction and Five-Year Development Plans

Until recently, disaster risk reduction in Pakistan had been basically revolving around flood disasters with a primary focus on rescue and relief. Hence, emergency response has been the predominant policy approach to dealing with disasters. After each disaster episode, the government has to incur considerable expenditure directed at rescue, relief, and rehabilitation. In the late 1960s, the need for incorporating the flood control program to the national developmental plan was realized. However, due to some institutional bottlenecks, only the first draft of the national disaster plan was prepared in 1974. The plan suggested various procedures, strategies, responsibilities, organizational setup, and standard procedures for disaster management operations. The plan, however, was neither finalized nor implemented (UNISDR, 2005). In 1993, environmental issues got high priority in various plans and policies for the first time. The eighth five-year plan (1993–1998) dealt with sustainable development and water resource management. Efforts were made for improving environmental legislation and enforcement of National Environmental Quality Standards (NEQS); initiating Environmental Impact Assessment (EIA) procedures for public and private investments; and incorporating more environment-related investments in all development expenditures (GOP, 1993). However, this strategy also failed and did not achieve the stated objective. Keeping in view the eighth five-year plan, a more comprehensive plan (the ninth five-Year Plan, 1998–2003) was prepared. However, it only focused on water resource management, and there is no visible section that has directly targeted disaster risk management and climate change adaptation. To achieve the Millennium Development Goals, the 10-year (2001–2011) perspective development plan was launched in 2001. The plan recognizes the sternness of drought in Sindh and Baluchistan and the consequent scarcity of water that has motivated and exacerbated the ongoing water crises in the country. The plan's main objectives included the reduction of poverty, increased growth rate, environmental stability, and the provision of basic social services and social safety nets (GOP, 2004). The plan laws, however, are too scanty and do not meet the standard requirements of providing proper patronage and protection to environment and sustainable development.

Environmental Protection Policies

To ensure environmental protection, the Pakistan Environmental Protection Act was issued in 1997. The act provided various rules for the conservation, protection, and rehabilitation to curb environmental pollution and promote sustainable development. The key feature of this act was the establishment of the Pakistan Environmental Protection Council, the Environmental Protection Authority, the National Environmental Coordination Committee, and the Provincial Sustainable Development Fund (GOP, 1987, 1997). The Karachi Port Trust Act (1994) keeps restrictions on the release of pollutants as well as waste, oil, and radioactive material, hazardous material, and bilge discharge from tankers and vessels within the limits of the Karachi Port (GOP, 1987, 1994). The EPA also drafted certain rules and regulations in 1998 and 1999, to implement the 1997 Environmental Protection Act. In its ninth meeting (February 2001), the Pakistan Environmental Protection Council (PEPC) approved a National Environmental Action Plan presented by the United Nations Development Programme (UNDP). High priorities were given to the construction of new dams and canals. The Pakistan National Conservation Strategy, Forestry Sector Master Plan, Poverty Reduction Strategy Paper, and Sarhad Provincial Conservation Strategy also gave special attention to sustainable development and environmental preservation.

Disaster Risk Reduction Policies

Until recently, the predominant approach to disaster management in Pakistan was emergency response. In 1958, the National Calamities (Prevention and Relief) Act was promulgated. The act focused on the maintenance and restoration of affected areas from various catastrophes, and for providing relief to the people during disaster. Provincial Relief Commissioners have been given the responsibility of relief and evacuation (GOP, 1958). This act only focused on relief activities, and no measures were suggested for mitigation and prevention. The local government ordinance of 2001 also provides some avenues for disaster management; however, its only focus again is on relief and recovery. The draft Emergency Services Ordinance was promulgated in 2002, to deal with emergencies in an effective manner and address threats to the public from modern warfare, terrorism, and disasters. This ordinance defines responsibilities at each and every level. Under the 2002 ordinance, the federal-, provincial-, and district-level governments shall set up emergency services, which shall be accountable for preparedness and quick response to emergencies and disasters of any extent (GOP, 2002b).

The situation of flood hazard management in Pakistan is comparatively better since the time of independence. An Early Flood Forecasting and Warning System was established soon after the creation of Pakistan in 1947. The devastating floods of 1973 aroused the feeling that the flood forecasting arrangements in the country were still inadequate and did not meet the country's requirements (GOP, 1987). These floods, thus, paved the way to full-fledged activities for the further strengthening of the Flood Forecasting System (FFS). The subsequent floods of 1988, 1992, 1996, and 1997 again revealed many serious system discrepancies pertaining to areas of flood forecasting and warning dissemination (Khan, 2002, 2003). Consequently, it was felt that the system still badly needed to be further enhanced. The system went through many improvements and is now sufficiently comprehensive as a number of government organizations work for floods mitigation: Provincial Irrigation and Drainage Authority (PIDA); Water and Power Development Authority (WAPDA); Provincial Relief Organization; Pakistan Army; Commissioner for Indus Water (CIW); Emergency Relief Cell (ERC); Federal Flood Commission (FFC); and Flood Forecasting Division (FFD). The efforts are still underway to make the system more self-contained (GOP, 1987, 2007a; South Asian Association for Regional Cooperation [SAARC], 1992).

In case of earthquake risk reduction management, the major improvement has been the establishment of the Earthquake Rehabilitation and Recovery Authority (ERRA) and the Provisional Earthquake Rehabilitation and Recovery Authority (PERRA) in 2006, after the earthquake 2005 (GOP, 2007a). For the risk of drought and desertification mitigation, various government institutions and authorities are working at different levels. Among these are the Division Emergency Relief Cell (ERC) within the Cabinet Division; Drought Emergency Relief Assistance (DERA) program; National Steering Committee (NSC); Provincial Relief Commission; Drought Crisis Control Centre (DCCC); Provincial Drought Management Committee (PDMC); Provincial DERA unit; Pakistan Meteorological Department; Pakistan Agricultural Research Council; Ministry of Environment, Local Government and Rural Development (MELGRD); Sarhad Developmental Authority (SDA); Sindh Arid Zone Developmental Authority; Punjab Land Utilization Authority (PLUA); Cholistan Developmental Authority; Thal Development Authority; and WAPDA (GOP, 1987, 2007a; SAARC, 1992).

Keeping in view the losses of the earthquake of 2005, the government of Pakistan promulgated the National Disaster Management Ordinance

(NDMO) of 2006. The implementation of NDMO has been ensured by the National Disaster Management Commission. The government of Pakistan through NDMO established the National Disaster Management Authority (NDMA) at the federal level. The NDMA is the only focus group for coordinating and facilitating the implementation of strategies and programs on disaster risk reduction, response, and recovery. On the provincial level, the Provincial Disaster Management Authorities (PDMA) are responsible for the disaster management operations. Currently, the PDMA exists only in two provinces (Punjab and NWFP). Efforts have been made by the INGOs and donor agencies to establish districts and Tehsil Disaster Management Authorities. For the technical guidance of the entire disaster risk reduction process, a National Disaster Risk Management Framework Pakistan (NDRMFP) has been formulated. The main purpose of the NDRMFP is to attain sustainable social, economic, and environmental development in Pakistan by mitigating risks and vulnerabilities, particularly of the poor and marginalized groups, and effective response to natural disasters (GOP, 2007a). The implementation of NDMO is not yet finalized due to some political and institutional problems. As a result, the ordinance has not yet been given the status of act through the parliament.

It is pertinent to mention here that even now, there has been no continuing, comprehensive, and sound institutional planning to deal with disaster issues with a long-term vision. Disasters are viewed in isolation from the processes of mainstream development planning and poverty alleviation programs. For example, disaster management, development, and environmental institutions function in isolation from one another, and integrated planning and proper coordination between these sectors is almost lacking. To conclude, the absence of a central authority for integrated disaster management and lack of coordination within and between disaster-related organizations is responsible for an effective, comprehensive, and well-organized disaster management program in the country. State-level disaster preparedness and mitigation actions are greatly skewed toward structural aspects as well as weak nonstructural elements such as the knowledge and capacities of local people, and the related livelihood protection issues, etc.

CONCLUSIONS AND RECOMMENDATIONS

Pakistan is a country of diverse physiographic, climatic, hydrological, and cultural conditions. The country's prevailing fragile environmental, economic, and sociopolitical conditions have increased the vulnerability of the country for climate change impacts. Pakistan is already experiencing the adverse impacts of global warming and climate change. Summers are becoming hotter, monsoons are more irregular, heavy rainfall occurs over a short period, causing severe flash floods and landslides, river flow and inundation during monsoon has increased, the frequency, intensity, and recurrence of floods has increased; as well, the country has experienced severe damages to crop failure due to flash and river floods, crop failure due to drought, prolonged cold spells, coastal erosion, riverbank erosion, glacial retreat, GLOFs, avalanches, etc. To adapt to the adverse effects for climate change and variability, the government of Pakistan needs to formulate effective policies, encourage research on the climate change related issues, mainstream disaster management into routine planning processes, and establish a central authority for coordinating between disaster-related institutions and other organizations in the relevant sectors.

In most cases, local people traditionally adapt to their environment. They know the risks in their surroundings and adopt mitigation measures to reduce their vulnerability. However, due to the climatic variability, the number, frequency, density, and intensity of extreme weather events are increased beyond the capacity of the communities. As a result, these communities are no more resilient to natural disasters. The worst aspect of the impact of climate change is that it increases the vulnerability of the poorest communities of the poorest nations. It is pertinent to mention that Pakistan is a small contributor of carbon dioxide to the atmosphere, but it is number 12 on the list of most affected countries due to climate changes. There is a need for various adaptation measures to mitigate the impact of climate change and disasters. Following are some recommendations and adaptation measures suggested to mitigate the impacts of climate change:

1. Effective policy formulation for climate change adaptation
2. Mainstream disaster management into routine planning process
3. Central authority for coordination between disaster-related institutions and other organizations working in relevant sectors, most particularly environment management
4. Credible and up-to-date scientific knowledge for development of climate policies
5. Improved efficiency of irrigation system
6. Construction of delay action and recharger reservoir in the upstream
7. Installation of an early warning and forecasting system
8. Use of more heat-/drought-tolerant crop varieties in areas under water stress

9. Government should undertake more robust tree plantation programs
10. Preservation of watershed and rangelands
11. Conservation of coastal zones
12. Protection of forest and reforestation in riverine forests
13. Government infrastructure at all levels that can withstand future climate conditions to ensure and maintain its long-term safety
14. The government undertaking of an awareness-building campaign through the media
15. Continued research on climate change impacts
16. A climate change cell under the Ministry of Environment
17. Sensitization of government agencies through training programs
18. More policies and new plans can be applied by the Environment Ministry, aiming at adaptation of climate change
19. Research on climate change issues in the context of Pakistan
20. Supporting community-led adaptation (the focus should be on empowering communities to monitor and take action to adapt to changing climate and environment, based on their own decision making and participatory technology)

ACKNOWLEDGMENTS

The author is grateful to Mr. Amjad Ali and Mr. Mushtaq Ahmad Jan, research associates in the Centre for Disaster Preparedness and Management, University of Peshawar, Pakistan, for their valuable help and support during the preparation of this chapter.

REFERENCES

Alam, M. (2009). Pakistan may face exceptional climate change. The NEWS. May 06.

Blood, P. R. (1995). *Pakistan: A country study*. Washington, DC: Federal Research Division, Library of Congress.

Dyurgerov, M. D., & Meier, M. F. (2005). *Glaciers and changing earth system: A 2004 snapshot* (p. 117). Boulder, CO: Institute of Arctic and Alpine Research, University Colorado.

EM-DAT. (2008). The OFDA/CRED International Disaster Database, Université catholique de Louvain, Brussels, Belgium. Data version: v11.08.

Georg, P.C. (2009). Earthquake and tsunami of 28 November 1945 in Southern Pakistan: Tsunami, earthquakes, hurricanes, volcanic eruptions, climate change, and other natural and man-made hazards and disasters. Disaster Archaeology, The Tsunami Page. Available at http://www.drgeorgepc.com/Tsunami1945Pakistan.html.

Government of Pakistan (GOP). (1958). *Pakistan National Calamities (prevention and relief) Act 1958*. Islamabad, Pakistan: Ministry of Interior.

Government of Pakistan (GOP). (1987). *Environmental profile of Pakistan* (p. 248). Islamabad, Pakistan: Environment & Urban Affairs Division.

Government of Pakistan (GOP). (1993). *Eighth Five-Year Plan, 1993–98*. Islamabad, Pakistan: Planning Commission of Pakistan.

Government of Pakistan (GOP). (1994). *Karachi Port Trust Amendments Act 1994*. Islamabad, Pakistan: Ministry of Environment.

Government of Pakistan (GOP). (1997). *Pakistan Environmental Protection Act 1997*. Islamabad, Pakistan: Ministry of Environment.

Government of Pakistan (GOP). (2002a). National Action Programme to Combat Desertification in Pakistan. Pakistan Agricultural Research Council, Ministry of Environment, Local Government and Rural Development (MELGRD), United Nations Environment Programme (UNEP), Economic and Social Commercial for Asia and Pacific (ESCAP). Available at http://www.pakistan.gov.pk/divisions/environment-division/media/napintro.pdf.

Government of Pakistan (GOP). (2002b). *Emergency service ordinance 2002*. Islamabad, Pakistan: Federal Relief Commission.

Government of Pakistan (GOP). (2003). *Pakistan's initial national communication on climate change*. Islamabad, Pakistan: Ministry of Environment.

Government of Pakistan (GOP). (2004). *Pakistan millennium development goals report 2004*. Islamabad, Pakistan: Centre for Research on Poverty Reduction and Income Distribution.

Government of Pakistan (GOP). (2007a). *National disaster risk management framework Pakistan* (p. 91.). Islamabad, Pakistan: National Disaster Management Authority.

Government of Pakistan (GOP). (2007b). *Earthquake-8/10: Learning from Pakistan experience*. Islamabad, Pakistan: National Disaster Management Authority.

Handwerk, B. (2006). Some glaciers growing due to climate change, study suggests. National Geographic, National Geographic News, Available at http://news.nationalgeographic.com/news/2006/09/060911-growing-glaciers.html

Haque, Islam-ul, Anum, & Ruhmma (2009). Effects of the climate change on planning and managing water resources in the Potohar region of Pakistan. *Electronic Journal of Environmental, Agricultural and Food Chemistry*, *8*(8), 647–656

Hewitt, K. (2009). Glaciers and climate change in the Karakoram Himalaya: Developments affecting water resources and environmental hazards. Outline of Presentation at China Environment Forum, Cold Regions Research Centre, Wilfrid Laurier University, Waterloo, Ontario, N2L 3Z9. Available at http://www.wilsoncenter.org/events/docs/resources_hewitt.pdf

Hyogo Framework for Action 2005–2015 (HFA). (2005). *World conference on disaster reduction*, January 18–22, 2005, Kobe, Hyogo, Japan, A/CONF.206/6.

ICRC. (2003). *World disasters report 2003*. Geneva, Switzerland: International Federation of Red Cross and Red Crescent Societies.

Institute for Social and Environmental Transition (ISET). (2008). *Climate adaptation in Asia: Knowledge, gaps and research issues in South Asia* (p. 46). Nepal: Design and type in Kathmandu.

Intergovernmental Panel On Climate Change (IPCC). (2007). *Climate change 2007 – the physical science basis contribution of working group I to the fourth assessment report of the IPCC*, ISBN 978 0521 88009-1 Hardback; 978 0521 70596-7.

Jianchu, X., Sharestha, A., Vaidya, R., Eriksson, M., & Hewitt, K. (2007). *The melting Himalayas: Regional challenges and local impacts of climate change on mountain ecosystems and livelihood.* ICIMOD, Technical Paper, p. 15, Kathmandu, Nepal.

Kahlown, M. A. (2002). Welcome address. In: *Proceedings of the SAARC workshop on drought and water management strategies* (pp. 5–7). Lahore, Pakistan: Pakistan Council of Research in Water Resources.

Khalid, S. (2009). *Pakistan facing severe climate effects* (June 06, 2009). Islamabad, Pakistan: The NEWS.

Khan, A.N. (1992). *Landslides and housing: A case study of Murree, Pakistan.* Unpublished Ph.D. thesis, Environmental Planning, Institute of Planning Study, University of Nottingham, UK.

Khan, A. N. (1993). Towards an appraisal of hazard-response theory with special reference to landslide hazard. *Journal of Rural Development and Administration, 25*(4), 42–63.

Khan, J. A. (1993c). *The climate of Pakistan* (Kingsolver, J.G.). Karachi, Pakistan: Rehbar Publishers.

Khan, A. N. (1994a). Extent and evaluation of the adverse effects of landslides on housing in Murree, Pakistan. *Journal of Rural Development and Administration, 26*(1), 119–140.

Khan, A. N. (1994b). Towards an appraisal of Landslide Hazard Reduction Programme (LHRP) in Murree, Pakistan. *Pakistan Journal of Geography, 4*(1&2), 15–30.

Khan, A.N. (1995). Landslide hazards and policy response in Pakistan: A case study of Murree. In: *Proceeding of the national symposium on "Geological Hazards: Prediction, Mitigation and Control."* Department of Geology, University of Peshawar, Vol. 2, Issues 2, pp. 15–37.

Khan, A. N. (2002). An evaluation of flood hazard reduction policies: A case study of Kabul-Swat flood plain, Peshawar Vale. *Peshawar University Teachers' Association Journal (PUTAJ), 9*, 1–14.

Khan, A. N. (2003). Impacts of flood hazard in Muzaffarabad, Azad Jammu and Kashmir. *Peshawar University Teachers' Association Journal (PUTAJ), 10*, 1–12.

Khan, A. N. (2004a). An evaluation of landslide hazard mitigation measures: A case study of Muzaffarabad, A.J.K. *Journal of Engineering and Applied Sciences, 23*(1), 79–88.

Khan, A. N. (2004b). Planning for flood hazard: A geographical approach to flood problem in district Jhelum, Punjab. *Peshawar University Teacher's Association Journal (PUTAJ), 11*, 185–199.

Khan, A. N. (2004c). Causes and environmental impacts of flash floods in the Hindu Kush region: A case study of sample villages along the Buni-Chitral Road, Chitral, Pakistan. *Pakistan Journal of Geography, XIV*(1&2), 43–59.

Khan, F. K. (2004d). *Pakistan geography, economy and people* (No. 38, Sector 15, Korangi Industrial Area, PO Box 8214, Karachi-74900, Pakistan.). Karachi, Pakistan: Oxford University Press.

Khan, A.N. (2005a). Planning for the reduction of landslide hazard: A case study of Muzaffarabad, Azad Jammu & Kashmir. In: *Proceeding of the international symposium on landslide hazards in orogenic zone from the Himalaya to Island Arcs in Asia, jointly organized by Nepal Landslide Society, Japan Landslide Society and International Consortium on Landslides in collaboration with Centre for Applied Research and Development*, IOE, Tribhuvan University, Nepal and International Landslide Centre, University of Durham, UK, September 25–26, 2005, Kathmandu, Nepal, pp. 291–300.

Khan, A.N. (2005b). Planning for flood hazard reduction: A case of Neelum-Jhelum valley, Muzaffarabad, Azad Jammu and Kashmir. In: *Proceeding of the South East Asia survey congress, 2005 on geomatics and the community: Spatial way to sustainable development*, Bandar Seri Begawan, Brunei Darussalam.

Khan, A.N. (2006). Extent and causes of earthquake hazard: A case study of Quetta, Pakistan. *Punjab Geographers: A Journal of the Association of Punjab Geographers*, India.

Khan, A. N., & Atta-Ur-Rahaman. (2006). Landslide hazards in the mountainous region of Pakistan. *Pakistan Journal of Geography*, *XVI*(1&2), 38–51.

Khan, A. N., & Atta-Ur-Rahman. (2008). Application of a checklisting technique for the assessment of impacts of the Chashma right bank canal on land use and cropping pattern of D. I. Khan district, Pakistan. *International Journal of Irrigation and Drainage*, *57*, 165–173.

Mason, K. (1934). The study of threatening glaciers. *The Geographical Journal*, *85*(1) January 1935, pp. 24–35.

National Snow and Ice Data Center (NSIDC). (2009). *All about glaciers, supporting cryospheric research since 1976*. 449 UCB University of Colorado Boulder, CO 80309-0449. Available at http://nsidc.org/glaciers/questions/climate.html

Pachauri, R. K. (2009). *Climate changes to adversely impact Pakistan* (January 14, 2009). New Delhi, India: The Indian News.

Pakwatan. (2009). *Its all about Pakistan*. Travel Guide, Glaciers of Pakistan. Available at http://www.pakwatan.com

Pomee, M. S., Zaheer-ul-Ikram, Khan, M. A., & Ali, I. (2005). Drought mitigation measures: An overview under Pakistan's perspective. *Pakistan Journal of Water Resources*, *9*(2), 2–3.

Regional Specialized Metrological Centre (RSMC). (2009). Historical records of severe cyclonic storms which formed over Arabian Sea and made landfall at the Western coast of India during the period from 1970–1999. Available at http://www.imd.ernet.in/section/nhac/static/cyclone-history-as.htm

Samjwal, R. B., Mool, P. K., & Sharesta, B. R. (2007). *Impacts of climate change on Himalayan glaciers and glacial lakes: Case studies on GLOF and associated hazards in Nepal and Bhutan* (p. 119.). Kathmandu, Nepal: ICIMOD and UNEP.

South Asian Association for Regional Cooperation (SAARC). (1992). *Regional study on the causes and consequences of natural disasters and the protection and preservation of the environment*. Kathmandu, Nepal, p. 212.

South Asian Flood (SAF). (2009). Pakistan: An introduction. Regional Cooperation for Flood Information Exchange in the Hindukush and Himalayan Regions. Available at http://www.southasianfloods.icimod.org

The NEWS. (2009). Pakistan: Rural poor more vulnerable to climate change. Daily: The NEWS. November 22, 2009.

UNISDR. (2005). Disaster management policies and systems in Pakistan. Available at www.unisdr.org/eng/country-inform/reports/Pakistan-report.pdf

Yousaf, M. (2007). *Preparing for large scale emergencies: Country perspective of Pakistan*. Manila, Philippines: Asian Development Bank.

CHAPTER 11

INTEGRATING DISASTER RISK REDUCTION AND CLIMATE CHANGE ADAPTATION: INITIATIVES AND CHALLENGES IN THE PHILIPPINES

Juan M. Pulhin, Maricel A. Tapia and Rosa T. Perez

INTRODUCTION

The Philippines is one of the most vulnerable countries in the world as far as climate-related and other forms of disasters (e.g., earthquake and volcanic eruption) are concerned (World Bank and NDCC, 2004; UNISDR, 2008a, 2008b). In 2004 alone, 25 weather disturbances hit the country, of which four occurred successively in November and the first week of December. These triggered massive landslides and flooding in Southern and Central Luzon, leading to damages in lives and properties (Duque, 2005). With this the country is considered as one of the most disaster prone, ranking 12th among the 200 countries most at risk to natural hazards in the 2009 Mortality Risk Index of the United Nations International Strategy for Disaster Reduction (UNISDR, 2009).

Climate-related and other natural forms of disaster in the Philippines stem primarily from its unusually high exposure to natural hazards due to its

Climate Change Adaptation and Disaster Risk Reduction: An Asian Perspective
Community, Environment and Disaster Risk Management, Volume 5, 217–235

ISSN: 2040-7262/doi:10.1108/S2040-7262(2010)0000005017

geographic circumstances (World Bank, 2005). The country, an archipelago with approximately 7,100 islands, is situated along a stretch of risk-prone areas. It lies near the most seismically active part of the earth, the Pacific Ring of Fire, which is located in the western rim of the Pacific Ocean. It also rests at the junction of two large tectonic plates – the Philippine Sea Pacific Plates to the east and the Eurasian Plate to the west, making the country also vulnerable to volcanic eruptions and earthquakes (ADPC, 2001). Likewise, the eastern side of the Philippines faces a wide open expanse of Pacific Ocean, considered as one of the world's major typhoon belts. The archipelagic nature of the country also increases the vulnerability of its coastal areas to storm surges, tsunamis, and sea level rise (Punongbayan and Tayag, 1999 in ADPC, 2001).

Of the above, climatic hazards lead to the greatest number of disasters in the country and results in larger accumulated damages. They bring about the highest economic losses, most number of people affected, and most number of people killed (CRED EMDAT 2007). On top of this is the threat caused by climate change posing increased and more intense occurrences of these hazards. It is therefore in order that risks from these hazards be reduced and appropriate adaptation mechanisms, particularly with the changing climate, be installed.

This chapter reports the current efforts in the Philippines in integrating disaster risk reduction and climate change adaptation. Particularly, it highlights local initiatives in the municipality of Guagua in Pampanga as part of the devolution program of the Philippine government. Prior to this, it gives an overview of the different disasters experienced by the country over the years and the impacts of these on the different sectors. Challenges of the integration initiatives are also discussed to further enhance future actions and resilience of local communities to different types of disasters.

PHILIPPINE DISASTER EVENTS

Disaster originating from natural hazards can be categorized into two groups: hydrometeorological and geological. Hydrometeorological hazards are water- and weather-related, which include intense precipitation events, floods, droughts, hail, windstorms, and tropical cyclones. Geological hazards include volcano eruptions and earthquakes. Meanwhile, other hazards may be set off by the above weather-related or geological events, which are categorized as secondary hazards such as landslides and tsunamis (Sperling & Szekely, 2005).

In the Philippines, hydrometeorological hazards have caused more than 80% of the total disasters from 1905 to 2009. These are storms and associated surges, floods, droughts, and rainfall-induced landslides. Because of its geographic location, the country is visited by at least 20 tropical cyclones a year on average. Of this number, an average of 5.97 led to a disaster each year based on 1980–2008 records. Weather-related events in the same period also brought an average of 2.45 flood occurrences and drought almost every four years, with an annual average of 0.24 (CRED EMDAT in PreventionWeb, 2010) (Fig. 1).

With climate change, such occurrences, however, are expected to be more intense. For instance, climate change in the Philippines is predicted to bring significant increasing trend in temperature and significant increase in the frequency of warm days and hot nights. While change in rainfall trends and total number of annual tropical cyclones brought about by the changing climate appeared to be statistically insignificant (Hilario, 2008 in DILG, GTZ, & DIPECHO, 2008), the intensity of extreme events is, however, expected to be heightened increasing flood and drought risks.

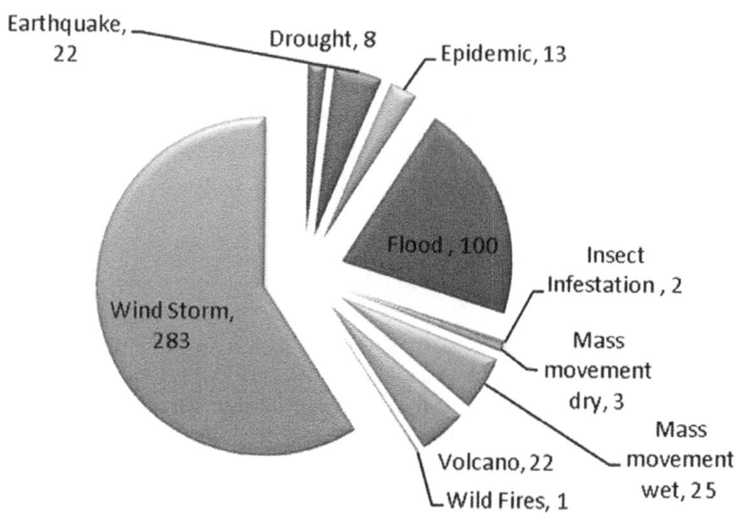

Fig. 1. Natural Disasters in the Philippines, 1905–2009 (CRED EM-DAT 2010-EM-DAT: The OFDA/CRED International Disaster Database – www.emdat.net– Université catholique de Louvain, Brussels, Belgium).

IMPACTS OF DISASTERS ON DIFFERENT SECTORS

Disasters have destroyed human, social, and physical capital, and they have derailed social and economic development, as funds are reallocated from ongoing programs to finance relief and reconstruction assistance. While the documented losses caused by disasters are formidable, the full impacts are not known and are probably much more significant (World Bank and NDCC, 2004).

In the Philippines, disaster events are one of the major threats to the country's economic progress and development process (Benson, 2009). They have had grave social and economic consequences, costing an average of PhP 15 billion (US$300M) per year in direct damages between 1970 and 2000 and killing about 866 people per year in that same period. Typhoons alone due to its high frequency in the country accounted for 65% of lives lost and 76% of total damage (World Bank and NDCC, 2004).

Poverty is closely linked to vulnerability to natural hazards due to the location of housing of poor people (e.g., on floodplains, riverbanks, steep slopes, or contaminated land previously occupied by industrial facilities); their level of access to basic services (e.g., garbage collection) particularly for squatters; sources of livelihood; and level of access to financial and other assets and resources, leaving limited recourse to intertemporal consumption smoothing (World Bank and NDCC, 2004). However, as demonstrated by tropical storm Ketsana (Ondoy) and typhoon Parma (Pepeng) that successively hit the country in September and October 2009, resulting in massive floods in Metro Manila and other provinces, the magnitude of the destruction did not spare even the affluent families living in exclusive villages nor the huge establishments of business tycoons.

Agriculture

For the agriculture sector, less rain or too much rain means less harvest. Changes in the timing of rain are also critical. A rise in carbon dioxide favors crops, but other limiting factors could catch up, for example, water unavailability, such that productivity is affected negatively. Temperature increases would cause the enhancement of activity of weeds and pests, and plant diseases; the reduced effectiveness of insecticides, pesticides, and herbicides; and the acceleration of soil degradation.

Crop yield data from the Bureau of Agricultural Statistics clearly indicated the influence of prevailing climate such as the El Niño-Southern

Oscillation (ENSO)[1] phenomenon. The 1997–1998 El Niño was by far the most intense one causing a large dip in both rice and corn yield (BAS, 2010).

Aside from production losses, it should also be noted that two-thirds of the Philippine poor are primarily dependent on agriculture for their livelihood, which is highly vulnerable to natural hazards (ADB, 2007 in Benson, 2009). In the period of 1990–2006 alone, the sector accounted for 63% of total reported disaster losses, with the typhoons in year 2006 accounting for PhP 10.5 billion (US$0.2 billion) in agricultural losses. The poverty incidence increase of 3% from 1997 to 1998 (from 25% to 28%) was also attributable to the El Niño event that brought devastating effects to the agricultural sector (Benson, 2009).

Infrastructure and Services

Infrastructure and services are among those damaged and disrupted when a disaster strikes. Typhoons, earthquakes, and volcanic eruptions have resulted in significant losses in terms of infrastructure, such as roads, bridges, buildings, and other properties. For instance, the typhoons in 2006 damaged 118 power transmission towers in South Luzon transmission system, leading to disruption of power services. Such destruction also derailed economic activities, as the blackout that resulted together with the continuing power deficit in the Bicol region amounted to US$250 million (PhP 10 billion) lost in just two days as of late 2007 (World Bank, 2007 in Benson, 2009). Furthermore, the recent typhoon Ketsana that hit the Philippines on 26 September, 2009, had resulted in an estimated PhP 3.684 billion (US$73.68M) infrastructural damage and about 42,556 totally or partially damaged houses (NDCC, 2009).

Disasters also have indirect impact on infrastructures and services as efforts for current improvement in transportation and other facilities only reached modest success as resources are redirected from such purposes to respond to the needs created by the disasters or calamities (Benson, 1997 in World Bank and NDCC, 2004).

Health

Disasters lead to deaths, injuries, and diseases. The impacts, particularly the outbreak of diseases, may have long-term effects on the human population.

For instance, floods may contaminate drinking water and provide sites for disease vectors (World Bank and NDCC, 2004). Sanitation is also another major problem when disasters strike, due to disruption in basic services such as water supply and garbage collection. Of particular concern is the situation in evacuation areas that are often overcrowded and wherein lavatory facilities, water, and other supplies may be lacking leading to diseases such as diarrhea, cholera, skin problems, among others. Further, damages and losses in health systems also aggravate the problem for providing immediate medical needs and so is the availability of food supplies for nutrition.

For the period of 1905–2009, eight of the top 10 natural disasters that led to the most number of deaths were weather-related (seven storms and one wet mass movement), bringing to a total of 15,464 killed just for these notable occurrences alone (CRED-EMDAT, 2010). Throughout the same period, storms also caused the most number of deaths (36,282), followed by earthquakes (9,589) and floods (3,003).

Environment

Environmental degradation is a cause and effect of disasters from natural hazards, particularly in the case of typhoons and its associated disasters such as floods, sediments, and landslides. Improper waste disposal and the absence of greeneries in urban areas cause the easy saturation of soil to rain waters, resulting in floods. On the other hand, deforestation contributes to increased runoff thereby resulting in landslides. Landslide events also erode the fertile part of the soil and therefore decrease its productivity. Effects of disasters can also change the complete landscape of an area. For example, the three tropical cyclones that successively hit the Philippines in 2009, tropical storm Ketsana (Ondoy), typhoon Parma (Pepeng), and typhoon Mirinae (Santi), left devastating effects that literally "erased" some areas in Luzon from the map. Several cities in Metro Manila and towns near Laguna Lake were submerged into a sea of mud and floodwaters that lasted for months, making some previously busy areas desolated. The disrupted environmental services caused by these disasters also have an impact on the economic activities, safety (both healthwise and related to physical structures), peace and order, governance, and cultural and other societal functions.

Social Impacts

While data on the social impacts of disasters are incomplete, disasters often lead to displacement of families from their homes and sources of livelihood, as well as job losses. This contributes to greater poverty, which further increases the vulnerability of the poor to natural hazards. For instance, the El Niño event experienced by the country in 1997–1998 has led to shift and diversification of activities of some for their livelihood as well as withdrawal of children from school and reduction in quantity and decline in quality of food intake (PCARRD/DA, 2001 in World Bank and NDCC, 2004).

Economy

Typhoons have brought about considerable economic costs in terms of direct damages. Twenty of the highest damages for the period of 1947–2006 occurred during the last three decades, including the present one (Fig. 2). Two of them, typhoons Ruping (1990) and Rosing (1995), also happened during the 1990 decade, which is still the warmest decade on record. Several typhoons have caused damages in the order of 1.21% of GDP (e.g., in 1984, 1988, and 1990) and 4% of agricultural production (e.g., in 1990 and 1996; Amadore, 2005). Table 1 provides information on other climate

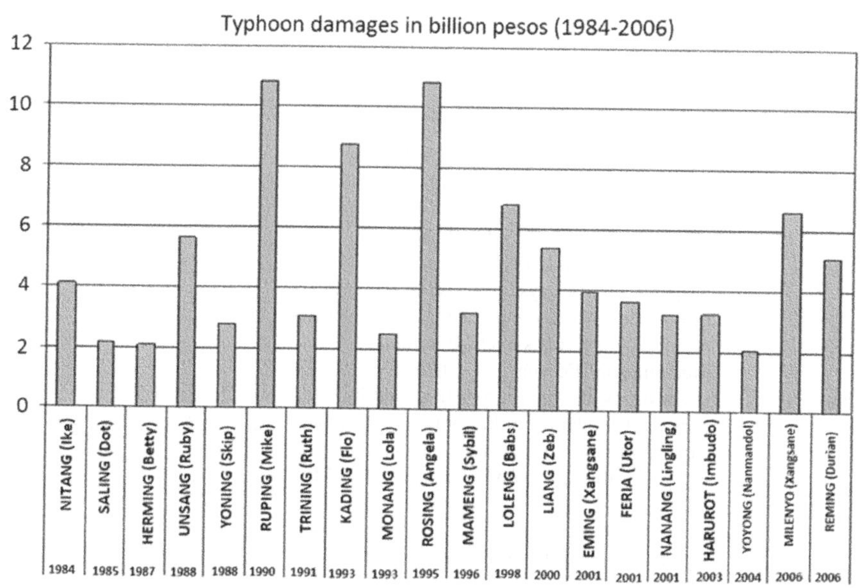

Fig. 2. Cost of Direct Damages of Typhoons, 1984–2006 (NDCC and Newspaper Reports). *Source:* NDCC and newspaper reports.

Table 1. Climate-Related Disasters in the Philippines.

Year of Event	Climate Change-Related Event (Specify Region)	Damage/Losses
1991	Ormoc flashflood – triggered by unusually heavy and continuous rains brought about by tropical storm Uring (region 8)	Deaths: 5,101–8,000+
1998	Multiple regions – El Niño phenomenon caused massive droughts and water shortages (April 1998)	Number of people affected: 2,600,000
1999	Cherry Hill landslide – triggered by three consecutive days of persistent moderate to heavy rains (region 4)	
2000	Payatas garbage slide (NCR) – triggered by continuous moderate to heavy monsoon rains over Metro Manila	
2001	Baguio-La Trinidad landslides (CAR) – triggered by record-breaking hour rainfall of 1,085.5 mm in Baguio City	
2001	Camuigin flashflood (region 10) – triggered by continuous light to moderate rains brought about by typhoon Nanang	Deaths: 157 Missing: 146 People affected: 35,3006 US$96 million
2003	Southern Leyte/Surigao disasters (region 8, ARMM) – triggered by record-breaking hour rainfall of 1,119.0 mm in Surigao and 699.0 mm in Leyte	
2004	Aurora/Infanta floods (regions 3, 4) – triggered by 20 days of persistent moderate to heavy rains brought about by typhoon Unding, tropical storm Violeta, tropical depression Winnie, and typhoon Yoyong, which also affected regions 1, 2, 5, CAR, NCR	
2006	Guinsaugon, Leyte landslide (region 8) – triggered by five days of persistent moderate to heavy rains, which is equivalent to almost three months of the area's average annual precipitation (February 17, 2006)	Deaths: 1,126
2006	Legazpi mudslide (region 5) – triggered by persistent moderate to heavy rains brought by super typhoon Reming, which also (November 30, 2006) affected regions 1, 2, 3, 4, 8, CAR, NCR	Deaths: 1,399 People affected: 2,562,517
2008	Tropical storm "Cosme" battered the northern Philippines, with powerful winds triggering floods and landslides. Wide areas remained without electricity as strong winds toppled electric posts and tripped power line	Deaths: 12 People affected: 35,000

Table 1. *(Continued)*

Year of Event	Climate Change-Related Event (Specify Region)	Damage/Losses
2009	Tropical storm "Ondoy" (Ketsana) swept Manila and neighboring regions of Luzon on September 26 with heavy rains causing flashfloods, landslides, and submerging many parts of these areas affecting 880,175 families, 4,320,699 persons, 1,902 barangays, 155 municipalities, 32 cities, and 25 provinces in 12 of the 16 regions of the country with a total damage of around PhP 10.5 billion in infrastructure and agriculture. Even before "Ondoy" had left, typhoon "Pepeng" (Parma) started to inflict similar destructions in eight regions of country particularly in Northern Luzon affecting more than 2.5 million persons and a total estimated damage of PhP 8.1 billion in infrastructure and agriculture (data as of October 13, 2009)	Deaths: 315 (Pepeng), 337 (Ondoy) Injured: 185 (Pepeng), 308 (Ondoy) Missing: 52 (Pepeng), 37 (Ondoy)

Sources: National Disaster Coordinating Council through the Office of Civil Defense.

change-related disasters in the Philippines with some estimates of deaths and cost of damages. It is generally believed that the more intense tropical cyclones and heavy rainfall episode causing large economic costs and loss of lives are already fingerprints of a changing climate.

INTEGRATING DISASTER RISK REDUCTION AND CLIMATE CHANGE ADAPTATION AT THE NATIONAL LEVEL

The institutional and disaster management system in the Philippines dates back in 1978 with the legislation of Presidential Decree 1566 creating the National Disaster Coordinating Council (NDCC). It relies on a highly reactive approach to disaster management, with emphasis on postdisaster relief and short-term preparedness (forecasting, evacuation, etc.) and with inadequate recognition of natural hazards as obstacles to long-term sustainable development (World Bank and NDCC, 2004).

The NDCC is the highest policy-making and coordinating body for disaster management in the Philippines. It is located within the Office of the

Civil Defense, chaired by the Secretary of the National Defense, and with 14 department secretaries as members. Its structure is also replicated at the regional, provincial, city, municipal, and *barangay* (district) levels, composed of representatives of national government agencies operating at local levels and local officials (Benson, 2009).

The comprehensive disaster management framework of the Philippines encompasses disaster risk reduction/mitigation and preparedness in the preevent and disaster response and rehabilitation/recovery in the postevent. This framework is supposed to be carried out by all stakeholders and levels in disaster management (Duque, 2005). However, disaster risk has been largely ignored in detailed economic planning, economic forecasting, and annual budgetary planning process (Benson, 2009). Furthermore, aside from the response-oriented approach noted above, information on disaster risk is lacking and measurement of socioeconomic impacts of disasters is inadequate. Limited risk reduction capacities and lack of coordination among stakeholders are also among the weaknesses of the disaster management in the Philippines (Duque, 2005). Clearly, there is a need to update the disaster management system and practice in the country, particularly including medium and longer-term visioning for mitigation and reducing risks to natural hazards.

Meanwhile, with the strong linkage between natural hazards leading to disasters and climate change, the need to integrate efforts on disaster risk reduction and climate change adaptation has been recently gaining greater recognition. This is also noted by Sperling and Szekely (2005) who highlight "a real opportunity to reduce current and future vulnerabilities to climate risks by building on and expanding existing disaster risk management." Integrating the two issues together through institutional, policy, and research coordination and development of joint strategies would ensure a more efficient use of efforts and resources and greater achievements (Benson, 2009).

The sections below discuss the initiatives in the country where disaster risk reduction and climate change adaptation are integrated, whether directly or indirectly, through policy, development plans/programs, and others.

Policy Initiatives

Various bills have been passed in the Philippine legislature that deal separately on disaster risk management and climate change. Proposed amendments on the disaster risk management of the country have also been noted to be filed in the last five years or so (Benson, 2009). Significant progress of late with regard to numerous bills submitted on the above issues is the recommendation of Senate Bill 3080 titled "An Act Strengthening Philippine Disaster Risk Capability by Institutionalizing the National Disaster Risk Management Framework, Appropriating Funds Therefor and for Other Purposes" (also known as Philippine Disaster Risk Management Act of 2009) and Senate Bill 2583 titled "An Act Mainstreaming Climate Change into Government Policy Formulations, Creating for This Purpose the Climate Change Commission, and for Other Purposes" (also known as the Climate Change Act of 2008).

The climate change bill had greater momentum with its enactment into a law already approved by the President on 23 October, 2009. The Act aims to mainstream climate change into policy formulation, development planning, and poverty reduction programs, creating a Climate Change Commission. Meanwhile, the disaster risk management bill, as of this writing, has already passed the second reading and has been sent to the House of Representatives requesting for concurrence. Nevertheless, this bill also addresses anticipatory responses to climate change as well as mainstreaming risk reduction in physical and land use planning, budget, infrastructure, health, environment, housing, and other sectors. Thus, it incorporates proactive and more comprehensive measures to disasters and acknowledges the relevance of these in poverty reduction and achieving sustainable development.

Major Plans and Programs

The Medium Term Philippine Development Plan (MTPDP) for 2004–2010 is the primary document that guides the country's national development programs. Its stipulations, however, have addressed responses to climate change and disasters separately. For instance, climate change was mentioned in terms of participation in the Clean Development Mechanism and the emerging carbon market. Disaster risk reduction, meanwhile, is mentioned under the agribusiness and environment and natural resources sections focusing on mitigation efforts for inherent risks in agriculture and mitigation of the occurrence of natural disasters through nonstructural (geohazard mapping, conduct of soil stability measures, and disaster preparedness and management strategies) and structural (flood control and drainage facilities) measures. Responses to disaster are also mentioned under the financial sector and respond to the basic needs of the poor but more from a reactionary perspective in the aftermath of disasters (NEDA, 2004).

While not clearly indicated, measures for disaster management in the MTPDP also relate to climate change adaptation, particularly as the same actions also address vulnerability and risks to climate-related phenomena. Future development programs could look at how the two can be integrated for a more effective and synergistic approach to these events.

Government efforts for increasing agricultural productivity, such as enhancing and expanding the efficiency of the country's irrigation network and development of hazard-resilient seed varieties (as stipulated in the Rice Self-Sufficiency Plan for 2008–2010), also address disaster risk reduction and climate change adaptation. Such measures enhance resilience of farmers to drought and enable them to plant ahead of the typhoon season (Benson, 2009).

Other Initiatives

Other initiatives that integrate disaster risk reduction and climate change adaptation are project-based in nature. For instance, the Global Environment Facility-funded "Philippine Climate Change Adaptation Project" aims to strengthen proactive disaster risk management with the NDCC, while a project carried out by the National Economic Development Authority (NEDA) and partner agencies produced a guideline on mainstreaming disaster risk reduction into subnational development, land use, and physical planning, implemented through the United Nations Development Programme (UNDP), with funding from the European Commission on Humanitarian Aid Department (ECHO) through its Disaster Preparedness Programme (DIPECHO). Another project entitled "Strengthening the Philippines' Institutional Capacity to Adapt to Climate Change" funded by UNDP/Spanish Millennium Development Goals Achievement Fund incorporates mainstreaming climate risk reduction into key national and selected local development plans. These include the MTPDP, National Physical Framework Plan and Medium-Term Public Investment Program, and regulatory processes such as environmental impact assessment (Benson, 2009).

LOCAL INITIATIVES: THE CASE OF GUAGUA, PAMPANGA

In 1991, the Philippine government embarked on a program of devolution through the Local Government Code, which assigned the local government units (LGUs; provinces, cities, municipalities, and barangays) major responsibilities in the areas of agriculture, environment and natural resources management, health, public works, and highways (World Bank, 2000 in World Bank and NDCC, 2004). A few have become political champions and embarked on mainstreaming climate change adaptation and enhancing resilience in climate-related disasters in their respective jurisdictions, such as the case of the LGU in Guagua, Pampanga.

The municipality of Guagua is located in the province of Pampanga, in Central Luzon. Guagua is beset with seasonal flooding discharge. This is largely due to the low topography of the municipality and the effect of the lahar flows due to the eruption of Mt. Pinatubo, which made Pampanga's major river channels (i.e., Pasig-Potrero River Control System and the Pasac-Guagua-San Fernando River) heavily silted. A high tide of 1 m already triggers flooding in the municipality. Furthermore, during wet seasons, rainwater from parts of the town with higher elevation drains toward the low-lying regions through secondary waterways and interior tributaries.

From 1991, the LGU regularly monitors typhoons in terms of its disaster potential, amount of rainfall, road closure, infrastructure damages, class suspension, flood level per barangay, evacuation of residents, disease outbreak, and damage to agriculture. The municipality was also hit by El Niño in 1995. Planting of rice, which is the major crop produced in the area, also shifted from May and June to June and July due to delays in the onset of rain. Consistent with the report of Benson (2009), the people in the municipality tend to view natural hazards as a fact of life, and so efforts to address these have been more of reactive in terms of response, relief, and rescue. Nevertheless, recent efforts are being undertaken to prepare and plan for the coming of hazards, such as typhoons.

Measures that were initiated by the LGU in response to the above hazards and disasters are the installation of small farm reservoir or small water impounding projects (SWIP) when the soil is not porous, as aid during the drought season. The municipal government also conducts 24-h monitoring of the amount of rainfall from mid-May to end of October over the last five years. This enables them to chart a pattern for local flood forecasting. For instance, an 80–100 mm rainfall already leads to 1–2 ft flooding in front of the municipal building within a 6-h period. Hence, when such amounts are registered by the rain gauges, the municipal government issues an announcement to the residents and businessmen for appropriate actions. A satellite market was also established as the central market easily gets flooded due to its low elevation (Cosio, 2009).

Another mechanism that both addresses disaster risk reduction and climate change adaptation set by the municipal government is the submission of soil boring test results and compliance with the standards of the local government prior to the issuance of building permits, especially for a three-storey or higher building. Other long-term responses of the local government include road improvement and the commencement of the westward development of the town proper that has higher elevation and thus does not easily get flooded.

Meanwhile, the residents of Barangays Betis and Poblacion initiated the construction of a secondary dike. They started soliciting voluntary contributions from students, residents, and banks, which eventually led to the creation of the Lahar Control Measure Fund. The collected contributions amounted to more than PhP 3 million (US$60,000). The dike constructed from the solicited funds serves as secondary protection of the residents, in case the primary dike collapses (Cosio, 2009).

While there is seeming confusion among the people in Guagua as to whether what they are experiencing are climate change or disaster-related, what is clear is that they understand the risks involved and are ready to act to reduce their vulnerability to these events. The availability of data through regular monitoring, effective dissemination of information, receptive and dynamic leadership, and well-informed and cooperative constituents are keys for effectively responding to disaster and climate change risks. Indeed, there are strong indications that the ability of the Guagua municipality to appropriately respond to these risks has started to pay off. The Municipal Disaster Coordinating Council (MDCC) estimated that the various disaster preparedness and adaptation measures employed by the municipal government have resulted in the reduction of annual flooding over the last several years from 90 days to 21 days, thereby minimizing the annual losses from the business sector (in terms of forgone business taxes, retails, services, transport, manufacturing, etc.) from PhP 20.7 million (US$460,000) to PhP 4.83 million (US$107,333) or a total reduction of about 76.67%. Moreover, appropriate disaster preparedness and effective adaptation measures have resulted in a significant downward trend in the house and infrastructure damages in the municipality as well as the expenditures of the municipal LGU in terms of repairs to these damages (Fig. 3).

CHALLENGES IN INTEGRATING DISASTER RISK REDUCTION AND CLIMATE CHANGE ADAPTATION

Given the high incidence of natural hazards in the Philippines leading to disasters, particularly those that are water- and weather-related, and the

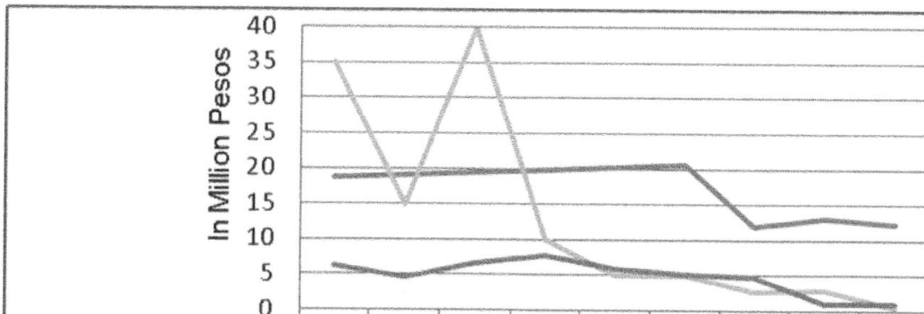

	2000	2001	2002	2003	2004	2005	2006	2007	2008
Houses	18.65	19.02	19.40	19.79	20.18	20.58	11.91	13.08	12.35
Infrastructures	35.00	15.00	40.00	10.00	5.00	5.00	2.75	3.00	0.50
LGU Expenditures	6.33	4.61	6.69	7.83	5.87	5.21	4.79	1.13	1.10

Fig. 3. Estimated Damages and Municipal LGU Expenditures Related to Typhoons, Guagua, Pampanga, 2000–2008 (Pulhin et al., 2009).

imminent effects of the changing climate on the severity and frequency of such hazards, it is important that the country has modern weather observation techniques and equipment to accurately forecast the path of typhoons, location, and amount of heavy rains to provide warnings that would save lives and reduce economic damages. However, the country is currently lacking in the needed weather surveillance systems and equipment, and if present, such equipment is already outdated or is already not working (World Bank, 2005). The effects of poor weather observation techniques were dreadfully felt recently when the country was hit by the tropical storm Ketsana, bringing rains equivalent to 30 days in regular rainy season in just 6 h. This resulted in more than PhP 10 billion (US$20M) damages in infrastructure and agriculture and almost 700 casualties. The Philippine Atmospheric, Geophysical and Astronomical Services Administration (PAGASA) does not have the much needed Doppler radar, which could have predicted the amount of rainfall that the typhoon brought with it and thereby issue the necessary warnings to the affected areas. The climate-related disaster was also compounded by the unpreparedness of the national and local governments and ineffective responses.

In view of the above, the obstacles to technological and capacity building for effective response strategies still need to be conquered for successful integration of disaster risk reduction and climate change adaptation. Both actions rely on the availability of data to make sound decisions that would eventually contribute to improved responses.

Political champions that would lead the trail toward disaster risk reduction and climate change adaptation are also needed to further this initiative. At the moment, concrete efforts are being undertaken by few LGUs. While they set good examples and demonstrate best practices in this regard, this does not give binding responsibilities to others to also embark on such undertakings. It is not also certain that these actions would be continued by their successors once their term in office has ended. Thus, the presence of enabling policies, such as the enactment of pending bills on disaster risk management and climate change, could greatly enhance actions for integration through institutionalization.

Last but not the least, awareness raising is a key to push for synergies between the two issues. Currently, a great deal of measures separately address disaster risk reduction and climate change adaptation because of limited understanding on avenues for integration. Likewise, if ever collaborative efforts exist, either this is done indirectly or one issue is taken nominally. Climate change adaptation has also gained greater political attention (Benson, 2009), with the equally urgent need for disaster risk reduction lagging behind. Hence, putting the two issues together and on equal platform through enhanced awareness and understanding is in order to improve future actions.

CONCLUSION

Disaster risk reduction and climate change adaptation are two urgent issues that the country needs to tackle being highly exposed to disasters originating from natural hazards that are further exacerbated by the changing climate. The most predominant of such hazards are typhoons, floods, and landslides, comprising more than 80% of disasters experienced in the Philippines. Such events have caused the country billions of pesos damages in agriculture, infrastructure, health, environment, and the economy; loss of lives; and greater poverty, which further increases the vulnerability of the people to these hazards and disasters.

While disaster risk reduction and climate change adaptation are closely linked, efforts to address these are undertaken separately, at the national level in particular, with the latter gaining greater political attention. Furthermore, responses are more reactive in nature focusing on rescue, relief, and other postevent measures. Measures in terms of policy, programs/ plans, and projects that seemingly integrate the two issues are also found to be more of a product of overlap rather than collaboration.

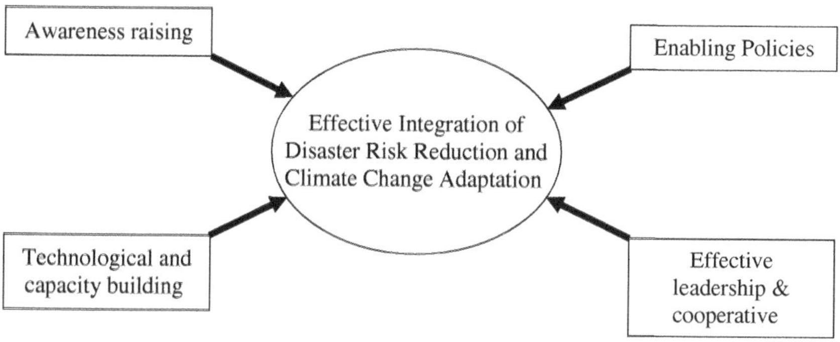

Fig. 4. Factors Leading to an Effective Integration of Disaster Risk Reduction and Climate Change Adaptation.

Integration initiatives at local government levels exist, but these are also highly driven by climate change or the expected risks from climate- or disaster-related events. Nevertheless, they demonstrate the key role of political champions in embedding disaster risk reduction and climate change adaptation for minimizing the impacts of natural hazards and increasing the resilience of the local communities.

Great opportunities are available both at the national and local levels for integrating disaster risk reduction and climate change adaptation that would lead to efficient use of resources and greater achievements in reducing vulnerability to hazards. However, this requires battling challenges such as raising more awareness on the two issues, technological and capacity building, effective leadership and cooperative constituents, and institutionalization through the formulation of enabling policies to ensure continued attention and action (Fig. 4).

Nevertheless, the experience of the people and the LGU in Guagua, Pampanga, showed that this is not difficult to follow. Indeed, when it is the safety and security of lives and properties that are at stake, inaction is a greater crime than the disaster itself.

NOTE

1. El Nino is the positive phase of ENSO or abnormal warming of the central equatorial Pacific Ocean, while La Nina is the negative phase or abnormal cooling of the central equatorial Pacific Ocean.

REFERENCES

Amadore, L. (2005). *Crisis or opportunity. Climate change impacts and the Philippines.* Greenpeace Southeast Asia in the Philippines: Quezon City, Philippines.

Asian Disaster Preparedness Council – ADPC. (2001). *The Philippine disaster management story: Issues and challenges.* Pathumthani, Thailand: ADPC.

Benson, C. (2009). *Mainstreaming disaster risk reduction into development: Challenges and experience in the Philippines.* Geneva, Switzerland: ProVention Consortium.

Bureau of Agricultural Statistics (BAS). (2010). Volume of cereals production, 1994–2009. Available at http://countrystat.bas.gov.ph/. Retrieved on 10 August, 2010.

Center for Research on the Epidemiology of Disasters (CRED) Emergence Events Database (EM-DAT). (2010). Philippines country profile – natural disasters. Available at http://www.emdat.be/result-country-profile. Retrieved on 4 April 2010.

Cosio, R. (2009). *Assessing the climate change vulnerability of the municipality of Pampanga. A vulnerability report produced from the APN CAPaBLE Project titled "Change Adaptation Capacity of Local Government Units and Scientists in the Philippines".* Laguna, Philippines: University of the Philippines Los Baños, College.

Department of Interior and Local Government (DILG), Deutsche Gesellschaft fur Technische Zusammenarbeit (GTZ) GmbH, and Disaster Preparedness European Commission Humanitarian Aid Department (DIPECHO). (2008). Mainstreaming disaster risk reduction in local governance. *Proceedings of the national conference on mainstreaming disaster risk reduction in local governance.* NCDRR2008, Makati City, Philippines, 4–6 March, 2008.

Duque, P. P. (2005). Disaster management and critical issues on disaster risk reduction in the Philippines [powerpoint presentation]. *International workshop on emergency response and rescue.* Available at http://ncdr.nat.gov.tw/iwerr/doc/pdf/S10%20PDF/s10-4%20ATTY.pdf. Retrieved on 5 October 2009.

National Disaster Coordinating Council – NDCC. (2009). NDCC update. SitRep No. 26 on humanitarian coordination on tropical storm "Ondoy" (Ketsana) Glide No. TC-2009-000205-PHL. NDCC, Quezon City, Philippines. Available at http://210.185.184.53/ndccWeb/images/ndccWeb/ndcc_update/TD_ONDOY/sitrep%20%2326%20re%20ts%20ondoy%20as%20of%2011%20oct%206am.pdf. Retrieved on 5 October 2009.

National Economic Development Authority – NEDA. (2004) Medium-Term Philippine development plan 2004–2010. Manila, Philippines. Available at http://www.neda.gov.ph/ads/mtpdp/MTPDP2004-2010/PDF/MTPDP%202004-2010%20NEDA_Chapter3_Environment.pdf. Retrieved on 5 October 2009.

Presidential Decree 1566. (1978). Strengthening the Philippine disaster control, capability and establishing the national program on community disaster preparedness. Available at http://www.wpro.who.int/NR/rdonlyres/1F5252CC-8DCC-4483-9406-54A59321C494/0/PDNo1566.pdf. Retrieved on 5 October 2009.

PreventionWeb. (2010). Philippine disaster statistics. Available at http://www.preventionweb.net/english/countries/statistics/?cid = 135. Retrieved on 3 April 2010.

Pulhin, J. M., Lasco, R. D., Delfino, R. J., Rangasa, M., Pulhin, F. B., & Cosio, R. (2009). Climate change adaptation and local governance in the Philippines. Paper presented during the fourth international conference ACT AHEAD with the theme disaster risk reduction: Investment for a safer future, February, 23–25, 2009, Jaypee Vasant Continental, New Delhi, India.

Sperling, F., & Szekely, F. (2005). Disaster risk management in a changing climate. Discussion paper prepared for the world conference on disaster reduction on behalf of the Vulnerability and Adaptation Resource Group (VARG). Reprint with Addendum on Conference outcomes. Washington, D.C.

UNISDR. (2008a). *Climate change and disaster risk reduction. Briefing note no. 1.* Geneva, Switzerland: United Nations.

UNISDR. (2008b). *Disaster risk and climate change [flyer].* Geneva, Switzerland: United Nations.

UNISDR. (2009). Global assessment report on disaster risk reduction. United Nations, Geneva, Switzerland.

World Bank. (2005). Natural disaster risk management in the Philippines: Reducing vulnerability. Follow-on study.

World Bank and Natural Disaster Coordinating Council. (2004). Natural disaster risk management in the Philippines: Enhancing poverty alleviation through disaster reduction.

CHAPTER 12

LOCAL ADAPTATION TO ENHANCE LIVELIHOOD ASSETS AND BUILD RESILIENCE IN ALBAY, PHILIPPINES

Noralene Uy, Yukiko Takeuchi and Rajib Shaw

INTRODUCTION

Compelling evidence in recent decades confirms that climate change is already happening as demonstrated by increasing mean temperature, changing precipitation patterns, rising sea level, and increasing frequency and growing intensity of extreme weather events in Asia. Regional climate studies have shown and projected that the worse is yet to come. According to the International Strategy for Disaster Reduction (ISDR), Asia registered the highest number of hydrometeorological disasters at 1,532 occurrences in the period 1991–2005 among all regions in the world.

By its location in the tropics along the western rim of the Pacific Ring of Fire and the Pacific typhoon belt, the Philippines is found to be naturally vulnerable to various disasters. The threat of natural hazards, such as earthquake, volcanic eruption, drought, tidal wave, landslide, flood, and tropical cyclones, brings constant risks to vulnerable populations year after year. Furthermore, most of the hazards that occur regularly are

Climate Change Adaptation and Disaster Risk Reduction: An Asian Perspective
Community, Environment and Disaster Risk Management, Volume 5, 237–259

ISSN: 2040-7262/doi:10.1108/S2040-7262(2010)0000005018

hydrometeorological in nature affirming the vulnerability of the country to climate change (Amadore, 2005). Tropical cyclones are the most destructive, as they bring about considerable loss of life and extensive damage to property. Studies in the Philippines reveal that climate variability and extreme event occurrences have increased in recent years. The country continuously faces the threat of extreme events as identified in the climate risk index (CRI) (Harmeling, 2009). The country was among the top 10 most affected countries in the years 1990–2008 and ranked fourth, first, and third in the CRI in 2004, 2006, and 2008, respectively. Of the major extreme events that occurred in the country between 1991 and 2006, typhoons and variability in precipitation triggered major disasters (Guerero, 2007).

Human beings have been adapting to the variable climate around them for centuries. Local climate variability influences peoples' decisions with consequences on their social, economic, political, and personal conditions, and effects on their lives and livelihoods (United Nations Framework Convention on Climate Change [UNFCCC], 2007). In this regard, local actors are critical in achieving real impact on the ground, as adaptation is highly site-specific. The local government especially plays a key role in building an enabling environment for adaptation. Decisions about adaptation are made at a range of levels from national down to the community (Carter & Raps, 2008). At the local level, the support of the local government in terms of policy can contribute to expeditious adaptation actions. Effective adaptation takes place through the dynamics of local governance, civil society engagement, and economic development building from the actions of local authorities, civil society organizations, and private businesses (International Federation of Red Cross and Red Crescent Societies (IFRC) and ProVentium Consortium, 2009). As indicated by other case studies in this book, critical factors to achieving a significant level of impact include the capacity of those taking actions, the information available at the local level, and funding support for adaptation initiatives.

This chapter looks at the case of the coastal municipality of Bacacay through an examination of the livelihood assets of six selected *barangays* (villages) to determine the appropriate adaptation strategies to build resilience in the coastal communities. It shows that microlevel variations among the *barangays* are important in decision making and planning for local adaptation. By identifying each *barangay's* individual strengths and weaknesses, resilience levels of livelihood assets can be enhanced and adaptive capacity and resilience to the impacts of climate change developed.

RESEARCH SCOPE AND METHODOLOGY

The study was conducted in three inland and three island coastal *barangays* of Bacacay in the province of Albay in the Philippines. The reasons for selection of the *barangays* included a large population of fisherfolks; other existing livelihoods, such as farming, handicraft making, and tourism; and coastal conservation initiatives, such as fish sanctuary and mangrove plantation. The inland coastal *barangays* are (1) Sogod, (2) Igang, and (3) Bariw, while the island *barangays* are (4) Uson, (5) Busdac, and (6) Misibis (Fig. 1).

The questionnaire used in the survey of Bacacay consisted of 68 multiple choice questions divided into five sections, namely household characteristics, climate change and adaptation, disaster risk management, livelihood assets, and livelihood strategies. There were a total of 187 respondents comprising at least 10% of total households in the six *barangays*. Specifically, there were 18 respondents in Bariw, 20 in Busdac, 39 in Igang, 19 in Misibis, 70 in Sogod, and 21 in Uson. The heads of households were targeted in the survey to obtain the best information on livelihoods. To deepen the understanding of the local context, focused group discussions were also conducted in three *barangays*, namely Bariw, Sogod, and Uson, with 9, 10, and 12 participants, respectively, composed mainly of members of the *barangay* council and representatives of fishermen and farmer

Fig. 1. Location of the Six *Barangays* in Bacacay.

associations, women and church groups, and community residents. The problem tree analysis was utilized as a tool for discussion.

The study used the Sustainable Livelihoods Approach developed by the Department for International Development of the United Kingdom as an analytical framework. Livelihood asset pentagons were constructed for each *barangay*, utilizing information on livelihood assets from the questionnaire survey. Among the indicators of the five livelihood assets included farmland and livestock ownership for natural; house and asset ownership for physical; sources of income and credit for financial; years in occupation and trainings for human; and membership in organizations and benefits derived for social. The responses to questions supporting the indicators were tabulated and weighted into a composite index to allow for comparison among the five *barangays*. These pentagons were developed to be able to depict the resilience to climate change impacts of each *barangay* in terms of its livelihood assets. Livelihood asset pentagon scores were categorized as 5 (very high), 4 (high), 3 (moderate), 2 (low), and 1 (very low).

PROFILE OF THE STUDY AREA

Sociodemographic Profile

The province of Albay lies between Camarines Sur on the north and Sorsogon on the south in Region V, also called the Bicol Region. It is bounded on the east by the Pacific Ocean, on the northeast by the Lagonoy Gulf, and on the west by the Burias Pass. It has 3 cities and 15 municipalities, 11 of which are in the coast. The capital of Albay, Legazpi City, is the regional center. Albay is an important location as it is the region's principal air and sea transportation hub. Legazpi Airport serves as the Bicol region's gateway to Manila and the Visayas islands. In addition, sea trans-shipment points can be found at Tabaco International, Legazpi National, Pio Duran Provincial, and the Pantao Regional Ports. The province has large reservoirs of geothermal steam in Tiwi and Manito, supplying a substantial amount of electric power to the Luzon Power Grid.

The province is generally mountainous with scattered fertile plains and valleys. About 40% of the land area is flat and cradled by the mountains of Masaraga, Malinao, Pantao, Catburawan, and Mayon Volcano. In the north, there are the islands of Rapu-Rapu, Batan, Cagraray, and San Miguel.

Albay's land area is 2,552.6 km^2, of which about 50% is used for agriculture. Agriculture is the main industry in Albay, which produces crops, such as coconut, rice, corn, vegetables, sugarcane, pineapple, and *abacá* (Hemp). The principal crop in Albay is coconut, which occupies 95,794 ha. Handicraft making is also a major source of income. The manufacture of *abacá* and *karagumoy* (Pandan leaf) products, such as mats, bags, hats, and slippers, supports small-scale industries in the rural areas. The province's 354 km of coastline sustains a variety of marine resources and is a source of high-value fish, including grouper, milkfish, *siganid*, mudcrab, and prawn. The majestic Mayon Volcano with its perfect cone draws tourists to the region and provides tourism income to Albay. Likewise, beautiful beaches and resorts contribute to tourism and support many coastal municipalities.

Based on census in 2007, Albay has a total population of 1,190,823, making it the 20th most populous province in the country. There are 208,640 households in the province, with an average size of 5.22 persons, significantly higher than the national average of 4.99.

Bacacay is one of the 15 municipalities of Albay. It is a second-class municipality located 23 km east of Legazpi City. It lies within 13° 11′ 05″ to 13° 28′ 12″ latitude and 123° 44′ 42″ to 124° 01′ 36″ longitude. Geographically, it occupies the eastern part of Albay, on the slope of Mayon Volcano, descending to the lowlands and facing the open sea. It is bounded on the north by Tabaco Bay; by the Municipality of Santo Domingo and Albay Gulf on the south; on the east and north by Lagunoy Gulf; and the Municipality of Malilipot on the west (Municipality of Bacacay, 2003). Some parts of the island of Cagraray fall under the jurisdiction of Bacacay, including Busdac, Misibis, and Uson.

The municipality has a total land area of approximately 12,220 ha of which 66.28% is devoted to agriculture and 3.35% is forestland. Of the forestland, 536.66 ha is production forest and 639.6 ha is protection forest. Bacacay's coastline measures 143 km. Also, there are 58.5 ha of fishponds located in different parts of the municipality.

Bacacay is one of the biggest fishing municipalities in Albay. It has a population of 61,574 people in 11,154 households (2007 census). It consists of 12 inland coastal and 24 island coastal barangays hosting a rich diversity of coastal resources and habitats (Bicol Cooperative Development Center, undated). The livelihood of its coastal communities is highly dependent on harvest from coastal waters, *karagumoy* and *abacá* for handicraft, and coconut plantations (Municipality of Bacacay, 2004) (Fig. 2).

Climate and Hazard Context

Albay is bounded by the Pacific Ocean on the east, Samar Sea on the southeast, and Sibuyan Sea on the southwest. Because of this location, there

Fig. 2. Main Livelihoods in Bacacay.

is a 16% frequency probability of the province experiencing typhoons. An average of 20 typhoons enter the Philippines every year, most of them often passing through the Bicol region en route from southern Philippines in a northwesterly direction. According to the Philippine Atmospheric, Geophysical and Astronomical Services Administration (PAGASA), there have been 86 tropical cyclones, which passed within 50 km of Albay in a period of 58 years, mostly during the months of October to December. The most recent typhoon disaster that the Bicol region experienced was in November 2006 when Typhoon Reming made landfall, resulting in 604 people dead, 1,465 injured, and 419 missing (Table 1). Albay was the worst affected because strong winds of up to 250 kph and torrential rains of a record high of 446 mm, coupled with earlier eruptions of Mayon Volcano, caused mudslides to bury villages along its path.

Albay's climate is categorized as Type II, which is described as no dry season with pronounced maximum rainfall from November to January. Annual rainfall trend for the period 1963–2008 shows a gradual increasing trend averaging at 3,416 mm as shown in Fig. 3. Average monthly rainfall is 285 mm,

Table 1. Damages by Tropical Cyclones in Albay from 1994 to 2006.

	Name of Typhoon	Year	Strength (kph)	Affected Population				Total Amount of Damages (US$)
				Number of People	Number of Dead	Number of Injured	Number of Missing	
1	Akang	1994	54	18,036	47	112	1	2,211,904
2	Garding	1994	80	6,799	1	2	1	1,546,644
3	Mameng	1995	97	10,126	0	0	0	1,588,884
4	Rosing	1995	137	440,372	44	20	2	11,991,106
5	Pining	1997	90	1,800	0	0	0	836,956
6	Loleng	1998	130	201,834	1	7	1	6,754,448
7	Sendang	1999	60	1,122	0	0	0	2,444
8	Reming	2000	110	27,547	12	1	2	7,188,989
9	Senyang	2000	150	22,882	0	0	0	91,111
10	Dindo	2004	170	33,892	0	6	1	5,038,046
11	Unding	2004	100	1,744	0	0	0	942,094
12	Yoyong	2004	100	18,372	0	10	1	1,124,229
13	Caloy	2006	65	47,065	0	5	0	2,207,708
14	Milenyo	2006	130	698,460	14	176		37,007,025
15	Reming	2006	209	1,060,875	604	1,465	419	71,787,460
	Total							153,419,031

Source: Albay Provincial Safety and Emergency Services Office (APSEMO).

with a maximum average of 533 mm in December and minimum average of 157 mm in April. Rainfall occurs at its maximum in the months of October to January. Maximum temperature in Albay occurs in May and June at 32 °C and minimum temperature occurs in January and February at 22 °C. The trend of annual average maximum temperature as depicted in Fig. 4(a) does not show a significant pattern. However, Fig. 4(b) illustrates a steady increase in the average minimum temperature, which can be construed as more hot than cold days.

Climate change is expected to worsen the impacts of weather-related events. Some of the potential climate change impacts to Albay include sea-level rise, increased rainfall, intense and frequent typhoons, and warmer temperatures. Hydrometeorological events will have profound effects to coastal communities in Albay, which are greatly dependent on farming, fishing, and coastal tourism for livelihood.

Manila Observatory (2005) undertook a study to identify areas in the Philippines that are at high vulnerability and risk to environmental disasters. Using geographic information systems (GIS), environmental modeling tools, and resulting spatial databases, hazards and disasters were mapped and analyzed. Albay was identified as the top province vulnerable

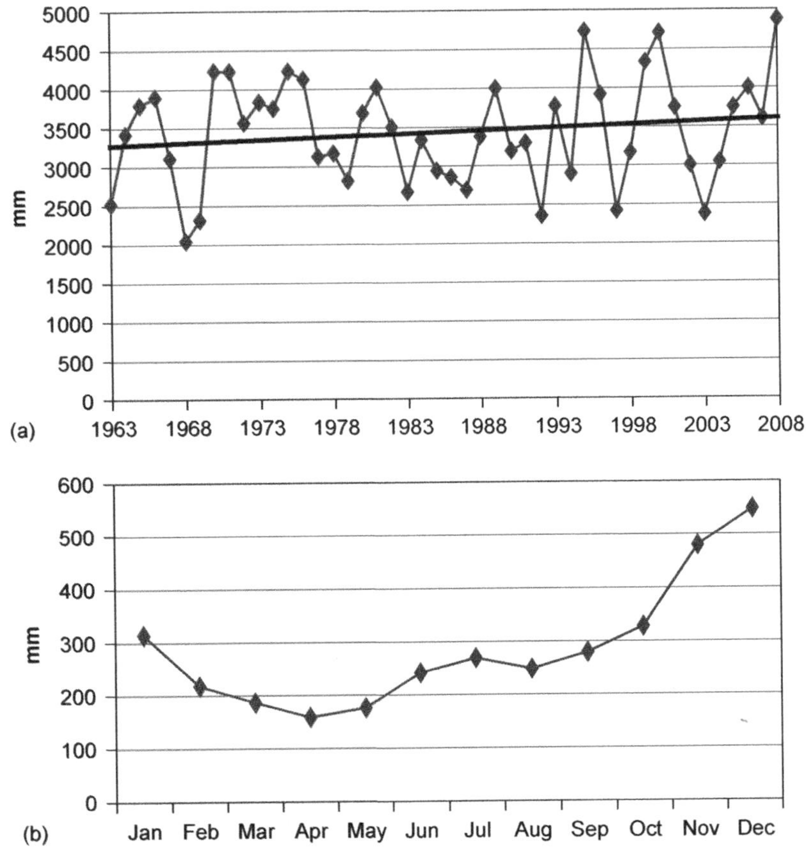

Fig. 3. Rainfall Trends in Albay for the Period 1963–2008. (a) Average Annual Rainfall. (b) Average Monthly Rainfall. *Source:* PAGASA.

to combined risk to climate disasters. This represents high scores in categories, such as risk to typhoons and risk to projected rainfall change.

In general, Albay is considered a high-risk province due to hazards brought about by volcanic eruption, landslide, earthquake, flood, and typhoon. Along with typhoons, volcanic eruption is a common risk, primarily because Mayon Volcano, located in Legazpi City, is highly active. It has recorded 48 eruptions since 1616. Mayon Volcano has an elevation of 2.46 km and a base circumference of 62.8 km, which encompasses the municipalities of Camalig, Malilipot, and Sto. Domingo. The hazards associated with Mayon Volcano are airfall tephra, lava flows, pyroclastic flows, and lahars.

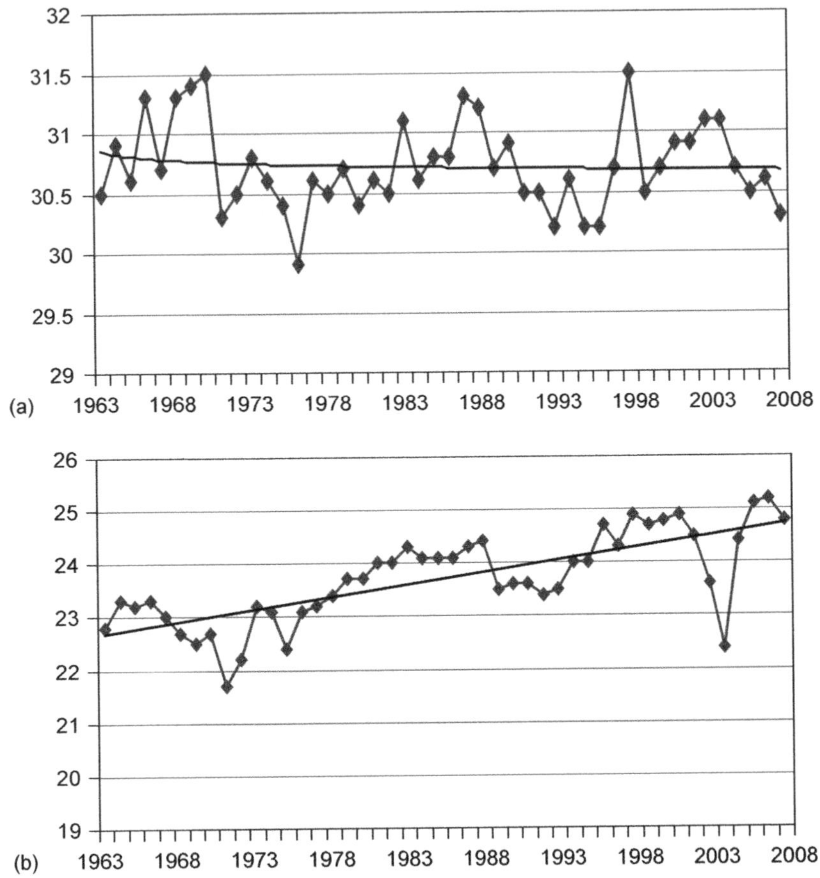

Fig. 4. Temperature Trends in Albay for the Period 1963–2008. (a) Annual Average Maximum Temperature. (b) Annual Average Minimum Temperature. *Source:* PAGASA.

CLIMATE CHANGE AND DISASTER RISK PERCEPTIONS IN BACACAY

As shown above, climate trends in Albay show increasing trends in precipitation and minimum temperature. These trends correlate with the perception of the communities in the six *barangays* in Bacacay, which have observed increasing rainfall intensity and number of rainy days as well as

increasing number of hot days. Similarly, they have observed the increasing occurrences of climate-related hazards, such as typhoons and torrential rains. The seasonality of these hazards in the months of October to December causes stresses during these lean months. In general, the identified main hazards in the mainland *barangays* are typhoon, severe heat, and flood, while the island *barangays* name typhoon, severe heat, and storm surge as significant hazards. The shocks the communities have experienced in recent years were mostly related to very strong typhoons, such as Milenyo and Reming in 2006, which severely affected Bacacay.

The provincial government of Albay has exhibited political will and great enthusiasm for climate-friendly development. Because of the province being highly disaster-prone, the local government has taken a proactive role in addressing disaster and climate risks. Disaster risk reduction (DRR) and climate change adaptation (CCA) are governing policies central to the provincial strategy of disaster- and climate-proof development. In addition, because climate change is anticipated to have far-reaching effects on sustainable development, including the ability to attain the United Nations Millennium Development Goals by 2015 (United Nations [UN], 2007), the provincial government of Albay has given adaptation action a high priority. In combating climate change, the Provincial Government of Albay has developed a program called Albay in Action on Climate Change (A^2C^2) implemented by the Centre on Initiatives and Research on Climate Adaptation (CIRCA). With regard to DRR, the provincial government has institutionalized the Albay Provincial Safety and Emergency Management Office (APSEMO) to undertake its DRR initiatives.

Moreover, the province has taken a high-profile stance in CCA advocacy by convening the National Conferences on Climate Change Adaptation (NCCCA). The first NCCCA in Albay in October 2007 produced the Albay Declaration on Climate Change, which was instrumental in pushing forward CCA into the national agenda. The second NCCCA, called NCCCA+2, was held in October 2009 in Manila. Moving forward on Albay Declaration 2007, it made an appeal to the Philippine president through the Manila Declaration 2009.

Despite these initiatives, 97% of the survey respondents attested to having no knowledge of the provincial program on CCA. This showed that the provincial government has little interaction with the local people and presently plays a small role in the local communities' lives. An adequate understanding of climate change and disaster risk has not been instilled in the minds of the people at the *barangay* level.

RESILIENCE OF LIVELIHOOD ASSETS

Resilience, as defined by ISDR (2009), is the ability of a system, community, or society exposed to hazards to resist, absorb, accommodate to and recover from the effects of a hazard in a timely and efficient manner, through the preservation and restoration of its essential basic structures and functions. It is determined by the degree to which the community has the necessary resources and is capable of organizing itself both prior to and during times of need.

The majority of the survey respondents articulated that their livelihoods are the most impacted by climate change. As their livelihoods are mostly in the climate-sensitive sectors of fisheries and agriculture, this impact is well substantiated. The impacts on the main livelihoods in Bacacay are shown in Table 2. Common in all four livelihoods are the impacts on income and food security, which are the major concerns of the communities.

Table 2. Climate Change Impacts on Livelihoods in Bacacay.

Type of Livelihood	Impact
Fisheries	• Income loss: low fish catch/less fishing days • Danger at sea/loss of life • Sickness/injury • Damage to fishing equipment • Unemployment • Household food insecurity • Loss of savings • Loan availment
Agriculture	• Income loss: loss/damage to crops • Reduced soil fertility • Sick or weak livestock • Household food insecurity • Loss of savings • Loan availment
Labor	• Income loss • Unemployment • Loss of savings • Household food insecurity
Handicraft making	• Income loss: loss/damage to raw materials and finished products and difficulty in processing raw materials during wet season • Unemployment • Household food insecurity

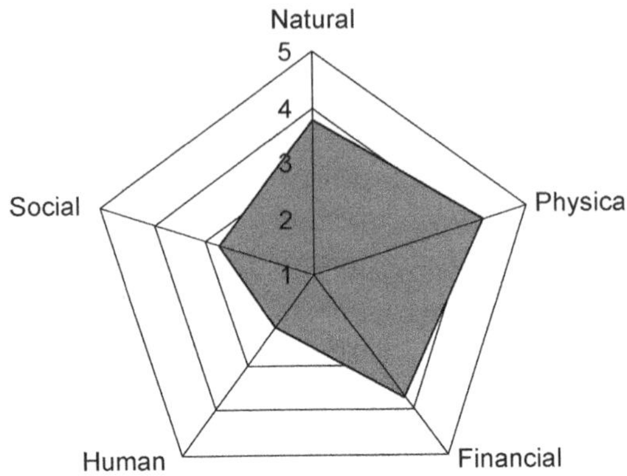

Fig. 5. Overall Resilience of Livelihood Assets in Bacacay.

Based on the livelihood asset pentagons developed, the strongest assets of the *barangays* in Bacacay are natural and physical assets, while the weakest assets are human and social assets (Fig. 5). As shown in Fig. 6, each *barangay* exhibits different levels of resilience in terms of individual livelihood assets. Among the six *barangays*, Bariw came out as the most resilient, while Misibis was the least resilient.

Looking at livelihoods from a sustainable livelihoods approach lens was effective as a means of assessing adaptation capacity, because it was able to (i) reveal local vulnerabilities, (ii) build understanding of macro- and micro-level enabling conditions for adaptation, and (iii) identify locally relevant resilience-building options (Spanger-Siegfried, Dougherty, Goutbi, & Osman, 2005). Given the analyses of livelihood assets as depicted in the pentagons, strategies to improve human and social assets in Bacacay would therefore support resilience building. Moreover, the understanding of the micro-level conditions in the *barangay* allows for the consideration of approaches to enhance the adaptive capacity of each *barangay* by addressing its unique context.

STRENGTHENING LIVELIHOOD ASSETS TO ENHANCE RESILIENCE

The case study of Bacacay demonstrated the coastal *barangays*' vulnerability to climate change and disaster risk and gave evidence of low

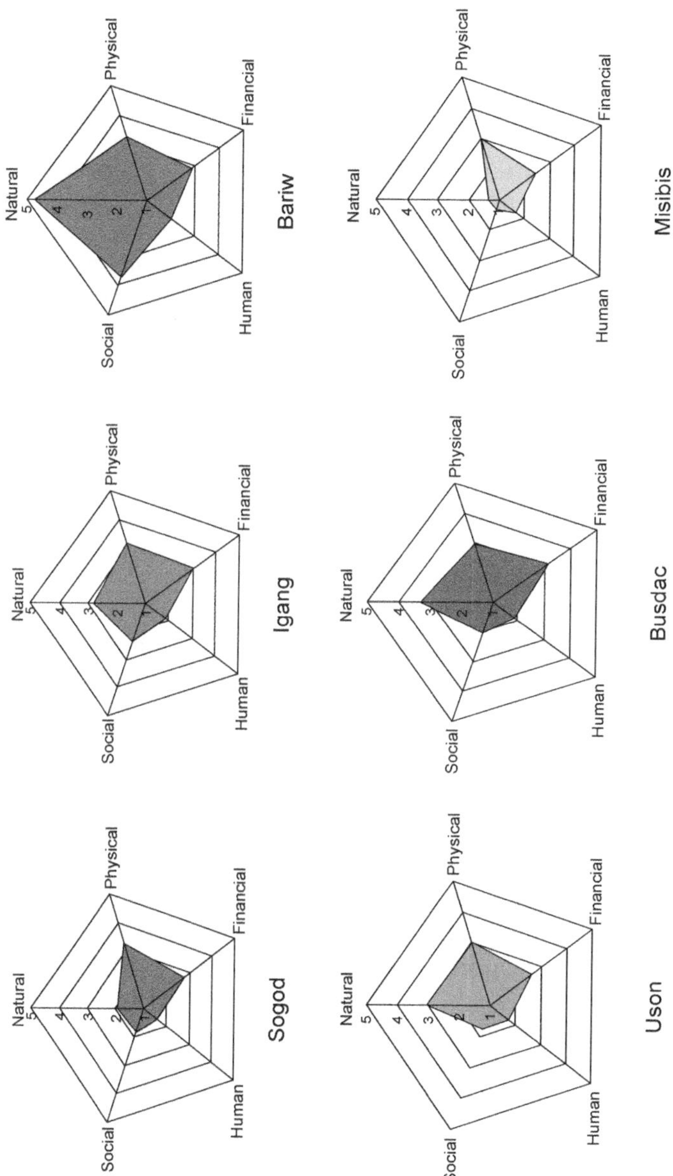

Fig. 6. Resilience of Each *Barangay's* Livelihood Assets.

human and social assets. Making livelihood assets one of the basis of resilience in the municipality, soft approaches through enhancement of its weak human and social assets will therefore be an appropriate strategy for Bacacay.

A comparison of the resilience levels of the *barangays* shows that Bariw is a representation of a resilient *barangay*, which other *barangays* in Bacacay can emulate. The important factors that account for its resilience are strong *barangay* leadership and solid community ties. The local governance practiced in Bariw demonstrates that dedicated leaders can guide actions that can gradually enhance the community's socioeconomic conditions, thereby developing resilience. Since the *barangay* council is constrained by its budget allocation, alternative ways to help the community are pursued through building social capital. The close relationships among the community residents form the basis for mutual and self-help. The trust and support they give to one another become a powerful impetus for action. With the community working together even out of the little they possess, activities to sustainably maximize the use of natural assets and increase physical and financial assets are undertaken. The example of Bariw reveals that the community can accomplish a lot on their own together and even more when supported by local leadership.

In the case of Misibis, which shows the least resilience among the *barangays*, existing conditions, especially its weak livelihood assets, dictate the consideration of alternative measures to improve the resilience of the community. Since there is little support from the *barangay* council in addressing its various vulnerabilities, the residents should strengthen their relationships with one another and develop a connection that is socially and economically valuable. The fact that the people in the community have weathered through regular stresses and shocks in the climate and economy on their own indicates that building social capital can be most advantageous for everyone. Community members' helping one another is not uncommon in the Philippines. The traditional *bayanihan* concept has evolved from merely community members carrying a house to a new location to a community's united effort to achieve a particular goal or resolve an issue. Depending solely on the local government for actions to improve socioeconomic conditions is not an option for Misibis, despite many believing that the municipal government is important in reducing their vulnerability. Since the people are directly affected by climate-related events and other hazards, they should take it upon themselves to act collectively to improve their livelihood assets and develop resilience.

In general, consideration of the following in planning for adaptation actions to reduce climate change and disaster risk vulnerability and enhance human and social assets in Bacacay would be most beneficial:

- Increasing awareness on CCA and DRR
- Increasing appreciation of the environment, especially coastal resources
- Promoting income diversification through enterprise development
- Building the capacity of municipal and *barangay* government officials
- Promoting and increasing multistakeholder participation
- Empowering communities to sustain CCA and DRR activities at the *barangay* level

The survey in Bacacay showed that the communities rely to a great extent on the municipal government and *barangay* council in reducing their vulnerabilities. These local government institutions must promote an enabling environment by formulating and programming adaptation strategies based on the local variations of each *barangay*, especially building on the *barangay's* strong assets. Maladaptation will be minimized when national and provincial frameworks are customized to each *barangay's* local context. Adaptation actions can be initiated by the municipal government or by the community and other stakeholders with support from the municipal government and *barangay* council.

Being a coastal municipality, Bacacay will benefit from "soft strategies" focusing on research, information, education, and communication (IEC), enterprise development, and training and technical assistance as discussed in the following.

Research

Gaps exist in knowledge and information sharing at the local level. Knowledge on climate change, disaster management, and coastal resources management differs among the different strata at the local level. Specifically, the mapping of coastal resources has not been fully undertaken and its values unaccounted for. In terms of adaptation, the roles of social networks and institutions on which the vulnerable populations rely on as well as social equity, fairness, and gender equality in the adaptation context need to be further studied. To resolve these important gaps, the following proposed actions can be undertaken:

- Building public–private partnerships (PPP) on knowledge building, especially with Bicol University and international organizations and

NGOs operating in the province, such as the Spanish Cooperation (AECID) and Bicol Consortium for Development Initiatives

- Actively participating in the activities of the Centre for Initiatives and Research on Climate Adaptation and the Albay Provincial Safety and Emergency Management Office
- Strengthening collaboration with relevant national government agencies, such as the Department of Environment and Natural Resources, Bureau of Fisheries and Aquatic Resources, Department of Interior and Local Government, Department of Transportation and Communication, Department of Science and Technology, and Philippine Council for Aquatic Research and Marine Development
- Involving the local community in research and monitoring activities to encourage participation and support

The stakeholders that play important roles in this strategy are the provincial and municipal governments, *barangay* council, Bicol University, relevant national government agencies, NGOs, and the local community.

Information, Education, and Communication

IEC combines strategies, approaches, and methods that enable individuals, communities, and groups to play active roles in protecting and sustaining the coastal environment. The essential aims of an IEC program should include raising public awareness, understanding and appreciation of coastal management, climate change, and disaster management issues. In addition, it should be able to promote the utilization of science and indigenous knowledge in decision making. As the process of learning is embodied in IEC, it contributes to human and social capital formation by increasing knowledge of local communities and fostering participation in community-based management. Consequently, it empowers people to make decisions, modify behaviors, and change social values.

Scaling up of IEC on CCA, DDR, and coastal resources management needs to be undertaken in Bacacay. As climate change is an urgent issue requiring immediate action, IEC must be prioritized. Appreciation of the environment, especially coastal resources, must be ingrained in the people's minds so that a change in attitude and behavior can happen. Information dissemination initiatives must also target capacity building of government officials, civil servants, NGOs, people's organizations, religious groups, teachers, and trainers. The local government can benefit by collaborating with relevant institutions, especially Bicol University, CIRCA, and the local media in education and awareness-raising programs. Training and technical assistance from academe, research institutes, and national government agencies can enhance the knowledge base as well as strengthen livelihoods. Heightened advocacies and high profiling of climate change, disaster risk, and coastal resources in the municipality can create an enabling environment for greater involvement of stakeholders. Activities focusing on coastal resources, climate change, and disaster can include (Chua, 2006; Shaw et al., 2009) the following:

- Presentation, short lecture, orientation session, open forum, interactive workshop, and focused group discussion
- Film and documentary showing
- Art competition (e.g., photography, painting, and essay writing) and exhibits
- Leaflet, pamphlet, brochure, information kit
- Mass media (e.g., radio broadcast, newspaper article, Internet [e-mail bulletin and online discussion group])
- Sports, games, and contests
- Coastal cleanup

Enterprise Development

Income diversification was identified in the Bali Action Plan as one of the actions to build resilience. Climate change poses significant challenges to economic activities and livelihoods. As with the livelihoods in Bacacay, economies in developing countries are highly susceptible to the impacts of climate change, since their economic activities often depend largely on natural resources. Where the economy relies heavily on natural resources, economic resilience depends on the availability of resources and their ecological resilience (UNFCCC, 2009).

Enterprise development and alternative livelihoods are essential to reduce pressure on coastal resources and should precede other climate-proofing and conservation initiatives in the municipality, especially as existing livelihoods are on the climate-sensitive sectors of agriculture and fisheries. Alternative livelihoods particularly on nonclimate sectors and for the lean months of October to December must be explored. Investments on the important economic activities in Bacacay, such as tourism and handicraft making, must be promoted. Viable livelihood options based on availability of resources and recommendations from survey respondents and focused group discussions include the following:

- *Abaca* and *karagumoy* handicraft making (e.g., mats, bags, slippers, and home decors)
- Livelihoods in coconut and its by-products; *pili* nuts and its by-products; and pineapple
- Mud crab farming
- Seaweed farming
- Sea cucumber culture
- Rice trading
- Food processing (e.g., meat processing such as corned beef, *longganiza* [sausage], and *tocino* [cured meat])
- Livestock raising (e.g., poultry, swine, cattle, and *carabao* [water buffalo])
- Tourism-related activities

Training and Technical Assistance

Strengthening scientific, technical, and indigenous education and training in sustainable management of coastal ecosystems, CCA, and disaster management should be the core objectives of this strategy. Activities to build capacity in these important areas include the following (Shaw et al., 2009):

- Marine ecology, climate change, and disaster management seminars
- Community mapping (e.g., most vulnerable households, hazard-prone areas, indigenous knowledge [signs and stories], evacuation sites, and escape routes)
- Preparation of an emergency plan (community and household), including emergency supply kit
- Evacuation drills and first aid, and cardiopulmonary resuscitation workshops
- Orientation on early warning systems
- Regular training on agriculture and fisheries (e.g., technologies, effective methods, and climate-resilient practices)

CONCLUSION

Climate studies in the Philippines give evidence that climate change is increasingly being felt and seriously affecting the country. Terrestrial and marine ecosystems are showing a decline in condition and diversity.

Temperature and precipitation changes and variability have grave implications, especially for a developing country, like the Philippines, which is still highly dependent on agriculture. In addition, the threat of sea-level rise makes coastal zones and low-lying areas highly vulnerable.

The Philippines is extremely prone to natural disasters, most of which are climate-related. It is in this context that it is important to relate climate change with vulnerability reduction. The most serious impacts of climate change in recent years have been the increasing frequency and intensity of extreme weather events, especially typhoon disasters. Typhoon disasters bring about extensive social and economic damage that set back poverty alleviation and development efforts. The adaptive capacity of the people must therefore be enhanced to enable them to face the impacts of climate change. For a country that is developing and highly vulnerable to natural hazards, climate change has the potential to create a vicious cycle of poverty and vulnerability, given that the poor are more vulnerable to natural disasters and that climate change tends to increase the frequency and intensity of many of these disasters (Ibarraran, Ruth, Ahmad, & London, 2007).

Climate trends and survey results in Bacacay indicate the need for adaptation strategies based on local context. In addition, the inherent conditions to which Albay is exposed to, such as various climate-related hazards, necessitate a proactive role in scaling up local adaptation actions. The foresight of the provincial government to tackle the issues on climate change has led to initiatives, which sensitized and encouraged people to take actions on climate change, whether individually or as a community, municipality, and province.

This study has identified the fundamental need of enhancing human and social capital as a prerequisite to building resilience in the coastal communities of Bacacay. Human and social capital enhancement is recognized as the most appropriate adaptation strategy as resilience is affected by weaknesses in these two aspects. Specific action plans considering the observed microvariations can strengthen weaknesses and build on strengths of livelihood assets in each *barangay*. The local government plays a key role in integrating these considerations into local governance. More importantly, the community can carry out adaptation actions on their own based on knowledge and understanding of their local context.

The "soft" adaptation strategy espoused in this study focusing on research; IEC; enterprise development; and training and technical assistance can empower the local community and create a change in attitude and a greater appreciation of the coastal system, thus promoting collective action.

Community participation is critical to ensuring buy-in and sustainability of initiatives at the local level. The success of initiatives is often dependent on the community. Most are successful because the community feels a sense of ownership of the activity. There is a common desire to protect marine resources and the environment. In addition, the exposure to hazards and recognition of vulnerability make the people responsive to efforts to enable them to cope with the impacts of climate change. The active involvement of the community, partner organizations, and the local government contributes to making CCA actions flourish.

Moreover, this study has shown that the coastal *barangays* in Bacacay are highly dependent on the *barangay* council and municipal government. However, the provincial government also plays a very critical role, especially as it has made climate- and disaster-proof development its provincial goal (Lasco, Rangasa, Pulhin, & Delfino, undated). The institutionalization of provincial entities to focus on CCA and DRR affirms the seriousness of the local leadership in addressing climate and disaster risks. The role of the government cannot be overemphasized. It is in the best position to demonstrate good governance and strong institutions in effectively undertaking actions and responses to climate change and disaster risk. Adaptation entails governance reforms toward more effective administrative and operational approaches by governments. Specifically, governance for adaptation will change existing conventional institutional arrangements and priorities. The dominant institutional cultures at various levels will strongly influence the shaping of policy and work on adaptation (Resurreccion & Sajor, 2008). Challenges would most likely relate to interagency coordination, sectoral and department separatism, and involvement of civil society. A framework, which addresses interagency competition and turf wars, and rigid separatism of functions and tasks between agencies as well as promotes involvement of civil society for a more participatory mechanism, will allow a successful integration of adaptation into the planning processes at all levels of government.

The fact that coastal communities are at the frontline of climate change necessitates appropriate interventions to reduce the risk to natural and human systems in coastal zones. Effective coastal management entails an integrated approach, including CCA and DRR, for it to have significant impact. In the case study of the municipality of Bacacay, climate trends and perceptions clearly showed the vulnerability of livelihoods of coastal communities. Since the main livelihoods in the *barangays* are fishing and farming, which are highly sensitive to hydrometeorological changes, climate change will significantly affect every household's goal to improve livelihoods and rise out of poverty. Through a sustainable livelihood analysis lens, the articulation of the local vulnerabilities of each *barangay* gave an understanding of microlevel enabling conditions for adaptation in Bacacay. The variations within the *barangays* in the municipality indicate that it is imperative to address each of the *barangay's* unique context for more meaningful and truly bottom-up actions. The assessment of vulnerabilities and strengths affords the local planning process to effectively consider locally relevant resilience-building options. Thus, looking from macro- and micro-perspectives, the local government can plan and strategize actions in a holistic and integrated manner. In addition, knowledge of individual vulnerabilities and strengths gives each of the *barangay* an understanding of the appropriate approaches and measures the community can undertake on their own.

CCA should be mainstreamed in all the functions of the local government, especially in development planning. Initiatives on CCA at the local level must also be undertaken in consonance with initiatives for DRR and development at the provincial, regional, and national levels. Promoting coherence in adaptation actions at all levels can lead to a systematic and holistic approach to scaling them up. With legislation in place for climate change in the country, it is hoped that an enabling environment for CCA be established to build the resilience of Filipinos to the impacts of climate change.

ACKNOWLEDGMENTS

Noralene Uy is thankful for a research scholarship grant from the Ministry of Education, Culture, Sports, Science and Technology (MEXT) of the Government of Japan, and support from EML and GCOE-ARS. The kind assistance of the Municipal Government of Bacacay and the Provincial Government of Albay – Centre for Initiatives and Research on Climate Adaptation during field work is highly appreciated.

REFERENCES

Amadore, L. A. (2005). *Crisis or opportunity. Climate change impacts and the Philippines*. Quezon City: Greenpeace.

Bicol Cooperative Development Center. (undated). *Barangay profile*.

Carter, L., & Raps, B. (2008). Adapting to climate change: It is just better planning. In: J. Tobey (Ed.), *Basins and Coasts, 2*(1), 29–31.

Chua, T. (2006). *The dynamics of integrated coastal management: Practical applications in the sustainable coastal development in East Asia*. 468pp. Global Environment Facility/United Nations Development Programme/International Maritime Organization Regional Programme on Building Partnerships in Environmental Management for the Seas of East Asia (PEMSEA), Quezon City Philippines.

Guerero, L. (Ed.) (2007). *The Philippines: A climate hotspot. Climate change impacts and the Philippines*. Quezon City: Greenpeace.

Harmeling, S. (2009). Global climate risk index 2010: Who is most vulnerable? Weather-related loss events since 1990 and how Copenhagen needs to respond. Germanwatch e.V. Available at http://www.germanwatch.org/klima/cri2010.pdf. Accessed on 4 January 2010.

Ibarraran, M. E., Ruth, M., Ahmad, S., & London, M. (2007). Climate change and natural disasters: Macroeconomic performance and distributional impacts. *Environment, Development and Sustainability, 11*(3), 549–569.

International Federation of Red Cross and Red Crescent Societies (IFRC) and ProVentium Consortium. (2009). *Climate change adaptation strategies for local impact: Key messages for UNFCCC negotiators*. Technical Paper for the IASC Task Force on Climate Change. Available at http://www.preventionweb.net/english/professional/publications/v.php?id = 12166. Accessed on 16 November 2009.

International Strategy for Disaster Reduction. ISDR disaster statistics occurrence. Avail able at http://www.unisdr.org/disaster-statistics/pdf/isdr-disaster-statistics-occurrence.pdf. Accessed on 8 January 2010.

International Strategy for Disaster Reduction. (2009). *2009 UNISDR Terminology on disaster risk reduction*. Geneva: UNISDR.

Lasco, R. D., Rangasa, M., Pulhin, F. B., & Delfino, R. J. (undated). The role of local government units in mainstreaming climate change adaptation in the Philippines. Available at http://www.albaycirca.org/pdf/THE%20ROLE%20OF%20LOCAL%20-GOVERNMENT%20UNITS.pdf. Accessed on 16 November 2009.

Manila Observatory. (2005). Mapping Philippine vulnerability to environmental disasters. Available at http://www.observatory.ph/vm/. Accessed on 10 January 2009.

Municipality of Bacacay. (2003). *Comprehensive land use plan*.

Municipality of Bacacay. (2004). *Municipal coastal environmental profile of Bacacay, Albay*.

Resurreccion, B. P., & Sajor, E. E. (2008). *Climate adaptation in Asia: Knowledge gaps and research issues in South East Asia*. Katmandu: ISET-International and ISET-Nepal.

Shaw, R., Takeuchi, Y., Shiwaku, K., Fernandez, G., Gwee, Q. R., & Yang, B. (2009). *1-2-3 of disaster education*. Japan: Kyoto University.

Spanger-Siegfried, E., Dougherty, B., Goutbi, N., & Osman, B. (2005). *Methodological framework. An internal scoping report of the project strategies for increasing human resilience in Sudan: Lessons for climate change adaptation in North and East Africa*. AIACC Working Paper No 18. Washington, DC: AIACC.

United Nations (UN). (2007). The millennium development goals report. United Nations, New York. Available at http://www.un.org/millenniumgoals/reports.shtml. Accessed on 8 January 2010.

United Nations Framework Convention on Climate Change. (2009). *Report on the technical workshop on increasing economic resilience to climate change and reducing reliance on vulnerable economic sectors, including through economic diversification.* UNFCCC. Available at http://unfccc.int/documentation/documents/advanced_search/items/3594.php?rec=j&priref=6000005427#beg. Accessed on 4 January 2010.

United Nations Framework Convention on Climate Change. (2007). *Climate change: Impacts, Vulnerabilities and adaptation in developing countries.* Germany: UNFCCC.

CHAPTER 13

CLIMATE CHANGE AND DISASTER RISKS: THE SINGAPORE RESPONSE

Bing Li and Vikram Rajola

INTRODUCTION AND BACKGROUND

The Intergovernmental Panel on Climate Change (IPCC) Fourth Assessment Report (AR4) (Intergovernmental Panel on Climate Change, 2007) of 2007 concluded that most of the warming of the climate is very likely driven by human activities that increase greenhouse gas (GHG) concentrations in the atmosphere. Activities such as burning of fossil fuels for power generation and in vehicles, as well as increasing deforestation, result in emissions of four long-lived GHGs: carbon dioxide (CO_2), methane (CH_4), nitrous oxide (N_2O), and halocarbons (a group of gases containing fluorine, chlorine, or bromine). The report projects that by the end of the 21st century global temperatures could rise by 1.1–6.4°C over 1990 levels, while global mean sea levels could rise by 18–59 cm, depending on future scenarios of varying global emission levels. This is likely to adversely impact ecosystem resilience, putting many plant and animal species at the risk of extinction. Sea level rise and coastal erosion coupled with temperature extremes, heat waves, and heavy precipitation events that are projected to become more frequent will affect the health and well-being of millions of people around the world.

In view of the aforesaid projections, and considering its small size and tropical climate, Singapore has recognized its vulnerability to climate change. Singapore's socioeconomic development has always been driven by judicious

Climate Change Adaptation and Disaster Risk Reduction: An Asian Perspective
Community, Environment and Disaster Risk Management, Volume 5, 261–283

ISSN: 2040-7262/doi:10.1108/S2040-7262(2010)0000005019

balance between environmental sustainability and economic growth. Being a small country with few natural resources, it is imperative for Singapore to optimize the use of available environmental and energy resources, and at the same time achieve synergies across the policy objectives of environmental sustainability, economic competitiveness, and energy security.

To achieve this, Singapore has based its policy to combat climate change on the three key factors of optimizing energy usage to reduce GHG emissions, using more efficient processes and fuels, and promoting research and development into technologies that help contain both the causes and effects of climate change. Toward this end, the National Climate Change Strategy was formulated in 2008, which details Singapore's current and future efforts to address climate change in vulnerability and adaptation, as well as mitigation of GHG emissions. The National Climate Change Strategy reiterates Singapore's commitment to do its part in the international effort to address climate change.

ENVIRONMENTAL SUSTAINABILITY AND CLIMATE CHANGE

Efforts to contain climate change have several benefits. The burning of fossil fuels in power stations, industries, and vehicles is the main source of both CO_2 emissions and air pollution in Singapore. Efforts to mitigate GHG emissions in the power generation, industry, and transport sectors can help to reduce air pollutants. Adopting cleaner fuels like natural gas and more energy-efficient technologies to reduce CO_2 emissions will contribute significantly toward cleaner air, as well as a better quality of life for Singaporeans.

Climate change actions can also generate economic opportunities, as in the clean energy and energy efficiency services sectors. Being more energy efficient can help both businesses and households reduce their energy consumption costs. At the national level, this helps to mitigate Singapore's dependence on imported fossil fuels, thereby enhancing energy security.

Singapore has therefore adopted an action-oriented approach to environmental sustainability that recognizes environmental resources are finite and valued. Development policies are underpinned by long-term vision, holistic urban planning, sound environmental policies, and high regulatory standards. Over the years, Singapore has strived to do the following:

(a) Maintain good ambient air quality;
(b) Manage waste to reduce the need for landfill;

(c) Conserve water resources; and
(d) Increase energy efficiency.

This practical approach has enabled Singapore to achieve economic prosperity, social progress, and a quality living environment.

VULNERABILITY TO CLIMATE CHANGE AND MITIGATION STRATEGIES

Singapore is a small and low-lying island nation located just north of the Equator, with a generally flat coastline and tropical climate. Almost 5 million people reside in an area of 700 km^2 most of which is only about 15 m above the sea level. Singapore faces a grave risk from aggravated impacts of climate change, particularly sea level rise and flooding, which could adversely impact its water supply resilience. Sea levels during high tides pose a challenge for drainage to the sea; this can be especially aggravated during rainstorms or storm surges, where the sea is pushed to the shore by the force of wind, leading to inland flooding. In the 1960s and 1970s, floods were a common occurrence in Singapore. In 1978, Singapore experienced its most severe flood when unusually heavy rainfall coincided with a high tide leaving roads submerged under a meter of water. Seven people died and more than 1,000 had to be evacuated from their homes.

To address the risk posed by urban flooding, Singapore has invested in construction of new drains and canals. These have helped reduce flood-prone areas by more than 95% since the 1970s. The drainage infrastructure development and improvement have reduced flood-prone areas from 3,200 ha in the 1970s to 124 ha, which will be further reduced to less than 66 ha by 2011. Additionally, the Public Utilities Board (PUB) of Singapore has a deliberate policy to raise low-lying areas in conjunction with redevelopment projects. Since 1991, PUB requires new reclamation projects to be built to a level 125 cm above the highest recorded tide level. This addresses the IPCC AR4's sea level rise projection of 59 cm by the end of the 21st century under the worst-case scenario assessed. The efficiency of this improved infrastructure was tested when Singapore experienced its third highest storm on record, with 366 mm of rain over a 24-hour period on December 19, 2006. Despite the intense rain, there were no major floods, and only few isolated spots (about 15 ha) were affected.

Singapore's National Climate Change Strategy

The Singapore government is committed toward addressing climate change in an environmentally sustainable manner that is compatible with economic growth. The nation became a Party to the Kyoto Protocol in 2006. As a further sign of commitment toward addressing climate change, the National Climate Change Strategy was formulated in 2008, which outlines local competency-building efforts and its participation in international climate change efforts. Singapore aims to improve the energy efficiency of major energy sectors, namely power generation, industries, transport, buildings, and households. The nation is also committed to the global research effort on climate change and energy technologies and is investing to develop technologies that can help meet the climate change challenge, in the areas of solar energy and water.

The strategy adopted by Singapore to combat the impacts of climate change is illustrated in Fig. 1.

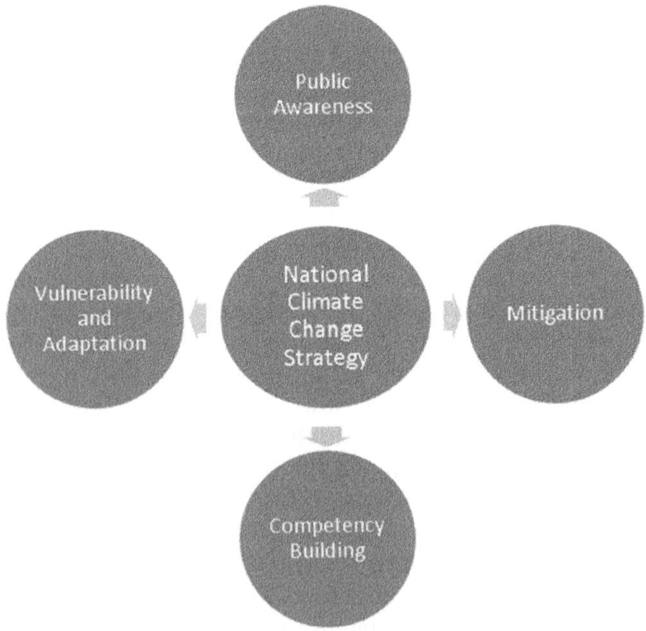

Fig. 1. Framework of Singapore's National Climate Change Strategy.

It broadly entails the following measures:

(i) *Vulnerability and adaptation*: This involves commissioning studies to better understand the detailed impacts of climate change on Singapore to facilitate further planning of adaptation measures. The study will look into the possible impacts of climate change on Singapore such as increased flooding and coastal land loss.
(ii) *Mitigation*: It refers to the reduction of GHG emissions in a manner that balances the economic and environmental goals of sustainable development, encouraging the improvement of energy efficiency with economic growth. To achieve this target, Singapore will focus on improving energy efficiency and using less carbon-intensive fuels in the power generation, industry, buildings, transport, and household sectors. Some examples of mitigation measures that will be explored further include the promotion of energy audits and purchase of energy-efficient equipment by companies, expanding the fuel economy labeling scheme for vehicles and promoting energy-efficient building designs.
(iii) *Competency building*: Competency building will entail promoting climate-related research and development in Singapore as well as helping companies to gain from the economic opportunities from climate change action, such as the Clean Development Mechanism under the Kyoto Protocol and increased demand for alternative energy and energy efficiency services.
(iv) *Public awareness*: The public will be informed of the effects of climate change, particularly the link between energy use and climate change, and simple habits they can adopt to save energy, save money, and help to mitigate climate change.

National Climate Change Committee (N3C)

To bring together efforts by all sectors on climate change, the Ministry of the Environment and Water Resources chairs a National Climate Change Committee (N3C) with representatives from various government agencies, industry representatives (e.g., from the Singapore Manufacturers' Federation and the Real Estate Developers Association of Singapore), academia, and nongovernmental organizations (e.g., the Singapore Environment Council). As illustrated in Fig. 2, the N3C comprises several subcommittees and workgroups. It is a platform for policy makers to engage the private sector and the general public on climate change issues and policies, gather

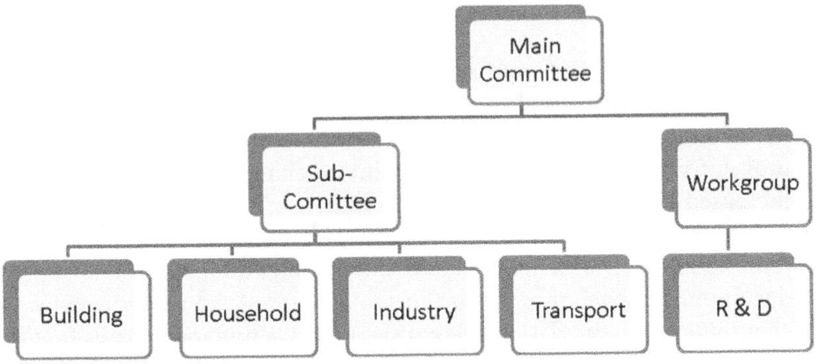

Fig. 2. National Climate Change Committee.

feedback, and provide the platform for key stakeholders to debate and share ideas on climate change issues.

The terms of reference of the NCCC are as follows.

To address climate change by

(a) Promoting greater energy efficiency and less carbon-intensive energy in key sectors;
(b) Raising awareness among the people and private and public sectors on the impacts and opportunities arising from climate change, and the actions they can take;
(c) Building competency in Singapore to better respond to climate change such as through promoting research and development of low-carbon technologies; and
(d) Understanding Singapore's vulnerability to climate change and facilitating the adaptation actions needed.

LONG-TERM IMPACT OF CLIMATE CHANGE ON SINGAPORE

In 2007, the National Environment Agency (NEA) commissioned a study to assess the long-term impact of climate change on Singapore up to 2100, as a first step toward building up local knowledge in climate science. Phase 1 of the study, covering the effects of climate change (e.g., temperature, sea level rise, and rainfall) on Singapore, up to 2100, has been completed.

The study projects that mean sea level around Singapore could rise by 24–65 cm by 2100. These findings are consistent with global projections by

Table 1. Findings of National Environment Agency (NEA) Commissioned Study to Assess the Long-Term Impact of Climate Change on Singapore up to 2100.

Climate Change Projections	IPCC AR4 Projections (Year 2100)	NEA Singapore Phase 1 Study Local Findings (Year 2100)
Increase in average temperature (°C, in 2100 relative to present)	+1.7 to +4.4 (A1B scenario, SE Asia)	+2.7 to +4.2 (A1B scenario)
% Increase in rainfall	−2% to +15% (A1B scenario, SE Asia)	No discernible trend Further studies needed
Increase in mean sea level (in meters, in 2100 relative to present, not including ice sheet melt)	+0.18 to +0.59 (All IPCC scenarios, global)	+0.24 to +0.65 (3 IPCC scenarios)

Source: Study on long-term impact of climate change on Singapore, Ministry of the Environment and Water Resources.

the IPCC. The findings,[1] as mentioned in Table 1, are comparable to the projections in the IPCC (AR4).

An interministry Adaptation Taskforce led by the Ministry of National Development (MND) has been set up to look into adaptation measures, to prepare Singapore for the effects of climate change. It has been established that the existing infrastructure will be sufficient to address the possible climate change effects over the short to medium term as indicated by the results of Phase 1 of the study. For example, the current land reclamation levels are required to be at least 1.25 m above the highest tide level, which can address the study finding of a 24–65 cm rise in sea level by 2100.

Further, the government will continue to monitor the effects of climate change and keep abreast of the state of understanding on climate science. In this regard, the NEA recently launched Phase 2 of the study to investigate changes in urban temperature profile and energy consumption of buildings, impacts on public health, and impacts on biodiversity in Singapore. This is expected to be completed by 2013.

RISKS, VULNERABILITIES, AND ADAPTATION EFFORTS

As a relatively low-lying, densely populated island in the tropics, with high uniform temperature and abundant rainfall, Singapore is vulnerable to climate change. Given these circumstances, the potential impacts of climate change on Singapore include (Ministry of the Environment and Water Resources, 2008):

(a) Increased flooding,
(b) Coastal land loss,
(c) Water resource scarcity,
(d) Public health impact from resurgence of diseases,
(e) Heat stress,
(f) Increased energy demand (see Annexure 1), and
(g) Impacts on biodiversity.

These potential impacts are not solely due to climate change but can be aggravated by adverse global climate change. The Intergovernmental Panel of Climate Change Annual Report 4 (2007) predicts the following climate change impacts on the Southeast Asian region in general:

(a) By the 2050s, freshwater availability in large river basins is projected to decrease.
(b) Coastal areas will be at greatest risk due to increased flooding from the sea and, in some megadeltas, flooding from the rivers.
(c) Endemic morbidity and mortality due to diarrheal disease primarily associated with floods and droughts are expected to rise due to projected changes in the hydrological cycle.
(d) Climate change is projected to compound the pressures on natural resources and the environment associated with rapid urbanization, industrialization, and economic development.

Recent examples of erratic weather events in Singapore include the longest dry spell of 40 days, which occurred between January 18 and February 26, 2005, and the third highest rainfall (366 mm) of previous 75 years recorded in a 24-hour period on December 19, 2006.[2]

Flooding

A higher sea level will make it more difficult for the rainwater to drain into the sea. This can aggravate inland flooding during storm surges and rainstorms. As mentioned earlier, the development of drainage infrastructure and raising of low-lying areas in Singapore has reduced the risk of submergence and Singapore's vulnerability to floods.

Coastal Land Loss

A sea level rise of up to 59 cm can result in some coastal erosion and land loss in Singapore, particularly as Singapore has a relatively flat coastline. Currently, about 70–80% of Singapore's coastal areas have hard wall or stone embankments, which help protect against coastal erosion. The rest are natural areas such as beaches and mangroves. Increased erosion could impact recreational areas at the coasts; therefore, the government will look at adapting to sea level rise through the protection of foreshore and coastal areas as the need arises. Existing revetments that protect against erosion may have to be strengthened and reinforced, while natural areas may have to be protected using different coastal defense systems.

Water Resource Scarcity

A rise in sea level can also result in saltwater intrusion wherein seawater could flow into some of Singapore's coastal reservoirs that will render stored water unfit for drinking. This is being contained by building reservoir dams that are much higher than the sea level rise projected by the IPCC AR4. The gate structures for the dams can be raised further if required.

Rising global temperatures may also change rainfall patterns and affect the amount of water stored in reservoirs. The unpredictability in rainfall can cause difficulties in capacity planning of water resources. This is catered to by introduction of NEWater and desalination, which are not rainfall dependent. NEWater is treated used water that has undergone stringent purification and treatment process using advanced dual-membrane (microfiltration and reverse osmosis) and ultraviolet technologies. NEWater could be mixed and blended with reservoir water as an additional source of raw water and then undergo conventional water treatment to produce drinking water.[3] These new water treatment technologies have diversified and increased the resilience of Singapore's water supply, even during prolonged dry spells.

Public Health Impact from Resurgence of Diseases

Singapore is situated in a region in which vector-borne diseases, particularly dengue, are endemic. Dengue patterns are affected by many factors, including climate. NEA is studying the link between climatic factors such as temperature, humidity, and rainfall with dengue cases. Preliminary results indicate that the number of dengue cases in Singapore is correlated with the ambient temperature. To address dengue, NEA has put in place a comprehensive mosquito surveillance, control, and enforcement system, which includes preemptive action to suppress the mosquito vector population and a review of building designs to reduce potential breeding habitats.

Heat Stress

Warmer temperatures due to both climate change and the urban heat island effect[4] can lead to greater use of air-conditioning and increase Singapore's energy demand. Higher annual temperatures may also mean more frequent and more severe episodes of warm weather, leading to increased occurrences of heat stress and discomfort, particularly among the elderly, the sick, and those without access to air-conditioning.

Measures that can lower ambient temperature include increasing the amount of greenery in the city (e.g., city parks, rooftop gardens, and vertical greening in buildings) and modifying building layouts and designs (e.g., using building materials with better thermal properties, lighter-colored building surfaces, designing building interiors and exterior building layouts for better ventilation, and maximizing the wind tunnel effect).

The Urban Redevelopment Authority (URA) and the National Parks Board (NParks) have been working closely to plan and provide greenery island-wide, such as providing parks and green open spaces, and planting along roads and around developments. URA and NParks have also been promoting rooftop and vertical greenery on residential and commercial buildings through planning guidelines and incentives. The Housing Development Board (HDB) is in the process of introducing rooftop greenery to multistorey car parks and residential buildings wherever feasible.

Impacts on Island and Marine Biodiversity

A rise in sea level can lead to loss of mangroves, reduced biodiversity, and aggravated coastal erosion rates. The coral reefs around the southern islands of Singapore help in sequestering carbon and mitigating storm damage and erosion. The NParks is looking into the role of nature reserves in carbon sequestration and is monitoring long-term tree diversity, tree growth, and survival in marked study plots. A coral nursery has been established off one of these islands, which will enable to proactively enhance existing marine habitats by maximizing the survival of naturally occurring corals. Mangroves help to protect coasts against erosion and NParks is developing preemptive management strategies to counter mangrove erosion at some coastal areas. Singapore cannot prevent global trends from affecting seawater temperatures locally but will seek to mitigate any additional influences on seawater quality through measures such as the release of cooling water or sedimentation.

FURTHER RESEARCH AND STUDY TO EVIDENCE AND DOCUMENT CLIMATE CHANGE

Singapore policy makers recognize the need to continuously improve understanding of the detailed effects and resulting impacts of climate change on Singapore. NEA, in consultation with other government agencies, has commissioned a study of Singapore's vulnerability to climate change. This study will project climate change effects such as temperature, sea level, and rainfall patterns in Singapore and the impacts of such effects like increased flooding and impacts on water resources. The results of the study will facilitate the identification of new adaptation measures as well as the review of existing adaptation measures.

Climate Change Mitigation Efforts

National Circumstances

The main contribution to Singapore's GHG emissions is CO_2 from the use of energy to meet development and human needs. Due to geographical constraints, large-scale adoption of alternative sources like wind and hydroelectric energy is unlikely in Singapore. The forms of renewable energy that will be more applicable besides waste-to-energy would comprise solar energy and biofuels. However, these sources of renewable energy are not yet cost-competitive with conventional fossil fuels.

Singapore is also an export-oriented, energy-intensive economy with much of the energy used by industry for exports rather than local consumption. Nonetheless, Singapore produces less carbon in the process of generating each dollar of gross domestic product (GDP) compared with other countries. Singapore's CO_2 intensity (kilograms of CO_2 produced per US dollar of GDP at purchasing power parity prices for year 2000) measured at 0.33 is below the world average of 0.47 (International Energy Agency, 2009).

Mitigation of Greenhouse Gas Emissions

The main contribution to Singapore's GHG emissions is CO_2 from the use of energy to meet development and human needs. Singapore's methane emissions are negligible, as Singapore has no agricultural base. Waste is incinerated, and the little methane emitted from the existing landfill is flared off.

Singapore's CO_2 emissions in 2005 were approximately 40 million tons. Table 2 shows the 2005 CO_2 emissions of Singapore in the year 2005 (Ministry of the Environment and Water Resources, Singapore, 2008).

In Singapore, GHG emissions are mostly a result of the combustion of fossil fuels for energy. With little possibility of increasing forest cover in the nation, the key strategies to mitigate GHG emissions in Singapore are:

(a) To support efforts to use less carbon-intensive fuels (e.g., natural gas, renewables);
(b) To improve energy efficiency in all sectors of the economy.

Increasing energy efficiency in Singapore would help to reduce Singapore's CO_2 emissions, improve air quality, reduce energy costs for companies and consumers, and help to improve energy security by mitigating energy demand growth. Energy efficiency is therefore Singapore's key strategy to reduce CO_2 emissions.

Table 2. Breakdown of CO_2 Contribution (Key Contributors in kilo tons) in 2009.

	Primary Consumption (Combust Fuels)	Secondary Consumption (Use Electricity)	Overall
Electricity generation	19,315 (48%)	↓	
Industry	13,465 (33%)	8,328 (21%)	21,793 (54%)
Transport	7,056 (17%)	930 (2%)	7,986 (19%)
Buildings	325 (1%)	5,910 (15%)	6,235 (16%)
Consumers/households	216 (1%)	3,415 (8%)	3,631 (9%)
Others		732 (2%)	732 (2%)
Total		40,377 kton	

Total CO_2 in 2005 = 40,377 kton.
Source: Ministry of the Environment and Water Resources (2008).

Carbon Intensity Improvement

By 2006, Singapore's carbon intensity reached 30% below 1990 levels, due to a rapid switch to natural gas for power generation and ongoing improvements in energy efficiency.[5] From 2000 to 2006, the electricity generated by natural gas has increased from 19% to 78%.[6]

Support Efforts to Use Less Carbon-Intensive Fuels

With the liberalization of the electricity market, Singapore has moved significantly toward using natural gas as an electricity source. As the government does not subsidize energy prices, generation companies have an inherent incentive to choose the most efficient technology. Electricity generation using gas led to significantly lower CO_2 emissions from the power sector, as natural gas emits 40% less CO_2 than fuel oil per unit of electricity generated. Singapore is also focusing on promoting renewable energy, by investing in R&D and test-bedding to improve their performance and cost-effectiveness. This includes the setting up of research institutes, provision of funds, and test-bedding platforms.

Waste-to-Energy

About 54% of solid waste is recycled with the remaining 43% incinerated and 3% landfilled directly.[7] As such, Singapore's landfills generate negligible amounts of methane, which is a GHG. The resulting ash from incineration is subsequently disposed off at the only landfill in Singapore, Semakau Landfill. As both incineration plants and landfill are capital-intensive infrastructures and require large tracts of land, it is not sustainable for land-scarce Singapore to continue building incineration plants and landfills to cope with the growing demand for waste disposal. To tackle the challenges of solid waste disposal in Singapore, Ministry of the Environment and Water Resources has identified several long-term objectives for sustainability and formulated core strategies to achieve these targets:

(a) Toward zero landfill – extend the life span of Semakau Landfill to 50 years; and
(b) Achieve 60% recycling rate by 2012.

Increasing Energy Efficiency

Singapore's policy of not subsidizing energy costs provides a market incentive for energy consumers to be more energy efficient. This policy, together with past energy efficiency efforts, has led energy intensity to improve by 15% between 1990 and 2005.

To drive future energy efficiency efforts in the various sectors of the economy, the NEA chairs the Energy Efficiency Singapore Programme Office (E2PO). The E2PO has developed a national plan to promote energy efficiency, also known as Energy Efficient Singapore (E2 Singapore), which comprises actions in the following areas:

(a) Promoting the adoption of energy-efficient technology and measures by addressing the market barriers to energy efficiency;
(b) Raising awareness to reach out to the public and businesses so as to stimulate energy-efficient behavior and practices;
(c) Building capability to drive and sustain energy efficiency efforts and to develop the local knowledge base and expertise in energy management; and
(d) Promoting research and development to enhance Singapore's capability in energy-efficient technologies.

Power Generation

The power generation sector is the single largest primary source of carbon dioxide emissions in Singapore, accounting for 48% of the carbon emissions in 2005.[8] Because of competition in the electricity market, power generation companies have an incentive to be energy efficient. Gross efficiency of power generation in Singapore increased from 38% in 2000 to 44% in 2006.[9] Further efforts to help improve the energy efficiency of power generation in Singapore include the use of cogeneration and trigeneration.

Industry

Singapore's industry sector accounts for 54% of Singapore's carbon dioxide emissions, mostly in export industries such as refining, petrochemical, pharmaceutical, and wafer fabrication. Improving the energy efficiency of industries lowers carbon emissions and increases cost-competitiveness. Going forward, industries will be encouraged to

(a) Design their facilities to be efficient;
(b) Use energy-efficient equipment; and
(c) Have trained personnel who can recognize and implement energy-efficient measures.

Energy Audits

In July 2002, an Energy Audit Scheme was introduced for major industrial consumers of energy, such as oil refineries. Although the scheme is voluntary, to date, three major oil refineries and two petrochemical companies have opted into the scheme, demonstrating commitment toward improving energy efficiency and mitigating the environmental impact of their operations. All industrial facilities, especially those whose operations are energy intensive, are urged to participate in the Energy Audit Scheme, as the collective efforts of industries will make a significant impact in improved energy use and reduced emissions of air pollutants and carbon dioxide.

In April 2005, MEWR/NEA launched the S$10 million Energy Efficiency Improvement Assistance Scheme (EASe) to help manufacturing companies and building owners/operators improve their energy efficiency by cofunding energy audits. As of August 2007, a total of 70 applications have been approved, with projected energy savings of about S$16 million annually (Ministry of the Environment and Water Resources, Singapore, 2008).

Awareness Raising

To promote the uptake of energy-efficient equipment and implementation of energy efficiency practices, NEA will develop an energy efficiency Web site that will provide industry with details on the available energy efficiency schemes and energy efficiency case studies. To facilitate the sharing of knowledge, expertise, and best practices in energy efficiency, NEA together with other agencies such as the Building Construction Authority (BCA) will periodically arrange energy efficiency seminars and workshops that bring together local and overseas experts and various stakeholders. NEA will also provide a Singapore Certified Energy Manager Training Grant to help equip facility owners and technical staff with the necessary knowledge and skills to manage energy services within their facilities. Besides, it will study the feasibility of making the appointment of energy managers mandatory for large manufacturing and building facilities in the coming years.

Transport

The transport sector in Singapore accounts for about 19% of GHG emissions. Improving the energy efficiency of the transport sector is achieved through the following key strategies:

(a) Managing vehicle usage and traffic congestion;
(b) Improving and promoting the use of public transport;
(c) Improving fuel economy; and
(d) Promoting green vehicles.

Managing Vehicle Usage and Traffic Congestion

Singapore employs various means of controlling vehicle usage in the city, which have helped to reduce energy use as traffic congestion results in fuel wastage. Car ownership in Singapore is discouraged through the imposition of taxes on vehicle ownership such as the Additional Registration Fee, and through the Certificate of Entitlement (COE) system, which limits the number of vehicles registered. The Land Transport Authority (LTA) aims to lower the vehicle population growth rate from 3% per annum (2006–2008) to 1.5% from May 2009.[10] Vehicle usage in Singapore is managed through integrated land use planning, congestion control measures such as electronic road pricing (ERP), and efficient public transport system, which is a much more energy-efficient way of moving the masses compared with private cars.

Improving and Promoting the Use of Public Transport

Ownership of private vehicles is well controlled in Singapore, and there is a widespread use of public transport. LTA will continue to improve public transport by making significant public transport infrastructure investments and planning the entire system as an integrated whole from the commuters' perspective, with bus and rail working in close partnership to provide more seamless and convenient transfers.

Promoting Green Vehicles

To encourage the purchase of hybrid and compressed natural gas (CNG) vehicles, a Green Vehicle Rebate has been in place since 2001. Green vehicles are more fuel-efficient and emit less air pollutants than their conventional petrol or diesel equivalent. The rebate aims to bring about a change in consumer behavior to support clean emerging technologies, by narrowing the cost differential between a green vehicle and the conventional equivalent model.

Buildings

The buildings sector contributes about 16% of Singapore's GHG emissions. Most of the electricity used by buildings in Singapore is for air-conditioning (40–50%), mechanical ventilation (about 20%), and lighting (15–20%). Results from energy audits cofunded by NEA have shown that there is room for the energy efficiency of buildings in Singapore to improve.

Green Mark

The Building and Construction Authority (BCA) of Singapore introduced the BCA Green Mark Scheme in January 2005. The scheme is a green building rating system to evaluate a building for its environmental impact and performance. It provides a comprehensive framework for assessing building performance and environmental friendliness. From early 2008 onward, all new buildings and existing buildings undergoing major retrofitting works with gross floor area above 2000 m^2 must meet the Green Mark Certified standard.

Buildings are awarded the BCA Green Mark based on five key criteria:

(a) Energy efficiency,
(b) Water efficiency,
(c) Site/project development and management (building management and operation for existing buildings),
(d) Good indoor environmental quality and environmental protection, and
(e) Innovation.

The assessment process consists of an initial assessment leading to the award of the Green Mark, followed by triennial assessments to ensure that the Green Mark building continues to be well maintained.

Higher Energy Demand

To mitigate the increase in the amount of energy used by air-conditioning, BCA and NEA are implementing measures to further improve energy efficiency in buildings. For example, under the Building Control Act, air-conditioned buildings must be designed with a high-performance building envelope that meets the prescribed Envelope Thermal Transfer Value (ETTV), currently set at $50\,W/m^2$. The value sets a limit on the amount of heat gained by an air-conditioned building through its roofs, external walls, and windows. BCA has completed a study to review the ETTV standards and to explore the possibility of extending the ETTV regulations to residential buildings. The findings from the study were taken into consideration in the proposal to stipulate minimum Green Mark standards for new buildings, leading to amendments to the Building Control Act. BCA also sets minimum energy efficiency standards for the four major energy-consuming equipments and systems, viz air-conditioning equipment, water heaters, electric motors, and artificial lighting system.

Building Labels

The Energy Smart Building Labelling Scheme was launched in December 2005 to accord recognition for existing office buildings with good energy performance. The Energy Smart Buildings Scheme has been extended to the hotel sector in 2007. In future, the label will be expanded to other building types including shopping complexes and hospitals.

Households

The households sector represents about 10% of total energy use, with the bulk of energy consumption being used for air conditioners and refrigerators, followed by appliances such as consumer electronics (e.g., TVs, DVD players, and computers), lighting, and water heating.

Mandatory Energy Labeling

Mandatory energy labeling informs consumers of the energy efficiencies of different models of an appliance, thereby helping them choose a more energy-efficient model. Starting from January 2008, all air conditioners and refrigerators sold in Singapore have to carry an energy label. The scheme will be extended to other appliances, including household lightings and water heaters in future.

Electricity Vending System

The feasibility of a scheme called the Electricity Vending System (EVS) is being studied. EVS enables consumers to buy electricity from any electricity retailer and allows consumers to monitor their electricity consumption figures by the half hour. By being more aware of electricity use patterns, consumers can reduce their electricity bills through more prudent use of electricity.

Electricity Consumption Tracking Device

NEA is also looking into introducing an electricity consumption tracking device that not only displays electricity consumption in dollar terms but also tracks the electricity usage of key energy-intensive household appliances such as air conditioners.

Public Awareness Programme

The Climate Change Awareness Programme (CCAP) was launched in April 2006 by the Singapore Environment Council and supported by NEA. It aims to raise awareness among households and motorists on how simple changes in daily habits can save money and energy, and help to address climate change. Moving forward, NEA will enhance the public awareness efforts to encourage households and motorists to move from awareness to adopting these simple energy efficiency habits and reducing their energy consumption.

COMPETENCY BUILDING

To facilitate Singapore's efforts to mitigate CO_2 emissions and adapt to climate change in the most environmentally sustainable and cost-effective way possible, there is a need to build up a broad range of competencies, in technology, skills, and expertise. Research into clean and renewable energy will decrease reliance on carbon-intensive fossil fuels. Developing energy-efficient technology will reduce the impact of growing energy needs. The objective of research is to improve the current state of technology and bring down production costs to a level that would make large-scale adoption commercially viable.

Research and Development

Academia

Research in energy and climate-related activities are already being carried out by several research institutions in Singapore. For example, A*STAR's Institute of Materials Research and Engineering, National University of Singapore and Nanyang Technological University are looking into novel materials for solar cells. Nanyang Polytechnic and Temasek Polytechnic have set up fuel cell R&D programs.

Industry

In the private sector, the world's top wind technology company, Vestas is setting up a global wind R&D center in Singapore. Rolls Royce is working with various research organizations in Singapore to develop stationary fuel cells. In October 2007, Renewable Energy Corporation (REC) from Norway announced its intention to build an integrated solar wafer, cell, and module manufacturing facility in Singapore. Slated to be the largest solar manufacturing complex when completed, REC is expected to also undertake research in Singapore related to manufacturing automation and improvement, as well as solar cells.

Government

Singapore Government is also taking an active role in driving R&D investment. In March 2007, the Research, Innovation and Enterprise Council (RIEC) endorsed Clean Energy as a key growth area for Singapore. The Economic Development Board (EDB) thereafter set up the interagency Clean Energy Programme Office (CEPO) to grow the Clean Energy industry in Singapore.

CO_2 Mitigation Technology

Research into clean and renewable energy will help decrease Singapore's reliance on carbon-intensive fossil fuels, while developing energy-efficient technology will reduce the impact of growing energy needs. The objective of research efforts (e.g., energy efficiency, solar energy, green buildings, etc.) is to improve the current state of technology and to bring down production costs to a level that would make large-scale adoption commercially viable.

Energy Management and Carbon Expertise
With the nationwide drive on energy efficiency and worldwide move to reduce CO_2 emissions, there will be a demand for energy management capabilities and services and carbon consultancy services. The establishment of the relevant training courses, accreditation schemes, and institutes will help to ensure a continued buildup of such skills and expertise in Singapore in the future.

Vulnerability Assessment and Adaptation Technology and Expertise
Research into better vulnerability assessments and adaptation technologies will help to better understand climate change effects as well as address vulnerabilities to climate change more effectively.

Research centers in Singapore will also undertake research in areas such as human health, future cities, environmental monitoring and modeling as well as clean water. To enhance Singapore's capability in water policy research, PUB is collaborating with the Lee Kuan Yew School of Public Policy to establish an Institute of Water Policy (IWP) that will engage in research, education, and consultancy work in water management and policy for public sector organizations.

Promoting Renewable Energy
Efforts in promoting renewable energy such as biomass and solar energy are focused on promoting R&D and test-bedding to improve their performance and cost-effectiveness.

CONCLUSION

Climate change is a long-term phenomenon, and its impacts will be felt for decades. The MND, Singapore, leads an interagency Taskforce to review existing infrastructural adaptation measures, as part of the government's overall approach to address climate change. Government agencies will continue to work closely to assess the possible impacts of climate change on Singapore, regularly review the sufficiency of Singapore's existing adaptation measures, identify new measures as necessary, and establish national systems to actively monitor and manage these impacts.

Singapore's climate change strategy is basically focussed on

(a) Assessing and studying the vulnerability to climate change and addressing it through requisite adaptation/mitigation measures;
(b) Reducing the GHG emissions produced in the country through energy efficiency measures.

Both the above initiatives are buttressed by the concurrent focus on competency building in Singapore that will enable it to develop and implement technologies and processes that support its efforts in minimizing the adverse impacts of climate change.

To meet the consequences it could face as a result of climate change, Singapore has initiated a range of programs that will enable it to override the challenges. It is ensuring reliability and resilience of water supply through integrated water resource management and also by tapping on R&D and technology to develop and test-bed new technologies. Singapore hopes to offset the adverse impacts of an increased sea level and changing weather patterns through land reclamation, improved drainage infrastructure, building of reservoir dams, and protection of coastal areas.

Singapore has identified improving the energy efficiency of major energy sectors, to do its part in containing factors that cause climate change. Despite having an export-oriented, energy-intensive economy, the measures initiated by Singapore have resulted in lower GHG emissions per dollar as compared with other economies. Its efforts to adopt cleaner fuels, increase energy efficiency and not to subsidize energy prices have proven to work for reducing climate change factors and improving the quality of life in the country.

The various government agencies, research institutes, and private sector enterprises in Singapore have introduced a range of initiatives and programs, which will build up manpower expertise and technical competencies in Singapore. These efforts, along with the available funding for R&D and test-bedding, will enable to mitigate CO_2 emissions in Singapore and adapt to climate change in an environmentally sustainable and cost-effective way.

NOTES

1. Factsheet: Study on Long Term Impact of Climate Change on Singapore, Ministry of Environment and Water Resources, 2010.
2. National Environment Agency (NEA), Singapore.
3. Public Utilities Board, Singapore.

4. The urban heat island effect refers to the phenomenon whereby urban areas are warmer than rural areas, largely due to the replacement of natural land cover with pavement, buildings, and other infrastructure.
5. Ministry of the Environment and Water Resources, Singapore (http://app.mewr.gov.sg/data/ImgCont/1342/1.AirClimateChange.pdf).
6. http://www.e2singapore.gov.sg/power-generation.html.
7. Ministry of Environment and Water Resources, Singapore (http://app.mewr.gov.sg/web/Contents/Contents.aspx?ContId = 26).
8. National Environment Agency, Singapore.
9. http://www.e2singapore.gov.sg/power-generation.html.
10. Land Transport Authority, Singapore.

REFERENCES

Intergovernmental Panel on Climate Change. (2007). *Climate change 2007: Synthesis report.* An Assessment of the Intergovernmental Panel on Climate Change. Geneva, Switzerland: IPCC.

International Energy Agency. (2009). *Key world energy statistics.* Paris: International Energy Agency.

Ministry of the Environment and Water Resources. (2008). *Singapore's national climate change strategy.* Singapore: Ministry of the Environment and Water Resources.

Ministry of the Environment and Water Resources. (2008). *State of the environment 2008 report.* Singapore: Ministry of the Environment and Water Resources.

APPENDIX. INCREASED ENERGY DEMAND

An increase in ambient temperature is expected to increase the demand for energy, especially for air-conditioning in homes, offices, and industries. To mitigate the increase in the amount of energy used by air-conditioning, the Building & Construction Authority (BCA) and the National Environment Agency (NEA) of Singapore are implementing measures to further improve energy efficiency in buildings. Under the Building Control Act, air-conditioned buildings must be designed with a high-performance building envelope that meets the prescribed Envelope Thermal Transfer Value (ETTV), which is a measure of heat gain through external walls and windows of a building. BCA has also developed a green building rating system to evaluate buildings for their environmental impact and performance, one of the key criteria for which is energy efficiency. The BCA has completed a study to review the ETTV standards to explore the possibility of extending the ETTV regulations to residential buildings.

CHAPTER 14

CLIMATE CHANGE ADAPTATION AND AGROFORESTRY IN SRI LANKA

Rekha Nianthi

INTRODUCTION

Throughout the world, at one period or another in its history, it has been the practice to cultivate tree species and agricultural crops in intimate combination in most of the countries. The history of cultivating trees and crops in home gardens, social tree planting, protecting and managing forests, appreciating wildlife, and sustaining the beauties of nature in Sri Lanka go back to more than about 25 centuries. In chronicles, there are some references on social tree planting practices, and home gardens planted with flowering and fruit-bearing trees in Sri Lanka. Because of the traditions, influencing factors of the existing environment, and nature of agroforestry, the numerous examples of agroforestry practices are found in all agro climatic and ecological zones of Sri Lanka. Today, the traditional knowledge of agroforestry is being developed and expanded with the objective of improving living standards, especially the rural communities in Sri Lanka.

In the context of global climate change and its impacts on the impoverishment of farmers' economy, the role of agroforestry has become of vital importance in terms of balancing the conflicting issues of conservation of

Climate Change Adaptation and Disaster Risk Reduction: An Asian Perspective
Community, Environment and Disaster Risk Management, Volume 5, 285–305

ISSN: 2040-7262/doi:10.1108/S2040-7262(2010)0000005020

natural resources and their usage in sustaining agricultural development of rural areas and livelihood, in most of the developing countries. On the other hand, it is recognized that agriculture is the human-based activity that is most vulnerable to global warming and future climate change. With the climate change issue, subsistence agriculture as well as commercial agriculture in the tropical regions is particularly vulnerable in many ways. In developing countries, smallholder farmers in the agricultural fields do not have sufficient resources and planning to adapt to the existing climate change.

The traditional systems of agroforestry practices may possibly play a significant role in mitigating the atmospheric accumulation of greenhouse gases (GHGs). It is well recognized now that the agroforestry system has high carbon sequestration capacity to minimize the carbon dioxide (CO_2) in the atmosphere. Agroforestry also plays a role in helping smallholder farmers adapt to the changing climate in different ways. With the scientific evidences, agroforestry also has an important role in climate change adaptation, particularly for smallholders. It is necessary to initiate research concerning the role of agroforestry in both mitigation and adaptation to climate change in Sri Lanka. At present, various development partners as well as international policy makers have increasingly recognized the potential role of agroforestry for addressing climate change vulnerabilities.

Climate change adaptation has gradually gained more recognition or appreciation as a major aspect in agricultural development in different countries and regions throughout the world. The Fourth Assessment Report of the Intergovernmental Panel on Climate Change (IPCC) emphasizes the potential risks and vulnerabilities in developing countries, where smallholder farmers are subjected to extreme climatic events such as drought conditions with dry spells, high intensity of rainfall and floods, delayed onset of the main rains during the cultivation seasons, and other natural disturbances to traditional and existing climatic patterns.

Several studies have shown that various agroforestry systems have a unique opportunity to increase carbon (C) stocks or storage in the terrestrial biosphere, which contain all the spheres as atmosphere, lithosphere, and hydrosphere. Some studies have shown that one of the main advantages of agroforestry systems is the ability of carbon sequestration; therefore, it has a unique opportunity to increase C stocks in the terrestrial biosphere more in tropical zones. Agroforestry could play an important role in reducing the atmospheric concentration of CO_2 by storing carbon in tree biomass and in the soil, encouraging the natural carbon sinks and storage through the improvement of land productivity and the provision of forest products such as firewood and timber (Alain, Kandji, & Verchot, 2004). All these process will enhance through the carbon cycle and the various other chemical cycles in the sub systems of the geo-system.

Climate change forecast models indicate that the average global temperature will rise by 1.5–4.5°C, when the effective CO_2 concentration is double in the atmosphere (United Nations, 2010). Carbon dioxide is the primary source of carbon for agricultural crops and tree growth. A primary effect of increased CO_2 is an expected increase of gross photosynthesis. With adequate management and conservation of trees in various cultivated lands, a significant portion of the atmospheric carbon dioxide could be captured and stored in tree plant biomass and in soils.

The agro-silviculture (tree and crop combinations) as one of the agroforestry system allow full tree growth and have an important component in the particular systems, and therefore can store considerable quantities of carbon in the living biomass. One of the most important comparative advantages of this system is that sequestration does not have to end with wood or tree harvests; carbon can continue to the tree stems, roots, and tree branches. Even this stock can process in any form of long-lasting product in various components in the system. Thus, if the necessary planning is made to promote small-scale wood and timber processing industries (for example, sawmills, wood carving, furniture houses, handicrafts, etc.), then the use of high-quality wood trees in agroforestry practices can contribute significantly to climate change mitigation while providing subsistence as well as satisfactory financial aids to smallholders. This system will enhance even the paddy farmers' income during the water deficit or dry period and will be a survival strategy for their livelihood.

Agroforestry positively influences several other areas related to the man–environment interaction. Agroforestry can increase deep infiltrations of water and improve the soil nutrition in the different soil structures. It also increases water use efficiency and captures wind-transported fine materials. Therefore, it increases organic matter and soil nutrient availability of the system. In addition to these healthy benefits, the agroforestry positively influences favorable microclimate conditions in the environment.

Over the past decades, with the increase of carbon dioxide in the atmosphere, as an adaptation measure to decrease the carbon in the environment, the agroforestry practice started to gain more popularity in the world. Based on this mixed production system, fauna and flora have the potential to minimize the GHGs as well as protect the system through interactions among trees, crops, and, in some cases, engaging animals (livestock) in a sustainable way. This chapter reviews and discusses the adaptation capabilities of climate change on various agroforestry systems and practices. In addition, it highlights some previous studies related to the main premise of this particular theme.

Following definition highlight the most appropriate opinion for the function of the agroforestry system as:

> Agroforestry is a sustainable land management system which increases the yield of the land, combines the production of crops (including tree crops) and forest plants and/or animals simultaneously or sequentially on the same unit of land and applies management practices that are compatible with the cultural practices of the local population. (Bene, Beall, & Cote, 1977)

Concerning this previous understanding, it is difficult to discuss the benefits of agroforestry without taking a closer look at the nature of agroforestry systems. In general, there are three components of any agroforestry system (Nair, 1989a, 1989b):

- Trees with crops ("*Agrosilvicultural*")
- Trees with livestock ("*Silvopastoral*")
- Trees with both crops and livestock ("*Agrosilvopastoral*")

Agroforestry has been practiced for a long time throughout the world, but has been recognized as a science subject for nearly three decades. The trees and the other components in the system can be separated in space or time, but they must interact for the system to be truly considered as an agroforestry system (Gordon, Newman, & Williams, 1997). Ideally, these interactions provide both ecological and socioeconomic benefits to the smallholders in the particular system.

Population increase and urbanization with urban growth, land degradation, and global warming are causing food insecurity in large parts of Asian and African regions, especially in developing countries. Sri Lanka would not be an exception to this situation. The traditional knowledge of agroforestry is being developed and expanded with the objective of improving living standards, especially in rural communities of Sri Lanka.

AGROFORESTRY-ASSOCIATED PRACTICES IN SRI LANKA

Because of the age-old agroforestry tradition in Sri Lanka, numerous examples of agroforestry practices are found in all agro-climatic and agro-ecological zones of the country. Today, the traditional knowledge of agroforestry is being developed and expanded with the objective of improving living standards of rural communities in the country. A wide range of agroforestry systems is found in Sri Lanka, and the following summarizes some of their characteristics and features.

Shifting Cultivation

Shifting cultivation (known as *chena* cultivation in Sri Lanka) is traditionally practiced in many tropical countries known by various names. However, the indigenous knowledge acquired in the evolution of chena has many important environmental qualities for conservation and protection of natural resources in the various systems. Chena cultivation in Sri Lanka is an age-old agroforestry system in which forest and food crops are grown alternately in temporal cycle in particular seasons. The system is sustainable as long as a reasonably long fallow period (at least 15 years) exists. However, with population increases and shrinking forest cover in the country, the fallow period has been shortened, resulting in severe land degradation, especially in the dry zone areas of Sri Lanka. Dry zone in the country had large tracts of forest that vanished in the past decades due to uncontrolled and disorganized shifting cultivation and incompatible land use practices. Formerly, the issuing of cultivation permits by the government regularized the shifting cultivation in Sri Lanka. In 1981, issuing of these permits was officially terminated. Today, only a few illegal chena cultivation practices are taking place, especially in remote dry zone areas.

In the dry zone of Sri Lanka, clearing of chena lands begins in July or August (the rainless dry months in the dry zone), and farmers prepare the cultivation land by burning the vegetation, which is cut and heaped for drying. They complete the land preparation before mid-September, when the first rains are expected. The chena lands are usually planted with a variety of crops such as millet, maize, mung bean, mustard, and vegetables, at the beginning of *Maha* cultivation season (the main cultivation season in Sri Lanka). During the Yala season, (with less rainfall) chena land is planted with sesame, millet and pulses. After few seasons of cropping, the land is left to fallow. In the dry zone, in places where there is no limitation on the availability of forestland, the intensity of land use in chena cultivation take place less. Identification of suitable land for chena cultivation (Fig. 1), selection of seed and crops/plants, sharing of the land block among a group of farmers, time of cultivation, crop management and protection, and all other activities taking place in chena land are merely based on some sound

Fig. 1. Harvesting Sesames on Chena Land at the Paranahalmillewa Cascade in Anuradhapura (a Dry Zone).

scientific methods. If shifting cultivation is practiced in an environmentally sustainable manner, this will be a good carbon sequestration method that can easily be promoted to rural communities by acceptable approaches through government supports in response to climate change adaptation in the Sri Lanka perspective.

Cooperative Reforestation

This agroforestry system was a modified form of the "Burmese taungya system." This system was being practiced especially to convert the degraded dry zone natural forests into teak plantations with participation of shifting (chena) cultivators in this scheme (see Fig. 2). Cultivators were also provided an opportunity to grow agricultural crops over a three-year period, and the farmers were paid a monetary reward in many circumstances. In this system, the farmers formed cooperative societies and they actively participated in this cultivation program. This system has produced good outcome for the farmers in the dry and intermediate zone. Though this system was originally applied for teak plantation, in a later stage, some *Eucalyptus camaldulensis* plantations were also introduced into this program. The shortcoming of this system has been the insecure land tenure

Fig. 2. Teak Plantation at Anuradhapura, Dry Zone of Sri Lanka.

(*bukthiya*) for the cultivators. The system also encouraged nomadism, which did not normally improve the life standard of cultivators.

Intercropping with Other Cash Crops

Tea, coconut, and rubber are considered the main plantation crops in Sri Lanka. Coconut is the most widely planted industrial tree crop in Sri Lanka. It covers about 25 percent of the total cultivated area on the island. A large number of tree crops such as teak and mahogany are grown in association with coconuts. In addition to this practice, crops like banana, pineapple, pepper, and papaya are also cultivated as mix crops in most of the coconut lands. Tea is the second largest industrial crop in the country and is grown mainly in the central highland country and considerably on the southwestern lowlands (wet zone). A large number of leguminous and other shade trees are also grown in this agroforestry system (intercropping with other cash crops). The fuel wood needs for tea estates workers and domestic users, energy needs for tea processing, and tea needs for estate workers are also met within this system. Rubber is the third largest industrial crop in the country and various types of fruit trees are often intercropped on the rubber croplands. Pineapple and cocoa are grown in areas where rubber canopy is not too dense. In addition to the previously mentioned plantation crops, Palmyra palms (*Borassus flabellifer*), which yield many products, and intercultivatied cashew nuts are other agroforestry systems found especially in coastal areas of the dry and semiarid areas of the northern and eastern regions of Sri Lanka.

Agroforestry Based on the Sloping Agricultural Land Technology (SALT)

Sloping agricultural land technology (SALT) is also widely adopted in most of the mixed cultivation practices in the central hilly areas of the country. In this system, hedgerows are grown along contour lines on sloping lands mostly in abandoned tea areas, and on some other suitable lands alternating with agriculture crops. Fuelwood species, such as *Gliricidia sepium,* are commonly grown in hedgerows in this system. Continuous mulching with pruned biomass from hedgerows improves the soil nutrient conditions and reduces the soil erosion. Since the mid-1970s, the SALT-1 technology in Sri Lanka has been utilizing a number of fast-growing nitrogen-fixing trees and shrubs, especially in the upland areas. These are planted in double hedgerows every 3–5 m apart, along the contours of the sloping land. These nitrogen-rich fixing hedges act as physical barriers to soil erosion by providing rich mulch that enhances the soil nutrients and provides

Fig. 3. Tea with Shade Trees and Nitrogen-fixing Trees.

erosion control for the system and a good source of organic nutrients for cultivation systems. The alley (passage) or strips (carpet) in between the double-contour hedges are planted to cultivate seasonal and permanent crops to give a balance of food and income generation to smallholders (see Fig. 3).

KANDYAN HOME GARDEN SYSTEM

The *Kandyan* home garden system is also an age-old traditional agroforestry system practiced widely in the central hilly areas and other districts in the wet zone. This is an ideal form of land use system combined with agricultural, forestry, and livestock. Trees are grown in a multilayer structure with mixed cropping of trees yielding timber, wood, fuelwood, fodder, mulch, food, fruits, medicines, etc. Maximum utilization of space, both vertical and horizontal, is ensured in this system. This system also intensively utilizes the belowground resources, both horizontally and vertically. The system has the potential for continuous increased production, environmental sustainability, and the returns to smallholders. Therefore, this system is widely accepted not only in the wet zone areas but also in areas where this system has been investigating in particular areas.

Farmers' Woodlots

The Forest Department has started to establish the farmers' woodlots, raising partnerships with farmers in Sri Lanka's rural areas since the 1980s. This system is practiced mainly in the dry zone and in drier areas in the central hilly areas. Relatively small blocks of degraded state lands have been leased out to local farmers on a long-term lease basis. The Forest Department provides seedlings of forest and horticulture species and technical advice to the farmers. In this system, farmers are entitled to harvest the final tree crops, leave a few mother trees in the plot.

Agroforestry with Afforestation Programs

Plantation forestry is considered to be of great value in reducing pressure on natural forests and, at the same time, providing more regular products as compared to natural forests. The forest policy of 1929 emphasized increased wood production to ensure self-sufficiency and to allow for some exports. In 1938, the policy of planting natural grasslands (*patanas*) was commenced, and planting exotic tree species such as pines and eucalyptus in up-country reflected the prevailing emphasis of the policy. The forest policy of 1953 emphasized timber and fuelwood production that was expected to meet the increasing demands. Plantations of eucalyptus as well as acacias were also established at that time. During the 1970s and 1980s, eucalyptus was established extensively in the dry zone, together with acacia. Most of the *Pinus caribaea* plantations in the country were planted between 1965 and 1984. In the latter part of the 1980s, with environmental concerns about monocultures, the planting of *Pinus* species gradually decreased. In the dry zone, large-scale planting of teak was carried out until the 1980s, and since then, the scale of planting has reduced because the successful establishment of teak depends on clearing and burning natural forests, which is now regarded as undesirable for the environment. These species were planted in response to the concerns about a potential deficit in fuelwood supply in the country. It is necessary to mention that some of these plantations were not successful due to damages by wild animals and fires.

In 1993, the Forest Department (FD) implemented the Participatory Forestry Project (PFP), which was mainly concentrated to increase wood production in home gardens, farmers' woodlots, protective woodlots, and public reserve lands. This program was carried out through the participation of farmers. Presently, the Forest Department is implementing reforestation activities through the Forest Resources Management Project, funded under the Asian Development Bank. FD is involved mostly in regeneration cutting areas of mature forest plantations, while afforestation of degraded areas is mostly implemented through local peoples' participation. Relatively large blocks of land (over 25 ha) are leased out to private companies to establish commercial plantations, and farmer participation is sought in establishing plantations in smaller blocks of land.

Tree Improvement Practices

In the dry zone, the most important species for tree improvement practices are teak, eucalyptus, and acacias. Eucalyptus trees were introduced to Sri Lanka in the 1960s. Eucalyptus trees have been the principal species planted in the hill country, the most important species being *E. Grandis*. During the 1980s, there was a widespread interest in new species of *Acacia phyllodes*. *Acacia mangium* was introduced in Sri Lanka in the 1980s, and was planted in the different areas in the wet and intermediate zones. During the 1970s, the Oxford Forestry Institute, in cooperation with National Forestry Institutions, initiated several international species cum provenance trials of tropical *Pinus* species in a number of countries in Asia, including Sri Lanka. These trials provide the basis for the large-scale establishment of plantations in degraded sites in the upland and in the low country of the wet zone of Sri Lanka.

INDIGENOUS LAND USE PATTERN IN THE VILLAGE TANK CASCADE IN SRI LANKA

Agroforestry could contribute to livelihood improvement in the dry zone of Sri Lanka, where people have a long history and accumulated local knowledge connecting to their livelihood. Dry zone communities are particularly distinguished for protecting the tank cascade environment with different agricultural practices. Indigenous land use patterns have enhanced microclimatic conditions of the traditional tank village (small irrigation reservoirs in villages) in the dry zone of Sri Lanka. Sustainability of the traditional tank village system had been maintained in the past simply, not only from structural maintenance but also because every component of the ecosystem in the cascade system was given due consideration. The attention was paid not only on macro-land uses such as paddy land, settlement area, chena lands, tank beds, etc., but also on micro-land components such as *godawala, iswetiya, gasgommana, perahana, kattakaduwa, tisbambe, kiul-ela, gasgommana* (lots of trees naturally grown in vegetation in the upstream land strip) and *vaan gilma* (accommodating water only when spilling). Large trees such as *Kumbuk, Nabada, Maila, Damba*, etc., and climbers such as *Kaila, Elipaththa, Katukeliya, Kalawel, Bokalawel*, etc., are found in this environment, and these combinations of trees/crops signify the different types of agroforestry practices in the cascade areas. Around the tank, the tree strip acts as a wind barrier, reducing evaporation from the tank and lowering water temperature. It gets close to the bund from either side where roots of large trees make water cages, creating breeding and living places for some aquatic and fish species. This strip of trees demarcates the territory between human and wild animals.

Perahana (strainer) is the meadow developed under *gasgommana* (naturally grown vegetation in the upstream land strip above the tank bed) and filters the sediment flow coming from upstream *chena* lands. *Godawala* is the man-made water hole for trapping sediment and also for providing water to wild animals. This might have been one of the strategies to evade man–animal conflict. *Iswetiya* or *potawetiya* are the upstream soil ridges constructed at either side of a tank bund to prevent eroded soil from entering upper land slopes. *Thawula* is known as the upper part of the tank bed, where a shallow water body is found on an almost flat area.

Wew-pitiya is the deep area of a water body, which accommodates the major part of the tank water and is covered with water for about 8 months a year. *Mada-kaluwa* is the portion of the water mass found during even the dry months of the year and is located closer to the tank bund. This water cannot be moved through the sluice as it is below the spill level. Especially for drought periods, farmers keep water in the *mada-kaluwa* (dead storage) for urgent situations. If there is a good sign of rain, they release this water through the sluices downstream and they clean the dead storage (de-silt it) and fertilize the paddies. At present, *mada-kaluwa* is filled with sediments in most of the tanks in the dry zone. *Kulu wewa* is the small tank constructed above relatively large reservoirs only to trap sediment and not for irrigation purposes. It provides water for cattle and wild animals. *Tis-bambe* is a fertile land strip found around the *gangoda* (settlement areas) and does not belong to any body. Tree species such as *Mee,* mango, coconut, etc., are grown in scattered style in the system. This area was used mostly for sanitary purposes, as the resting place of buffalo. Buffalo were also used as a protection mechanism from wild animals.

Kiul-ela is the old natural stream utilized as a common drainage. Tree species such as the *karanda,* mee, *mat grass, ikiri, vetakeya,* and few rare small fish species are also found in water holes along the kiul-ela. Most importantly, it removes salt, iron, and polluted water, and improves the sanitary drainage condition for the paddy tracts. The *kattakaduwa* (interceptor) is reserved land below a tank bund. It consists of three microclimatic environments like the water hole, wetlands, and dry upland and, therefore, diverse vegetation is developed (Rekha & Dharmasena, 2009) in this land strip. This strip of land prevents salts and ferric ions from entering into the paddy field. The water hole referred to as "*yathuruwala*" minimizes bund seepage by raising the groundwater table. Villagers plant vetakeya along the toe of the bund to strengthen the bund stability. It appears to be a village garden, where people utilize various parts of the vegetation for different purposes such as fuelwood, medicine, timber, fencing materials, household and farm implements, food, fruits, vegetables, etc. (Dharmasena, 1995). Specifically, they harvest row materials from this vegetation for cottage and other domestic industries. In the present situation, the interceptors of the tank cascades are deteriorating, either

Fig. 4. Interceptors (Kattakaduwa) with Different Types of Tree Species.

because the new generation of the village community is unaware or they have neglected the system (see Fig. 4).

BENEFITS OF AGROFORESTRY

Agroforestry is now considered an important practice in modern agriculture and forestry. This practice is useful in minimizing environmental degradation, such as desertification and land erosion. In addition, the function of agroforestry systems has a greater potential for storing and absorbing carbon from the atmosphere. Furthermore, agroforestry systems produce timber and firewood that will bring profit to farmers and reduce the CO_2 concentration in the atmosphere. Management of trees in agroforestry can mitigate greenhouse gas emissions, supporting the aims of the Kyoto Protocol. Agroforestry systems are a better climate change mitigation option than other terrestrial options because they provide more benefits to farmers such as:

a. helping them to attain food security,
b. securing land tenure, specially in developing countries,
c. increasing farm income, especially in rural families,
d. restoring and maintaining aboveground and belowground biodiversity,
e. making corridors between protected forests,
f. decreasing greenhouse gas concentrations, as CO_2 and methane (CH_4) sinks,

g. maintaining watershed hydrology,
h. conserving soil,
i. mitigating the demand for wood,
j. reducing pressure on natural forests,
k. promoting the wood-carving industry, which facilitates long-term locking up of carbon in carved wood,
l. using local knowledge for the better development of systems,
m. encouraging trade and industry, and
n. creating equity and promoting livelihood security.

PROPOSED STRATEGIES ON AGROFORESTRY AS CLIMATE CHANGE ADAPTATION IN SRI LANKA

Environmental issues like climate change have become increasingly important, not only in a global context but also in a local aspect related to various sectors. One of the main sectoral concerns in the country is the long-term productivity and sustainability of the agricultural sector. In the climate change point of view, the exact reasons behind the changes in weather patterns are still not known. What is most important is to introduce agricultural systems that could withstand such adverse weather/climatic conditions, thereby protecting the country's agricultural system.

Perera and Sirapatham (2004) have examined that the coconut cultivation is the best practice as a measure of adapting and mitigating climate change in Sri Lanka. As far as the climate change is concerned, the coconut tree is the ideal choice for coastal areas because coconuts can be grown well using seawater in Sri Lanka (see Fig. 5).

The coconut husks, coconut shells, and coconut leaves, which can be harvested annually, contain a considerable amount of biomass. These products are to be used as an alternative fuel for electricity power generation, if coconut cultivation is properly promoted in the country. Various other economic benefits could be derived by engaging in large-scale coconut cultivation in the country. It is necessary to introduce more profitable combinations to coconut farmlands that can enhance the sustainability of the system as well as provide income to the coconut cultivators.

Coconut trees can fix more and more CO_2 and also can increase fuels production. Because coconut is an oil-producing crop, coconut oil cake or *poonac* could be used as cattle feed, which in turn would help to increase milk production for the country. Excess quantities of coconut oil cake could be used to feed wildlife, to maintain their well-being. Timber from mature coconut

Fig. 5. Coconut-growing Areas Along the Sea Coast of Tsunami-impacted Areas.

Fig. 6. Structure of Agroforestry: Coconut, Rubber, and Tea as Mixing Cultivation at Kuruvita (Wet Zone).

trees can lighten the pressure on natural forest resources in the country. Coconut treacle can also produce sugar, increasing local supply (see Fig. 6).

Rohinton (2004) has introduced the urban climate change mitigations through urban vegetation strategies for the Colombo metropolitan region.

Global climate change has been influenced mainly due to urban air pollution by greenhouse gases emissions, types of aerosols and various types of soot in the atmosphere. Air pollution is in many ways generally a problem for city dwellers, architects, and planners in the city of Colombo. With urbanization, Colombo's air pollution is rising rapidly with the growing level of motorization. The increase in vehicular population is the main cause of air pollution in Colombo, and more than 80 percent of air pollution is due to this vehicular emission. The University of Colombo conducted studies of polluted air particles; the studies showed high levels of suspended particulate matter (SPM) within Colombo at 10 different locations in 1995. High levels of this suspended SPM is now a significant threat to the public health of Colombo city dwellers.

Urban vegetation designed for Colombo city increases the favorable environment and human comfort in city areas. The mitigation program targets the landscape between pavement areas and attempts to control traffic emissions along selected streets by planting urban vegetation because some urban vegetation species are capable of combating air pollution. Therefore, these urban plants mitigate global climate change to some extent in the local context. The proper planning of urban vegetation guidelines will help to protect urban dwellers' health and comfort. For example, urban agroforestry practices have been introduced (in urban home gardens) most of cities in the Pacific Islands. Gradually, the urban gardens can be introduced and can contain a wide range of food trees and nonfood plants.

Climate change enhances rainfall and temperature variability and the frequency of extreme events leading to accelerated rates of degradation of soil and scarcity of water resources upon which agricultural farming communities depend for their livelihoods. These changes are threatening development advances such as the achievement of the Millennium Development Goals, especially in agricultural countries. Agricultural systems most vulnerable to climate change are those already affected by improper management, and land and resource degradation. Developing countries are going to bear the brunt of climate change and suffer most from its negative impacts on agriculture. Therefore, the trees play an important role in reducing vulnerability to climate change, increasing resilience of farming systems, and buffering households against climate-related shocks and risks. Agroforestry systems are therefore ideally placed to provide additional resilience of agro-ecosystems to future climate change.

Agricultural systems most vulnerable to climate change are those already affected by unsustainable agricultural management, incompatible land use systems, and resource degradation. Trees, crops, and animal combination agroforestry practices play an important role in reducing vulnerability of climate change by increasing of different farming systems' resilience against climate change–related risks. Removing atmospheric carbon and storing it in the different types of ecosystems in the terrestrial biosphere is one of the adaptation options, which have been proposed to minimize greenhouse gas emissions. Agricultural lands are believed to be a major potential sink and could absorb large quantities of carbon in the atmosphere, if carbon sequestration trees are reintroduced to these systems and wisely managed together with crops and/or animals.

To improve the resilience of farming systems and livelihood on agroforestry of smallholder farmers to current climate variability as well as long-term climate change, the following concerns need to be addressed in future programs in Sri Lanka:

a. Assessing the social and economic impacts that interact with agriculture/agroforestry practices on climate change–related shocks
b. Understanding the potential impacts of the different dimensions of climate change (water scarcity, temperature increases or decreases, high intensity of rainfall, increase of dry spells, etc.) at a number of scales of the main components of agroforestry
c. Using stakeholders' knowledge to enhance climate change adaptive capacity of smallholder farmers
d. Conserving existing carbon pools such as various types of forests, plantations, and other farming systems
e. Sequestering carbon by increasing the size of the carbon pool through reforestation and afforestation of degraded forests, croplands, and wastelands
f. Identifying and quantifying the carbon-fixing capability of various trees
g. Listing the multipurpose trees (MPT), that is, their management and multiple uses for characterizing their socioeconomic and ecological relevance with regard to different types of farmlands
h. Focusing research on understanding the emissions of GHGs, especially CO_2, nitrous oxide (N_2O), and CH_4 in agroforestry systems
i. Introducing more powerful methods for assessing cost/benefit analyses for agroforestry-based agriculture

j. Documenting baseline information on MPTs and tree-based systems by farm level to enable researchers and development actors to design adaptive tree management practices as well as adoptive agroforestry technologies
k. Providing models for sustainable development to promote tree-based agroforestry practices
l. Encouraging the adaptation of agronomic practices in rubber cultivation (for example, by mulching around rubber plants at the time of planting to minimize the effects of moisture stress when growing leguminous creeper as a ground cover, and identifying high-yielding rubber clones that respond to moisture stress)
m. Introducing selected vegetation species as an urban design for reducing urban emissions
n. Creating reforestation programs and proper forest management, because the anticipated change in rainfall in the hill country would have a differential impact on water yield, depending on type of vegetation and covers
o. Introducing rice varieties that produce minimum productive tillers with high yields to reduce methane emissions from paddy fields in the country, because paddy cultivation in Sri Lanka currently contributes 68 percent of the methane from the agricultural sector
p. Determining the agroforestry-related adaptive capacity not only on its ability to respond biophysically but also on the economic circumstances and institutional infrastructures

Agroforestry systems can improve farmers' and farming communities' resilience for coping with current climate variability, reducing net carbon emissions, and adapting to long-term climate change. This can be done through the use of trees/crops and animals by intensification, diversification, and buffering of farming. Fig. 7 shows the suggested future agroforestry approaches, of which some are already being implemented in Sri Lanka.

LIMITATION OF THE AGROFORESTRY PRACTICES

Rotational agriculture systems such as home gardens and boundary plantings can sequester substantial quantities of carbon in trees and plant biomass and in long-lasting wood products. Soil carbon sequestration constitutes another realistic option achievable in many agroforestry systems. The potential for agroforestry systems to mitigate CO_2 is well recognized

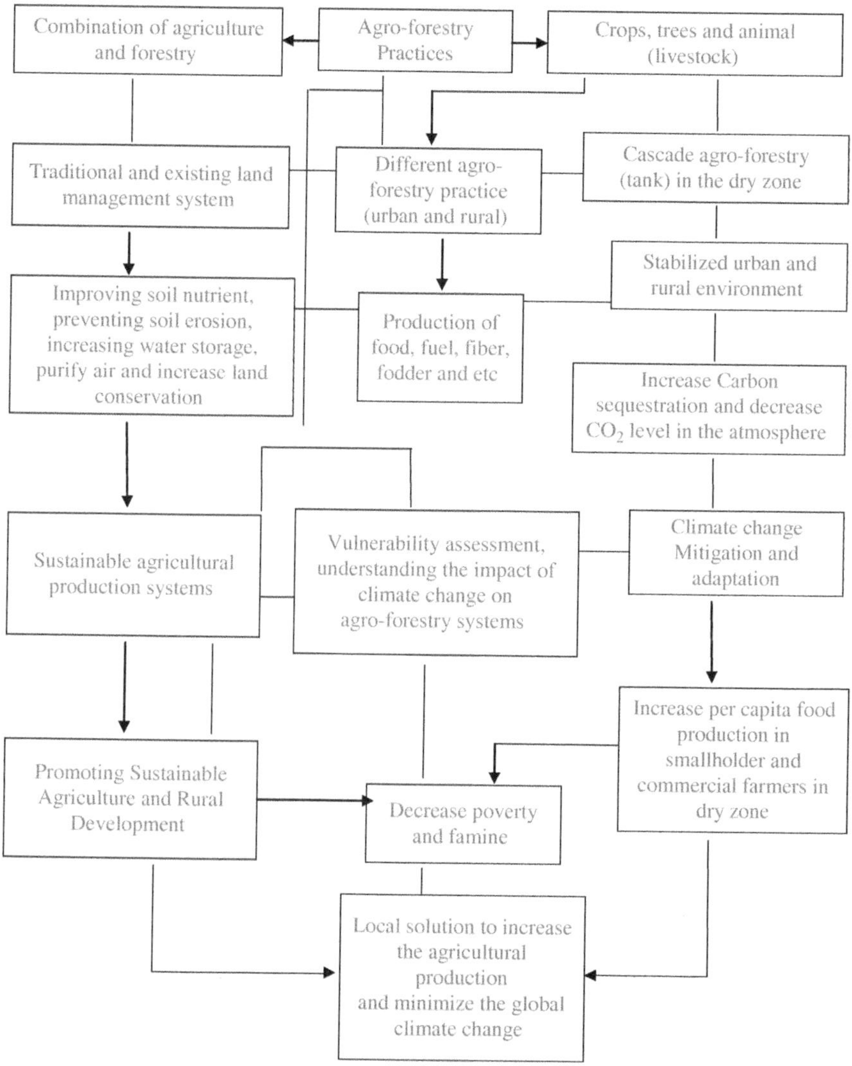

Fig. 7. Future agroforestry approaches in Sri Lanka.

in most countries now. However, the following are some shortcomings that need to be emphasized when we consider the climate change issues in agroforestry practices:

a. The uncertainties related to future shifts in global climate, which are based on different model forecasts and scenarios, will have different influences on agroforestry.
b. Land use systems and land cover are decreasing natural resources (as are water and soil resources, etc.) and an increasing population of humans is putting pressure on the resources.
c. Trees and crops grow poorly on substandard soils and in dry environments where water is scarce for tree/crops plantings.
d. Pests and diseases increase in different time periods/seasons that may further develop due to future climate change.
e. More efforts are needed to improve methods for estimating carbon stocks and trace gases such as CO_2, N_2O, CH_4, etc., to determine net benefits of agroforestry in the climatic system.

CONCLUSION

Agroforestry will continue to be a profitable and sustainable way of using land, as it has been (almost unnoticed) for thousands of years. It shows the potential for great benefits, for soil and water conservation, increased yields, enhanced biological diversity, reduced need for pesticides and fertilizers, and greater economic stability of for farming communities. Removing atmospheric carbon and storing it in the terrestrial biosphere is one of the important options, which have been proposed to compensate greenhouse gas (GHGs) emissions in the world. Agricultural and forest (green) lands are believed to be a major potential sink of carbon, if different trees are reintroduced to agricultural systems and wisely managed together with crops and/or animals. Thus, agroforestry as a land use system is receiving wider recognition, not only in terms of agricultural sustainability but also as an adaptation measure against climate change. This chapter discussed the numerous benefits of agroforestry systems in Sri Lanka and identified the usefulness for the absorption of CO_2 as another adaptation strategy for climate change. Appropriate policy responses should combine "crops/trees/livestock" as key assets that will strengthen adaptation and help to build the resilience of farming communities and households to local and global climate change. Conservation of existing carbon pools such as forests and other tree/crop farming practices are compulsory for continuing the existing level of carbon emission. Agroforestry systems as carbon sinks and a land management scheme that enhance the uptake of CO_2 or reduce its emissions. Therefore, the system has the potential to remove a significant amount of CO_2 from the atmosphere. However, those benefits are not always realized in practice, therefore underscoring the need for more research and experiments. Agroforestry is still a very young science, and as with all sciences, we can expect greater returns as we learn more and practice.

REFERENCES

Alain, A., Kandji, S., & Verchot, L. (2004). *Carbon sequestration in tropical agro forestry systems.* World Agro-forestry Centre (ICRAF), PO Box 30677, Nairobi, Kenya, February 2004.

Bene, J. C., Beall, H. W., & Cote, A. (1977). *Trees, food and people: Land management in the tropic.* Ottawa, Canada: IDRC.

Dharmasena, P. B. (1995). *Kattakaduwa*: A potential land for agro-forestry system development in Sri Lanka. In: H.P.M. Gunasena (Ed.), *Proceedings of 6th regional workshop on multipurpose trees*, University of Peradeniya, Peradeniya, Sri Lanka, August 17–19, 1995 (pp. 96–105).

Gordon, A. M., Newman, S. M., & Williams, P. A. (1997). Temperate agro-forestry: An Overview. In: A. M. Gordon & S. M. Newman (Eds), *Temperate agro-forestry systems* (pp. 1–8). CAB International: Wallingford, UK.

Nair, P. K. R. (1989a). Agro-forestry and biomass energy/fuel wood production. In: P. K. R. Nair (Ed.), *Agro-forestry systems in the tropics* (pp. 591–597). Dordrecht, The Netherlands: Kluwer Academic Publishers.

Nair, P. K. R. (1989b). Classification of agro-forestry systems. In: P. K. R. Nair (Ed.), *Agro-forestry systems in the tropics* (pp. 39–52). Dordrecht, The Netherlands: Kluwer Academic Publishers.

Perera, K. P. S. P., & Sirapatham, M. (2004). *Climate change of Sri Lanka: Project terminal report on climate change enabling activity phase II.* Colombo, Sri Lanka: Environmental economics & global affaires division, Ministry of Environment & Natural Resources.

Rekha Nianthi, K. W. G., & Dharmasena, P. B. (2009). Indigenous knowledge of farming practices and water management in the dry zone of Sri Lanka. In: R. Shaw, A. Sharma & Y. Takeuchi (Eds), *Indigenous knowledge and disaster risk reduction, from practice to policy* (pp. 329–348). New York, USA: Nova Science Publisher.

Rohinton, E. (2004). *Climate change of Sri Lanka: Project terminal report on climate change enabling activity phase II.* Colombo, Sri Lanka: Environmental economics and global affairs division, Ministry of Environment & Natural Resources.

United Nations. (2010). UNFCCC Executive Secretary: Governments make progress towards deciding shape of result at UN Climate Change Conference in Mexico, but need to narrow down number of negotiating options. Available at http://www.un.org/climatechange/bg.shtml

CHAPTER 15

COASTAL ECOSYSTEMS AND CLIMATE VULNERABILITY IN SRI LANKA

Deepthi Wickramasinghe

INTRODUCTION

Sri Lanka is an island described as a "pearl" situated between latitudes 5.55° 9.51′ N and longitude 79.41° 81.54′ E in the Indian Ocean. It has a coastline of 1,585 km (Coast Conservation Department [CCD], 1986a, 1986b). From the coastline, the exclusive economic zone (EEZ) extends 200 nautical miles. This is 6.7 times the country's land area, occupying 437,400 km^2. Coastal waters extend from the continental shelf to the other limits of the EEZ. The coastal area forms a dynamic interface of land and water and is of special significance in the country's economical, ecological, and social fabric. Thus, a "coastal zone" has been defined where activities are regulated by the government (Coast Conservation Act of Sri Lanka [CCD], 1981). Sri Lanka's coastal zone is defined in the Coast Conservation Act as the area lying within a limit of 300 m landward of the mean high waterline and, in the case of rivers, streams, lagoons, or any other body of water connected to the sea, either permanently or periodically, the landward boundary shall extend to a limit of 2 km perpendicular to the strait baseline drawn between the natural entrance points thereof and shall include waters of such water bodies.

Climate Change Adaptation and Disaster Risk Reduction: An Asian Perspective
Community, Environment and Disaster Risk Management, Volume 5, 307–326

ISSN: 2040-7262/doi:10.1108/S2040-7262(2010)0000005021

Coastal areas of Sri Lanka are relatively highly populated and demographically important. At present, out of the country's total population of 19.19 million, 11.36 million people live in coastal districts, amounting to about 59 percent (Coast Conservation Department [CCD], 1997). A large number of people visit the coastal region daily for a variety of economic, social, and commercial activities, thus making this area highly populated. The coastal area plays an important role in the country's economy. Around 44 percent of the country's gross domestic product (GDP) originates in the coastal region. Of these important sectors were the manufacturing, trade and tourism, banking, insurance and real estate, and forest sectors. Many trade- and service-related activities are concentrated in the western region around Colombo, the commercial capital. The growth and improvement of the tourism industry are facilitated by access to the sea in the coastal areas. The coast plays an important role in food security. Around 80 percent of fish production is carried out in the coast, and a significant extent of agricultural land lies in this area.

Due to multiplying developmental and population pressure, coastal areas are increasingly being exposed to several stress factors (Samaranayake, 2000). In developing countries like Sri Lanka, climate change could represent an additional stress on ecological and socioeconomic systems that are already facing tremendous pressures due to rapid urbanization, industrialization, economic development, and, above all, the impacts of three decades of war. With its growing population, a densely populated and low-lying coastline, and an economy that is closely tied to its natural resource base, Sri Lanka is considerably vulnerable to the impacts of climate change.

In this chapter, the current status of coastal resources of Sri Lanka is reported with likely impacts of climate change. Anthropogenic activities that cause stress on these resources and possible mitigatory actions through developing adaptive capacity are also discussed.

COASTAL HABITATS : SERVICES AND USES

Sri Lanka is gifted by nature with a wide array of coastal habitats that include coral reefs, mangroves, estuaries and lagoons, sea grass beds, mud flats, and sand dunes. Biodiversity is the measure of variety of life on earth. Article 2 of the Convention on Biodiversity Diversity (CBD) defines biological diversity as "the variability among living organisms from all sources, inter alia, terrestrial, marine and other aquatic ecosystems and the ecological complex of which they are a part; this includes diversity within species, between species and of ecosystems" (Convention on Biological Diversity (CBD), 2009). Sri Lanka is endowed with rich biodiversity and is considered one of the biodiversity hotspots in the world.

Mangroves

The mangrove cover on the island is about 12,500 ha , which is as little as 0.1–0.2 percent of the total land area. Mangroves have patchy distribution because all they are associated with the lagoons and estuaries of the country (Information Brief on Mangroves in Sri Lanka [IUCN], 2007). Twenty species of true mangroves have been reported in Sri Lanka. The mangrove ecosystems, which are very diverse, play a significant role by serving as a link between marine and terrestrial ecosystems (Choudhruy, 1997). This important link maintains the stability in the transient zone of sea and land, and supports many other equally important habitats, including estuaries and lagoons (see Fig. 1).

Mangroves provide various services for the community by providing locations for fishing, and for collecting timber, charcoal, fuelwood, and so on. They provide breeding habitats for commercially important fish, mollusks, and crustaceans.

Fig. 1. Mangroves Exposed to Low Tides in Puttalam Lagoon, North West Coast. *Source:* Photo taken by Deepthi Wickramasinghe.

Importantly, mangrove habitats serve as living buffers against disasters and reduce impacts. These habitats have been identified as a natural defense against the impact of wave action because they can absorb a major part of wave energy (Othman, 1994). A recent study carried out in South Asia reports that mangrove cover acts as an effective protection, especially to mitigate damage from storms and tropical cyclones (Anonymous, 2005; Dahdouh-Guebas et al., 2005). Mangroves act as a living barrier to trap sediments and stabilize the coastline, and therefore they protect the shoreline from erosion (Miththapala, 2008). Mangrove roots and organic matter in mangroves function to trap sediment, and also act like a sponge to absorb floodwater (McLeod & Salm, 2006). On the other hand, mangrove vegetation acts as a carbon sink, which directly reduces carbon dioxide in the atmosphere, thereby contributing to combating climate change.

Coral Reefs

Coral reefs, which are among the oldest ecosystems and the largest living structures on earth, are found on the coasts of Sri Lanka. The island's coast consists of coral reefs, both living as well as calcareous substances, sandstone, and rocky habitats. As per reports, about 2 percent of the coastline contains near-shore fringing reefs (Pathirana, Kamal, Riyas, & Safeek, 2008). Extensive coral reef habitats are restricted to areas with low sedimentation and semidry climates found in the northwestern and eastern coastal areas.

A total of 183 species of stony corals have been recorded in Sri Lanka (Rajasooriya, 2005). According to a recent survey, a total of nearly 1,000 known reef and reef-associated species have been recorded on the coast. Corals reefs harbor a variety of biodiversity and support many commercially important species of spiny lobsters, shrimps, crabs, and marine flora such as sea grasses and algae. Dolphins, whale sharks, and sea turtles have also been recorded along inshore and offshore reef habitats. Other key economic and social benefits associated with healthy coral reefs include high fishery yields, high tourism-related incomes, and good nutrition for coastal communities.

Coral reefs also play a vital role as a form of protection to the shore (Fernando, Samarawickrama, Balasubramanian, Hettiarachchi, & Voropayev, 2008). If a reef is located at a distance from land and if huge waves pass it, there is a tendency for the wave to actually break before it reaches the shore, because the reef reduces the depth of the ocean, increasing the wave height and causing the wave to break and lose its energy before reaching the shore (Conservation International (2008)). Some studies have reported the value of living coral reefs as a defense against tsunami waves in the Indian Ocean tsunami hit of 2004, suggesting that less damage occurs when coral reefs exist (Kunkel, Hallberg, & Oppenheimer (2006)).

Estuaries and Lagoons

Estuaries and lagoons act as entry points of inland freshwaters and sediments to the sea. There are about 45 estuaries and 40 lagoons along the coastline of the country, and they rank high among the most productive ecosystems (Kotagama & Bambaradeniya, 2006). One key reason for this is that they receive nutrients from inland activities. Thus, they support lots of microbial populations, attracting invertebrate as well as vertebrate components of food chains. Due to their quality of water being intermediate to sea and fresh water, they support unique biota too.

An abundance of fin fish and shellfish flourish in these waters (Samarakoon, 1986). In addition, they serve the special needs of migrating near-shore and oceanic species that require shallow protected habitats for breeding or as sanctuary for their larval stages. Lagoons and estuaries are famous for their fisheries, especially small-scale industries, and they provide a productive source of fish protein for the nation. Fishing in Negombo, Puttalam, and Rekawa lagoons are good examples (see Fig. 2).

These wetlands act as sponges or storage to absorb floodwaters from landslides and storm surges from the seaside, thereby reducing the risk of damage. During the tsunami hit in 2004, these coastal wetlands played a major role in accommodating intruding seawater, storing it for several days, and decreasing the severity of damage to the adjacent landscape.

Other Habitats

In certain coastal regions around Sri Lanka, sea grass beds occur, although not continuously along the coast (Amarasinghe & Pahalawattarachchi, 1996). Sea grasses are rooted, seed-bearing marine plants that grow in near-shore low-energy coastal waters. They are most extensive along the northwestern and northeastern coasts, especially around the small islands of Kalpitiya and Puttalam. They serve to stabilize sediment and provide a nursery/feeding area for a number of juvenile fish and shrimp. These

Fig. 2. Fishery Activities in Negombo Lagoon. *Source:* Photo taken by Deepthi Wickramasinghe.

underwater grasslands provide habitats for endangered species, including sea turtles and dugong.

Another important habitat is salt mashes. Because of the harsh environment under which they survive, all species that live in this ecosystem show adaptations for salt and heat tolerance. Very little attention has been given to salt marshes, compared with mangroves, as a component in the coastal ecosystem. The people living in the region do not understand the importance of this valuable resource, because the output is not direct, as in the case of mangroves.

During the 1990s, large numbers of shrimp farm developers acquired state land, mainly consisting of salt marshes around Puttalam Lagoon, to develop shrimp farms. And most of the state lands consisting of salt marshes were allocated to refugees for building their houses.

STRESSES ON COASTAL HABITATS

Mangroves

In recent years, the island's mangrove habitats have been drastically degraded in quality and quantity due to various illegal activities (Coast Conservation

Department [CCD], 1986a, 1986b). The increased human activities such as illegal felling, encroachment and clearing for infrastructural development, landfilling for housing development, cutting for firewood, and converting to other uses have caused serious pressure on the mangrove environment. Mangrove habitats, such as in Negombo and Chilaw, have been largely converted to areas of human settlement, where pressure of urbanization is high. In northwestern parts of Sri Lanka, the major threat to this ecosystem is clearance for the construction of aquacultural ponds, which have been a key element in the fishery export trade. Vast tracts of mangrove forests have also been cleared to make way for the establishment of coastal shrimp farming facilities. This has intensified pollution loads, flooding, and erosion in the lagoon and coastal areas. It is also evident in the recent past that the mangrove habitats have been polluted by indiscriminate solid waste dumping, including of household and municipal waste.

Coral Reefs

The main threats to coral reefs include rapid population growth in the coastal areas, pollution, coastal-based tourism development, mining and collection of corals, and fishing from mainland fishing vessels (Nayanananda, 2007). Recently introduced highly efficient techniques such as the bottom-set nets to catch spiny lobsters and reef fish cause severe damage to corals. Blast fishing using explosives is also commonly practiced in many parts of the country, being most prevalent in the southern coastal waters in the Galle District. In addition to fishing activities, collection of reef fish, invertebrates, and live coral for the ornamental fish export industry is of considerable importance. In fisheries export products, ornamental fish export is rated as the third highest in volume and value after prawns and lobsters.

Estuaries and Lagoons

Sri Lanka's coastline has become an important location for tourism development in the country. Swimming, snorkeling, scuba diving, and viewing corals through glass-bottom boats are popular activities that generate substantial income for the local populations. These result in numerous environmental problems that threaten the healthy function of reefs. Such threats include effluent and solid waste disposal, causing acute degradation of coral habitats, especially in some locations such as in Hikkaduwa.

It is also evident that estuaries and lagoons are threatened by urban encroachment, filling for infrastructural development, siltation, and over-fishing. Domestic sewage, garbage, and waste fuel from fishing crafts are major causes of the decline in water quality in several lagoons (Pahalawattaarachchi, Jayamanne, & Amarasinghe, 2005). Most of these waters have been degraded in such a way that water-quality parameters exceed maximum permissible levels for healthy aquatic life. Industrial effluents, agricultural runoff, and increased sedimentation from poor upstream land and water management schemes are also contributing factors. In certain lagoons, such as Negombo and Rekawa, frequent fish kills have been reported due to decreased dissolved oxygen levels. Unsustainable fishing practices, such as the collection of polychaete worms for brood-stock feed for shrimp hatcheries, are destroying portions of the habitat in the Negombo Lagoon. In addition, expansion of infrastructural facilities, pollution by effluents and solid wastes, and introduction of diseases and alien species have also created ecological imbalance (De Alwis, Ariyaratne, Azmy, & Dassanayake, 1995). This is clearly evident in the Chilaw and Puttalam areas.

Other Diminishing Habitats

Another recent development is mechanized and uncontrolled sand mining at river mouths (Gunaratne & Jayasooriya, 2006). Although sand is mined for commercial purposes in all major rivers, it is reported largely in the river mouths of the Nilwala River (in the Southern Province), and Deduru Oya and Ma Oya Rivers (in the North Western Province), where illegal miners remove tons of river sand even though these activities have been banned. This not only damages the riverbed and pollutes the water, but also greatly reduces the amount of sand being carried to the seashore. Consequently, the sea sand washed away due to wave action is not replenished by the inflow of sand from rivers, and results in severe coastal erosion. Excavation of sand causes deepening of river mouths, thereby allowing saltwater intrusion that ultimately affects the groundwater table in the area. Unstable, destroyed riverbanks exacerbate the impacts of weather hazards such as floods.

IMPACTS OF CLIMATE CHANGE ON COASTAL AREAS

It is understood that the overall change in climate is generally slow and unnoticeable, and the impacts are also not easily quantified and predicted.

In most cases, the integrated changes are complex and far from easy to interpret. Fundamentally, however, coastal areas are among the most threatened sectors on earth due to change in climate.

Natural coastal ecosystems in Sri Lanka increasingly are degrading in quality and quantity and are confined to smaller and more isolated fragments, and population sizes of species have generally declined. These constrictions have limited the options available to natural systems to contend with the predicted rapid changes in climatic extremes or in the frequency and intensity of disturbances.

Impact of Increased Temperature

A recent study modeled the potential impacts of climate change on species extinction risks for sample regions across the globe and found that, under mid-range climate warming scenarios, between 15 percent and 37 percent will be headed for extinction by the year 2050 (United Nations Framework Convention on Climate Change, [UNFCCC], 2009). Such extinction rates are unprecedented in historical times. Natural coastal ecosystems in Sri Lanka increasingly are degrading in quality and quantity and are confined to smaller and more isolated fragments, and population sizes of species have generally declined. These constrictions have limited the options available to natural systems to contend with the predicted rapid changes in climatic extremes or in the frequency and intensity of disturbances.

For any change, most species appeared to shift their distributions as though tracking the changing climate, rather than staying stationary. Within a species, there may be significant variation among individuals in their tolerance to any change or disturbance. Most commonly, species may change their life history patterns. Unfortunately, however, sensitive species might not be able to withstand unfavorable conditions and adapt their life histories accordingly, and the ultimate result would be loss of species. Similarly, some species would show decrease in population size that again causes threats to biodiversity.

Reduced population sizes often result in diminished genetic variation, which could limit potential for local adaptation. The increased separation between natural habitat fragments diminishes the ability for successful dispersal, thereby hampering simple shifts in species' distributions. Increased fragmentation also lowers the probability of successful recolonization of disturbed areas after catastrophic events. This was evident in coral reefs on the southern coast after the tsunami in 2004.

Biodiversity responds to variation in climate. For example, the onset of growth in both plants and animals is frequently triggered by environmental conditions that exceed critical temperature or precipitation thresholds. Similarly, the timing of migration and breeding is often driven and influenced by temperature and precipitation patterns. This is especially true for some aquatic species, including fish. Consequently, if temperature regimes change, the timing of seasonal events may also change. In addition, extreme weather events such as floods and drought are no doubt harmful to species. Many coastal areas would be flooded due to extreme weather events such as storms and heavy rains that would in turn affect wetlands and related biota. Similarly, the sediment loads carried by the floodwater are quite harmful to the coal reefs that are sensitive to even a small change in quality of seawater.

Mangrove species have demonstrated different tolerances to changes in sea level, salinity, and storms. One important prediction is that mangrove communities are not much affected by the increase in atmospheric temperature. Nevertheless, deposition of silt and mud pose severe threats to mangroves and water quality in estuaries and lagoons. At sites that are projected to experience rising seas, mangrove ecosystems on low relief areas and those deprived of sediment are especially vulnerable. Mangrove ecosystems with ample sediment supplies and/or room to move inland are likely to survive projected rates of sea level rise. In contrast, mangroves have shown to be damaged with impaired functioning when exposed to heavy storms in certain cases.

Any water temperature changes can be very disturbing to the coral reefs. This was seen during the 1998 and 2004 El Niño weather phenomena, in which sea surface temperatures rose well above normal, bleaching or killing many coral reefs. High seas surface temperatures (SSTs), coupled with high irradiance (light intensity), trigger the loss of live corals, causing coral bleaching. Warming, which is reported to be the key cause of coral bleaching, may also be the basis of a new emerging problem: increasing coral diseases.

Ecosystem processes and functions are also affected to different degrees, due to variation in environment. For example, the breakdown of dead organic material in the nutrient cycling process hinges on the ability of bacteria to chemically degrade and assimilate the material. The rates at which these chemical changes occur are affected by temperature and water regimes. Climatic changes that modify these limiting factors can, in turn, accelerate or decelerate the rates at which the ecosystem processes occur.

Impact of Sea Level Rise

According to a recent study, one of the major losses associated with sea level rise in Sri Lanka, is loss of land resources due to (1) inundation and (2) increased coastal erosion resulting from increased sea level (Fernando, 2000). On the assumption that 30 m of beach retreat will occur in 2010 and considering a 285-kmr study area, it has been estimated that the total value of lost land in 1990 would be about US$28 million. Nianthi and Shaw (2006) report that there would be significant economic impacts due to climate change on the coast, which include agriculture, tourism, and industries (Nianthi & Shaw, 2006).

Rise in sea level causes saltwater intrusion to connecting water bodies such as estuaries and lagoons. This poses negative impacts on the lifestyle and distribution of aquatic species. Resulting population decline would cause severe impacts on coastal fisheries and island residents' livelihood. Changes in abundance are likely to be more pronounced in sensitive estuaries and lagoons. The rate of climate change may prove a major determinant of the abundance and distribution of new fish and crustacean populations. Rapid change due to physical forcing will usually favor production of smaller, low-priced, opportunistic fish species that discharge large numbers of eggs over long periods.

BUILDING RESILIENCE AND DEVELOPING ADAPTATIONS

It is clear now that, to reduce the risks of disasters related to climate change, ecosystems play a significant role; nevertheless, human and natural causes hamper their ability to perform this invaluable service. However, one of the most effective ways of securing healthy ecosystems is to minimize various stressors outlined elsewhere in this chapter. The most important factor that generates continued pressure on coastal resource is the common impression that it is a "common property resource." Although the resource serves common ecological functions and benefits the community at large, no individual takes responsibility, or is accountable, for its well-being.

It is clear that anthropogenic impacts on natural resources add to the adverse effects that climate change would cause (Shaw & Krishnamurthy, 2009). The issue we need to address is how we minimize the short-term and long-term impacts and costs from these changes. The answers lie in response measures that bring together effective conservation, and development concepts that consider intergraded approaches. This is how mitigation and adaptation measures could be linked together with other management aspects.

Adaptation is the action of responding to experienced or expected impacts of changing climatic conditions to reduce impacts or to take advantage of new circumstances (Ministry of Forestry and Environment [MFE], 2000). Adaptation is not about returning to some prior state, because all natural systems are evolving and dynamic. It has been clearly shown that building resilience, which involves increasing the ability of a system (social and ecological) to withstand shocks and surprises and to regenerate itself if damaged, represents a vision of a sustainable response. Some natural and social systems process enhanced abilities to bounce back from adverse circumstances, whereas others have to learn how to become resilient (Tompkins & Neil Adger, 2003).

In the context of adaptations to the impacts of climate change, the following objectives should be met:

a. Maintain and support healthy functioning of coastal habitats
b. Reduce risk of vulnerabilities to disasters
c. Promote good governance that addresses social, economic, and environmental issues
d. Support community-based ecosystem conservation and restoration agendas

Thus, in this chapter, suggestions are included with respect to integrated institutional, ecosystem, and community approaches for adaptation (see Fig. 3).

Institutional Approach

The main institute that has regulations over the management of coastal resources in the country is the Coast Conservation Department. Under the Coast Conservation Act and amendments, the Coast Conservation Department is responsible for overall management of living and nonliving coastal resources, permit issuance for any developmental activity in the coastal zone, implementation of a ban on coral mining, and formulation and implementation of coastal zone management plans. The Coast Conservation Department acts in collaboration with the other agencies, including the Ministry of Natural Resources, the Central Environmental Authority,

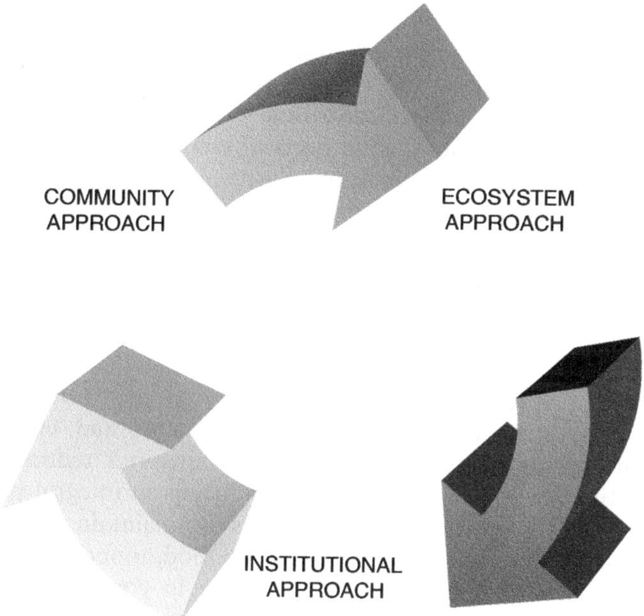

Fig. 3. Integrated Approaches of Adaptations to Climate Change for Disaster Risk Reduction.

the National Aquatic Resources Agency, and the Department of Wildlife Conservation. There are numerous policies, laws, action plans, and institutions involved in the management of coastal resources. Although more than 50 laws relate directly or indirectly to coastal resources, implementation has been sluggish.

Due to their complex nature, coastal resources cannot be managed merely by enacting laws and regulations. Awareness and participation of key stakeholders in their management are central to their sustainability. Ongoing initiatives such as the master plans, management plans, and policy statements are noteworthy. Contradictory and conflicting areas in the existing policy and regulatory enactments need to be examined, reviewed, and revised in light of climate change.

For long-term sustainable management of coastal resources, a strong research capability must be developed, research priorities identified and implemented, information gathered and analyzed, and databases developed, maintained, and upgraded. The capability of researchers and the availability of research facilities, both on the field and at respective institutions, need improvement. Links with national and international expertise should be sought to strengthen and promote research, especially with institutes of similar background in Asia and elsewhere. In certain universities, namely Universities of Colombo, Peradeniya, and Moratuwa, there are recently introduced research programs on impacts of climate change. Disaster risk reduction education is also gaining popularity. It is necessary to evolve an integrated, holistic approach to planning and plan implementation in interrelated sectors and interest groups.

Ecosystem Approach

As recognized in the Hyogo Framework for Action priority 4, "Reduce the Underlying Risk Factors," healthy ecosystems and environmental management are considered key actions in disaster risk reduction (DRR) (Sudmeier-Rieux & Ash, 2009). Ecosystems, being an integral part of any local environment, provide free-of-charge, easy-to-maintain natural barriers against disasters. To add to this, they provide food, timber, medicine, fuel, and other services that are of immense value in coping with disasters, especially for the well-being of the poor. An ecosystem approach to building capacity with immediate implications for conservation planners and managers is essential for reduction of risk in disasters. Ecosystem management approaches that emphasize ecosystem stability rather than control of the environmental disturbances should thus be promoted as a means of increasing ecological resilience. A number of actions could be taken to mitigate climate change impacts:

Risk assessment of different habitat sectors
A species' or habitat's level of endangerment and vulnerability may differ under a stable climate as compared with a changing climate. Risk assessments are of prime importance to evaluate the enhanced threats of climate change so that habitats that are likely to face more threats could be conserved accordingly.

Design new natural reserves to allow shifts in the distributions of target species
It is important to conserve the vanishing wealth of biodiversity by allocating more protected areas for their healthy survival. This allows controlled impacts of ever-increasing anthropogenic pressure. To ameliorate the disturbances caused by habitat fragmentations and to link existing habitats, possible links such as corridors could be created. Native corridors, such as

those along streams, waterways, and other minimally used land, could aid the redistribution of species between habitats.

Practice realistic conservation planning
Because the impacts of climate change are complex and evolving, conventional conservation planning is less effective. To address the changing nature of problems, flexible and dynamic plans are best suited but should be carried out with close monitoring and evaluation.

Alleviate the effects of other stressors
Climate change is occurring along with already existing human stressors. The fate of a species lies in the net effect of all stressors combined. Practically, it may be easiest to control the overall stress on an ecosystem by mitigating some of the non-climate stressors. For example, if both climate change and impacts of unsustainable developmental projects along the coastline are evident, it is more convenient to reduce the impacts of direct anthropogenic impacts such as pollution prevention.

Ex situ conservation
Certain species could be conserved outside of their normal habitat ranges. For example, turtle hatcheries could be operated on the coast to supplement declining populations. This is of importance because hatching turtle eggs are sensitive to variation in temperature, which might alter sex ratios. There are also reports of successful culture of corals in controlled conditions and replanting them in suitable coastal areas.

Community Approach

Adapting to climate change will entail adjustments and changes at every level – from community to national. Local coping strategies and traditional knowledge need to be used in synergy with government and local interventions.

During the last two decades, the importance of community participation in coastal resources conservation has gained much recognition. Participatory forest management is being increasingly considered in the protection of conservation areas, particularly in buffer zone development activities (see Fig. 4).

The question is then, how can communities enhance their networks of association, most importantly, their commitment of engagement?

Fig. 4. Coastal Communities and Estuaries: Livelihood at Risk? *Source:* Photo taken by Deepthi Wickramasinghe.

Local groups and individuals often feel their powerlessness in many ways, although none so much as in the lack of access to decision makers. In Sri Lanka, community-based adaptation has moved rapidly to the forefront of the climate change agenda. At present, there are several active community organizations in the coastal areas that are engaged in different activities, ranging from public awareness to planting mangroves. Various sectors in the stakeholder community including schoolchildren, youth groups, clergy, grassroots organizations, and specific groups such as fisheries communities funded by the government. International and local agencies are also actively involved in these activities.

Adopt Local Agendas

Climate change adaptations are more effective when driven by local people with a local agenda because the residents own the best knowledge and experience about a particular problem. For example, coastal communities residing in flood-prone areas very often possess know-how and skills for preparedness. Similarly, they are well aware about plants that can restore eroded riverbanks. On the northwestern coast of Sri Lanka, community organizations generate and maintain plant nurseries with trees that not only support ecosystem restoration but also provide local income at some stage. One such plant is bamboo, which protects riverbanks but also can be used in construction industry when fully grown.

Exploit Existing Capacities to Adapt

Disaster risk reduction approaches are more appropriate when they build on existing capacities of the group and allow ongoing group learning and adaptations. It should be understood that all systems are evolving, even the capacity of a community to adapt. For instance, when the domestic water supply or fisheries in an estuary are affected due to heavy pollution loads, the community should be made aware of the adverse impacts of their actions. Thus, they would gradually develop mitigation actions such as minimizing pollution inputs, reducing and reusing solid wastes, and so on.

Build New Capacities

Capacities could be built by disseminating knowledge, fostering training, and mobilizing action in any community, as long as they possess positive attitudes. Mobilizing the community to act on information and knowledge in support of adaptation requires that members have access to relevant information from credible sources. In this context, the government, non-government, and media sectors can play a key role. In certain cases, when the coastal communities were trained to culture brackish water fish to cope with diminishing lagoon fishery resources, new industry became more popular.

Promote Social Coherence

For actions to be effective and result-oriented, they should promote collective social integration and responsibility. This is very true for climate change adaptations also because the impacts experienced are local although the causes are global. Thus, high levels of social coherence should always be promoted and supported, which in turn would be beneficial in different programs ranging from management of natural resources to mangrove tree planting in destroyed habitats. For instance, a mangrove plantation aiming at lagoon restoration would generate benefits of several folds if all units of the community – children, youth, and adults – get together for a common goal.

CONCLUSIONS

It is clear that coastal ecosystems play a key role in maintaining ecological functions while providing economical, social, and cultural services to island nations. However, the damaging impacts of anthropogenic activities and climate change would cause severe stressors on coastal resources. The means of managing vulnerabilities involve general and specific mitigation measures to be adopted with effective institutional, ecosystem, and community approaches.

ACKNOWLEDGMENTS

The author acknowledges Professor Rajib Shaw of Kyoto University, Japan, for his valuable support and guidance. Special thanks also go to Chandana Seneviratne, Saminda Fernando, and Pubudu Wewalwala for assisting in information gathering.

REFERENCES

Amarasinghe, M. D., & Pahalawattarachchi, V. (1996). Mangals, salt marshes and sea grass beds on the northwestern coast of Sri Lanka. In: *Proceedings of the regional workshop on "Research for the Management of Coastal Resources in the Tropics."* Sri Lanka.

Anonymous – World Wildlife Fund. (2005). Mangroves shielded communities against tsunami. *ScienceDaily*, October 28. Available at http://www.worldwildlife.org. Retrieved on October 2010.

Choudhruy, J. K. (1997). Sustainable management of coastal mangrove forest development and social needs. In: *Proceedings of XI world forestry congress*, Turkey.

Coast Conservation Act of Sri Lanka (CCD). (1981). Government of Sri Lanka, No. 57. Colombo, Sri Lanka.

Coast Conservation Department. (CCD). (1986). *Coastal zone management plan*. Colombo, Sri Lanka: Ministry of Fisheries and Aquatic Resources.

Coast Conservation Department. (CCD). (1986). *Master plan for coastal erosion management*. Colombo, Sri Lanka: Ministry of Fisheries and Aquatic Resources.

Coast Conservation Department. (CCD). (1997). *Revised coastal zone management plan*. Colombo, Sri Lanka: Ministry of Fisheries and Aquatic Resources.

Conservation International. (2008). *Economic values of coral reefs, mangroves, and seagrasses: A global compilation*. Arlington, VA: Center for Applied Biodiversity Science, Conservation International.

Convention on Biological Diversity (CBD). (2009). Available at www.cbd.int. Accessed on March 2010.

Dahdouh-Guebas, F., Jayatissa, L. P., Di Nitto, D., Bosire, D., Lo Seen, D., & Koedam, N. (2005). How effective were mangroves as a defense against the recent tsunami? *Current Biology, 15*(14), 1337–1338.

De Alwis, P., Ariyaratne, S., Azmy, S. A. M., & Dassanayake, N. H. (1995). Environmental pollution and its impact on fishery management in Sri Lanka. In *Proceeding of Sri Lanka/FAO national workshop on development of community based fishery Management*. Colombo, Sri Lanka.

Fernando, H. J. S., Samarawickrama, S. P., Balasubramanian, S., Hettiarachchi, S. S. L., & Voropayev, S. (2008). *Journal of Hydro-Environment Research, 1*(3–4), 187–194.

Fernando, T. K. (2000). *Key vulnerabilities, information needs and capacity needs*. Sri Lanka: Ministry of Environment & Natural Resources.

Gunaratne, L. H. P., & Jayasooriya, K. A. S. P. (2006). *River sand mining in Sri Lanka: Trade-off between private profitability and environmental costs*. Sri Lanka: University of Peradeniya.

Information Brief on Mangroves in Sri Lanka (IUCN). (2007). *International Union for Conservation of Nature*. Sri Lanka: IUCN.

Kotagama, S. W., & Bambaradeniya, C. N. (2006). *An overview of the wetlands of Sri Lanka and their conservation significance*. IUCN Sri Lanka and the Central Environmental Authority. National Wetland Directory of Sri Lanka. Colombo, Sri Lanka: International Union for Conservation of Nature.

Kunkel, C. M., Hallberg, R. W., & Oppenheimer, M. (2006). Coral reefs reduce tsunami impact in model simulations. *Geophysical Research Letters, 33*.

McLeod, E., & Salm, R. V. (2006). *Managing mangroves for resilience to climate change* (p. 64). Gland, Switzerland: IUCN.

Ministry of Forestry and Environment. (2000). *Initial national communication under the United Nations framework convention on climate change*. Sri Lanka: Ministry of Forestry and Environment.

Miththapala, S. (2008). Mangroves. Coastal Ecosystems Series, Vol. 2, pp. 1–28, iii. Colombo, Sri Lanka: Ecosystems and Livelihoods Group Asia, IUCN.

Nayanananda, O. K. (2007). *The study of economic significance of the coastal region of Sri Lanka in the context of environmental changes of pre and post tsunami*. Sri Lanka: Coast Conservation Department and The Ministry of Environment and Natural Resources.

Nianthi, K. W. G., & Shaw, R. (2006). *Climate change and its impact on coastal economy of Sri Lanka: The global challenge*. India: Research Publishing.

Othman, M. A. (1994). Value of mangroves in coastal protection. *Hydrobiologia, 285*, 277–282.

Pahalawattaarachchi, V., Jayamanne, S. C., & Amarasinghe, M. D. (2005). Changes in seagrass ecosystems in Negombo Lagoon with special reference to past development activities. In: *Proceedings of the international workshop on fisheries and aquatic resources*, Sri Lanka.

Pathirana, K. P. P., Kamal, A. R. I., Riyas, A. I., & Safeek, A. L. M. (2008). *Management of coastal resources in Puttalam Lagoon* (Dubai, UAE). Sri Lanka: COPEDEC VII.

Rajasooriya, A. (2005). *Coral reefs of Sri Lanka: Current status and resource management. National Aquatic Resources*. Sri Lanka: Research and Development Agency.

Samarakoon, J. I. (1986). Fisheries and aquaculture in estuaries, lagoons and mangroves in the context of coastal zone management in Sri Lanka. In: *Proceedings of workshops on strategies for the management of fisheries and aquaculture in mangrove ecosystems*. Thailand.

Samaranayake R. A. D. B. (2000). Sri Lanka's agenda for coastal zone management. *The review of advanced technologies for the integrated management of EEZs and coastal zones worldwide*. ICG Publishing Limited, London.

Shaw, R., & Krishnamurthy, R. R. (2009). *Disaster management: Global challenges and local solutions*. India: Universities Press.

Sudmeier-Rieux, K., & Ash, N. (2009). *Environmental guidance note for disaster risk reduction: Healthy ecosystems for human security*, Rev. ed. Gland, Switzerland: IUCN, iii + 34pp.

Tompkins, E. L., & Neil Adger W. (2003). *Building resilience to climate change through adaptive management of natural resources*. Working Paper no. 27. Tyndall Centre. Available at www.tyndall.ac.uk. Accessed on March 2010.

United Nations Framework Convention on Climate Change (UNFCCC). (2009). http://www.unfccc.int. Accessed on March 2010.

CHAPTER 16

THE EXPERIENCES OF DRR THROUGH CCA IN TAIWAN

Jet-Chau Wen, Shao-Yang Huang,
Chia-Chen Hsu and Kou-Chiang Chang

INTRODUCTION

Taiwan is located between the world's largest landmass, the continent of Asia, and its largest ocean, the Pacific Ocean. The Tropic of Cancer passes through the island of Taiwan, giving it a subtropical and tropical oceanic climate. High temperatures and rainfall and strong winds characterize the climate. Because of Taiwan's position in the Asian monsoon region, its climate is greatly influenced by monsoons as well as by its own complicated topography. The annual mean temperatures in the lowlands are 22–25°C, and the monthly mean temperature exceeds 20°C for eight months starting with April each year. The period from June to August is the hottest season with mean temperatures of 27–29°C. Temperatures are cooler between November and March; in most places, the coldest monthly mean temperature is above 15°C. The climate is mild rather than cold and temperatures only fall dramatically when a cold front affects the region. Average annual rainfall in the lowlands of Taiwan is in the range of 1,600–2,500 mm. Due to the influences of topography and the monsoon climate, the rainfall differs greatly with different areas and seasons. In mountainous areas, average rainfall may exceed 4,000 mm/yr. Rainfall is generally higher in mountainous areas than in lowland areas, higher in the east than in the

Climate Change Adaptation and Disaster Risk Reduction: An Asian Perspective
Community, Environment and Disaster Risk Management, Volume 5, 327–347

ISSN: 2040-7262/doi:10.1108/S2040-7262(2010)0000005022

west, and higher on windward slopes than on the leeward side. The northeast monsoon prevails during the winter; this is the rainy season in the north though rainfall is not intense. But the same winter period is the dry season in the south. During the summer, the southwest monsoon prevails, often giving rise to convective thunderstorms and bringing intense and copious rainfall. With added downpours brought by typhoons, this season often accounts for over 50% of annual rainfall in the south so that central and southern regions often suffer greatly. Relative humidity on the island of Taiwan, surrounded by ocean, is high, usually measuring in the range of 78–85%. In the north, relative humidity is higher during winter than during summer. The situation in the south is the opposite. Over the past 100 years, the rainfall in the north has increased, while the rainfall in the south has decreased. The trend is not as consistent as that of the temperature change (Environmental Protection Administration, Executive Yuan, R.O.C. (Taiwan), 2002).

Because of the characteristics of the location, climate, and geology of Taiwan, there are seven kinds of disasters that have frequently occurred in the past. These are land erosion, earthquakes, scarcity of water resources, environmental change, landslides, land subsidence, and floods. Most of these have annually caused the loss of human life and cost the nation great expense. These disasters then affect the advancement of society by requiring restructure of governmental policies and efforts in Taiwan. These natural disasters are a constant threat to Taiwan.

IMPACTS OF CLIMATE CHANGE ON TAIWAN

People's activities have enlarged the emission of greenhouse gases since the Industrial Revolution, and in turn, the greenhouse effect has resulted into global climate change. According to the Intergovernmental Panel on Climate Change (IPCC), the concentration level of CO_2 in the atmosphere would at least double since the Industrial Revolution (280 ppm). The global average temperature from 1990 to 2100 may rise from 1.4 to 5.8°C, while the average temperature of the last 100 years has risen to 1.1°C in Taiwan (Hsu & Chen, 2002).

Climate change directly affects hydrology, water resources, agricultural production and agricultural water demands, ecosystems, and so on. With the natural environment of a subtropical island, Taiwan is very vulnerable to the impacts of climate change including sea level rise, damages of primary industries, public health, among others. The direct impacts of sea level rise cause the flooding of coastal lands, coastal erosion, and coastline recession. Coastal communities face the possible problems of relocation and subsequent social adaptation. The impacts on water resources include probable increase in frequency and intensity of droughts, lack of water resources, and impact on people's livelihood and industrial development. On the other hand, the increase in CO_2 concentration would enhance photosynthesis of plants and promote forest and agricultural growth. However, it would also promote the growth of pests and propagate diseases. Concerning the impact on the ecoenvironment, the long-term trend shows that some negative impacts would occur like reduction of inhabited areas of biological species and reduction or vanishment of biological groups (Environmental Protection Administration, Executive Yuan, R.O.C. (Taiwan), 2002).

Based on the studies of the Global Change Research Center at the National Taiwan University, in the last 100 years (1900–2000), the temperature of Taiwan increased 1.6°C/yr, which is about double of that of the northern hemisphere, and the sea level rising rate of the last 50 years is about 2.51 cm/yr, which is about 1.4 times of the global average rising rate. According to IPCC, the sea level will rise 1 m by the end of this century; at that time, 10% coastal or low-lying areas of Taiwan will be flooded.

Based on the studies of the Research Center of Environmental Change in Academia Sinica, winter and spring are the dry seasons and summer is the rainy season in Taiwan. Such weather pattern influences agricultural development and results in the lack of food and the insufficiency of storage water and, moreover, enlarges the frequencies of floods and droughts. In the most recent years, the intensity of abnormal climate has increased and the frequency is higher. According to the Institute of Earth Science in Academia Sinica, there were three typhoons with extreme rainfalls in Taiwan in August 1994, ten in October 1998, and one in September 2001. The interval during this period is about three to four years. But there were three typhoons in August 2007, two in September 2008, and even Typhoon Morakot, an extremely strong typhoon, appeared in August 2009. The interval currently has shortened to one to two years. Such situation shows that the interval of extreme weather has shortened, but the intensity has become stronger. If the temperature rises by 2°C by the end of this century, heavy rainfall in the top 10% scale will double. If energy conservation cannot be implemented, the temperature will rise by 4°C, and the heavy rainfall intensity ranked at the top 10% will increase 5.6 times. Typhoons will bring extreme rainfall that will cause debris flows; therefore, the environmental risks in mountain and coastal areas will eliminate available space to live and threatened people's livelihood and safety. The impacts of climate change in different sectors of

Table 1. The Impacts of Climate Change in Different Sectors of Taiwan.

Sectors	The Impacts of Climate Change
Sea level rise	Sea level rise may seriously influence the use of coastal lands. The major economic impacts resulting from damages regarding social cost includes some direct damages involving land and capital loss, cost of mitigation measures to reduce damages, and loss of structure and development opportunities near the coastline.
Water resources	The uneven rainfall of rainy and dry seasons may increase the extent of floods and droughts and increase the difficulties of disaster prevention. Related problems that may occur during drought is difficulties of allocating water resources and during the rainy season is insufficient capacity of storage structures to store excess rainfall, thus creating insufficient water supply during droughts.
Industries	The climate conditions of rising temperature and uneven rainfall may reduce crop production. The El Niño–Southern Oscillation affects the growth and decline of the Kuroshio main current and its branching process and reduces fishery production. The change of water temperature also changes the geological distribution of fish. The rise of temperature may also influence livestock and poultry propagation and growth.
Public health	The rapid temperature increase may result in an increasing death rate and in propagating the disease vector more easily, thus inducing infectious diseases. Dengue fever in Taiwan has spread to northern areas and the propagation of rats, the host of hantavirus, has also rapidly increased.
Ecological systems	Climate change may lower the production ability of aquatic life forms, reduce biodiversity, and influence oceanic temperature and current flow to affect the ecology of fish and other organisms.

Source: Environmental Protection Administration, Executive Yuan, R.O.C. (Taiwan) (2002).

Taiwan can be simplified in Table 1. The sectors that are discussed include sea level rise, water resources, industries, public health, and ecological systems.

International studies have indicated that climate change might change the available water demand and rising temperature might result in the increase of evapotranspiration (Tung & Haith, 1995), while domestic studies have indicated that climate change causes an extreme difference between rainy and dry seasons (Tung & Li, 2001). Therefore in order to understand the impacts of climate change on hydrology and the climate of Taiwan, this chapter analyzes some possible factors from climate change impacts and provides domestic researches in recent years that support the analysis of such factors.

Sea Level

Based on the long-term sea level variation data of 14 domestic tidal monitoring stations, the sea levels of Keelung and Kaohsiung have a rising trend in the most recent 90 years at the rising rate of 0.035 cm/yr in Keelung and 0.061 cm/yr in Kaohsiung, respectively. However, Taichung Port has the descending rate of 0.364 cm/yr. Overall, data from the tidal stations indicate that sea levels in parts of the northeast, northwest, and south have an increasing trend; sea levels in central Taiwan have a decreasing trend (Environmental Protection Administration, Executive Yuan, R.O.C. (Taiwan), 2002).

An analysis of long-term changes at the coastal lines of Taiwan shows that coastal lines in the north currently remains stable overall, but erosion is more serious in some parts. In the past 20 years, coastlines in the southeast have retreated by 20–50 m. Furthermore, coasts in central and southern Taiwan show signs of erosion in recent years and a decrease in the shorelines (Environmental Protection Administration, Executive Yuan, R.O.C. (Taiwan), 2002).

Rainfall

Based on the climate data of the last few years, the average annual rainfall in Taiwan is 2,500 mm, which is about 2.6 times of the world average value. However, 78% of the rainfall occurred during the rainy season (from May to October), and most of the rainfalls are from typhoons; on the other hand, only a small amount of rainfall occurred during the dry season (from November to April). Therefore, the hydrology and water resources conditions are full of much uncertainty. The average annual rainfall has changed little, but the number of rainfall days is continuously decreasing and the rainfall intensity is persistently increasing. According to rainfall data, the rainfall intensity of central, southern, and eastern Taiwan has an increasing trend, and it shows that the opportunity for heavy rainfall has apparently increased also (Lee, 2009).

The study of the Research Center of Environmental Change in Academia Sinica discusses the impact of climate change on the rainfall characteristics of Taiwan. The research divides Taiwan into northern, southwestern, and southeastern parts and has calculated the rainfall from 1941 to 2005. It indicates that the rainfall intensity, the ratio of annual rainfall amounts to the number of rainfall days, is 16.2 mm/day in Taiwan. Statistically, the individual regions indicate 16.4 mm/day in northern Taiwan, while southwestern Taiwan has the largest value of 17.7 mm/day, and southeastern Taiwan has the smallest value of 13.1 mm/day. These districts all have an increasing rainfall intensity trend. The study shows that the changes of rainfall patterns are different for northern and southern Taiwan. In recent years, northern Taiwan received more rainfall in fewer days, and this phenomenon may result in creating less reliable surface water resources due to the limited capacity of storage facilities. The trends of rainfall amount and the number of rainfall days are decreasing in southwestern Taiwan, but rainfall intensity has slightly increased. According to the rainfall data, southern Taiwan receives less rainfall amounts with comparatively more uneven seasonal distribution than northern Taiwan, and this is unfavorable for water resources (Hsu, Wang, Chen, Chen, & Ma, 2007).

Flow Rate

Based on past data, the river annual average runoff did not have any apparent increasing or decreasing trend in Taiwan. In northern Taiwan, when the northeastern monsoon prevails, approximately from October to January, the river runoff had a decreasing trend. But the river runoff increased at the onset period of the plum rainy season, from April to May, especially in central Taiwan (Juang & Chen, 1994). There were nine rivers impacted due to climate change, and the flow rates had an apparent decreasing trend in winter and spring and had an increasing trend in summer, but not so apparent in fall (National Science Council, 2005a).

There are researches that have simulated the impacts of climate change on the watershed flow rates in Taiwan. One research indicates the watershed flow rates will decrease as time increases, and the decreasing range is more apparent in northern Taiwan. Overall, the watershed flow rates will have a decreasing trend (National Science Council, 2008). Another research indicates that in the future Taiwan will be affected by climate change so much so that the flow rate will increase in the rainy season and will decrease in the dry season. This result means that the frequencies of floods and droughts may increase (National Science Council, 2005b).

POSSIBLE DISASTERS AS A RESULT OF CLIMATE CHANGE

Global warming will increase the number and severity of extreme events such as storms, floods, droughts, typhoons and related landslides, fires, and so on. In the last 100 years, Taiwan, in different locations and seasons, has had different warming trend variations and different abnormal climate conditions like storms, typhoons, droughts, and sandstorms. The occurrence frequencies and times of these conditions in Taiwan also have changed. More than 70% of the land in Taiwan is mountains with an average elevation of 600 m, so that the rivers in Taiwan have characteristics of steep slopes and rapid flows. As large-scale land development and urbanization have begun in recent years, the ability to contain water has worsened. Therefore the abnormal climate changes like storms and typhoons also have weakened the resistance of mountain soil and have caused severe damages due to debris flows and landslides. Besides the damages of basic livelihood structures and threats of the people's safety and property, abnormal climate changes would therefore result in subsequent effects on the ecosystem and environmental health (National Science Council, 2005b).

The annual rainfall amounts differ greatly during rainy seasons and dry seasons; therefore, abnormal rainfall in rainy seasons usually brings flood disasters. The floods and droughts that have occurred from 2003 to 2008 have indicated that the water resources supply and demand are very unstable in Taiwan due to climate change, such as Typhoon Nari in 2001 and the severe drought in 2002 (National Science Council, 2005a).

The surface water resource use in Taiwan is inefficient because precipitation differs considerably during rainy and dry seasons and rivers in Taiwan are steep and short in length. The lack of surface water resource is more significant because there are not enough water storage structures such as reservoirs in southwestern Taiwan. In addition, agriculture and fishery are the major economic productions that consume great quantities of water. If surface water supply is insufficient, groundwater will become the primary water resource and the overpumping of groundwater will cause serious land subsidence. Because of the low elevation in coastal areas, flood control and drainage systems lose their preventable potency in land subsidence regions. For this reason, serious flooding occurs during typhoons or torrential rains. Furthermore, when the Chi-Chi earthquake shook the entire island of Taiwan, it also loosened soil and rock on the hillsides in 1999. After that, the colluviums piled up in most valleys and have resulted in serious debris flows or landslides during heavy rains.

As a result of climate change, the intensity of typhoons has increased. The common typhoon paths in Taiwan are shown in Fig. 1. From 1897 to 2003, the paths statistically experienced came from nine different directions with the fifth and sixth directions being the predominant pattern with the fifth path going northwest off of the southern coast of Taiwan and the sixth

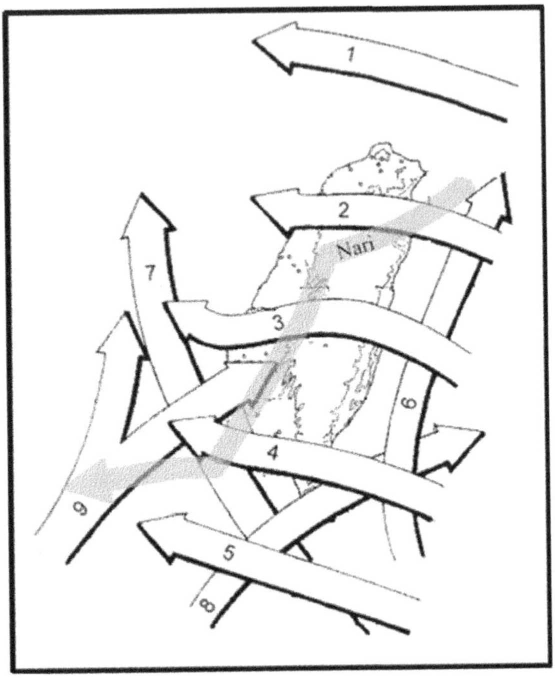

Each typhoon path:

(1) Nari in 2001: special zigzag path shown on the map

(2) Rammasun in 2002: did not directly invade Taiwan (path type is similar to 6, but very far away from Taiwan)

(3) Aere in 2004: **1**

(4) Mindulle in 2004: **6**

(5) Morakot in 2008: **3**

Fig. 1. Common Typhoon Paths in Taiwan (100 Questions about Typhoons, Central Weather Bureau, 2004, Modified).

path going northeast off the eastern coast of Taiwan. Paths of several specific typhoons mentioned in this chapter are also illustrated in this figure.

The following listed items are disasters highly relative to Taiwan as a result of climate change with supportive data.

Coastal Disasters

Damages from sea water level rise affect people's livelihood and coastal ecosystems; such disasters include sea water intrusion, tidal surges, coastal area floods, and so on. More specifically, there are some townships in Taiwan classified as severe land subsidence areas, and the effect of the rise of sea water level has greatly impacted these areas due to land subsidence. The primary damage is caused by tidal surges that are distributed mainly in low-lying lands of southwestern Taiwan. In recent years, flooding has occurred in these areas due to typhoons or heavy rainfalls. Furthermore, these areas have severe land subsidence, and the intensity of flooding may become even more severe because of the combined effects of seawater rise and land subsidence.

In recent years in Yunlin County, for example, the range of coastal areas being flooded due to heavy rainfalls or typhoons correlate with the land subsidence areas (Fig. 2). It presumes that land subsidence may be one of the factors contributing to disasters, such as the heavy rainfalls that occurred from 2004 to 2007. On July 2, 2004 heavy rainfall resulting from typhoon Mindulle (Fig. 1) flooded the western coastal areas of Yunlin County in Kouhu, Tuku, Taisi, Dabi, Yuanchang, Douliou, Shuelin, and so on. The following year on June 12, 2005, heavy rainfall with a daily precipitation amount of more than 350 mm occurred in southern Taiwan, which resulted from the plum rainy front and southwestern airflow. This rainfall caused flooding in these areas and threatened people's safety and damaged property. Then on June 9, 2006, the Central Weather Bureau continually announced torrential rainfall warnings posed by a stationary front. In Yunlin County, the persistent rainfall also resulted in severe flooding in low-lying areas of Taisi, Kouhu, and Sihhu. Severe flooding took place again on August 19, 2007 when Yunlin County received large amounts of rainfall that severely flooded the coastal areas of Lunbei, Mailiao, Taisi, Sihhu, Kouhu, and Sheilin, among others. In these areas, some classes were suspended due to flooding. According to the rainfall statistics of the Central Weather Bureau, the rainfall amounts were between 160 and 300 mm in Sihhu, Mailiao, Baojhong, Beigang, Gukeng, and Douliou.

Flood Disasters and Droughts

According to some researches (National Science Council, 2005b), the rainfall intensity in recent years (1941–2005) has an increasing trend in Taiwan. Therefore flood disasters will increasingly cause more disasters. In addition, the water resource supply in Taiwan mainly comes from stationary rainfalls (plum rains) from May to June and tropical cyclones (typhoons) from July to October, while the rainfall amounts during the other months amount to only 20–30% of the annual rainfall. Therefore, if plum rains and typhoons bring very little rainfall amounts, drought may occur in Taiwan. Based on the data of the monthly and annual average rainfall from 1953 to 1990, the annual rainfall in northern and eastern Taiwan had increasing

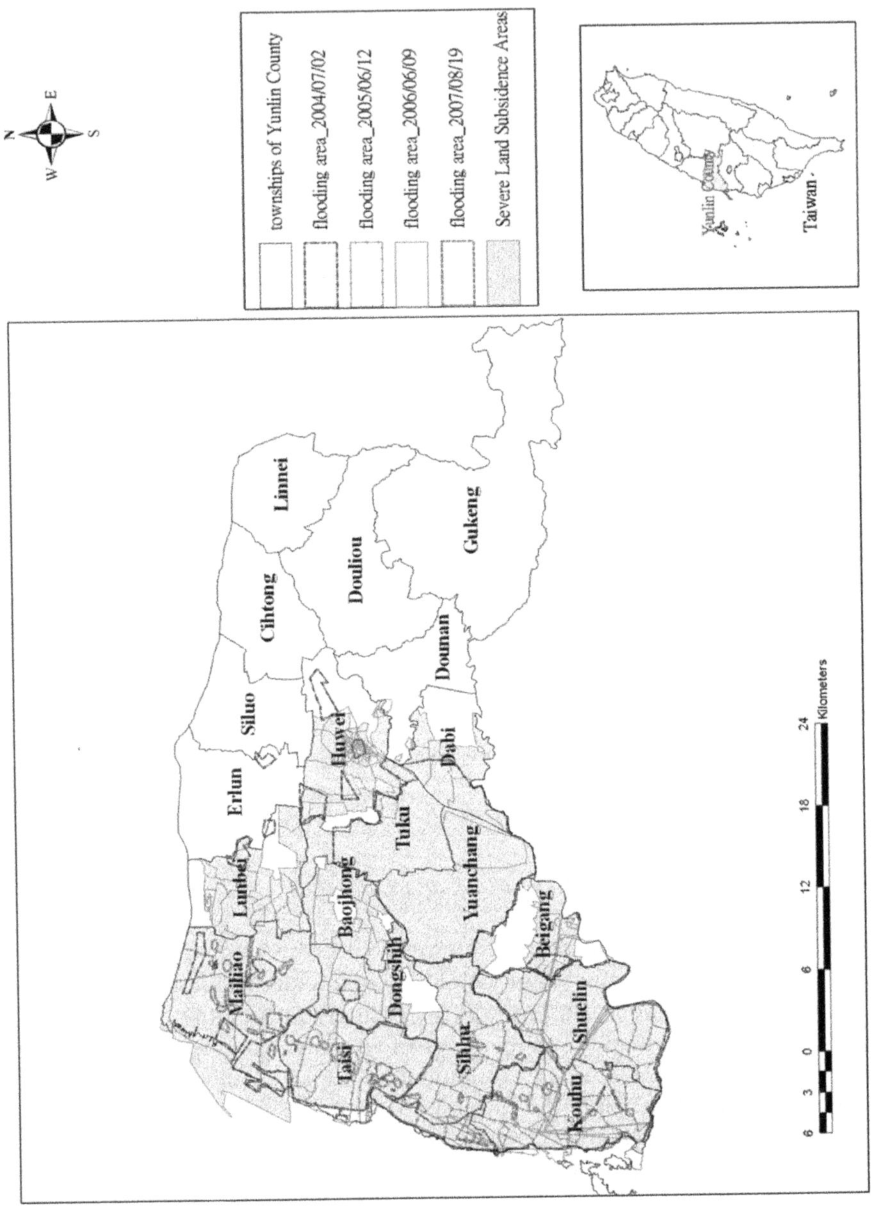

Fig. 2. Correlation of Previous Flood Ranges of Yunlin County with Severe Land Subsidence Areas.

trends, but central and southern Taiwan had decreasing trends. In addition, days lacking rainfall in southern Taiwan had increased.

The Shihmen reservoir water supply system, for example, had observed changes from floods and droughts that occurred in Taiwan during the years 2002 and 2004 (National Science Council, 2005b). The drought, which occurred in northern Taiwan in 2002, had an accumulated rainfall amount of only 216.66 mm in the Shihmen Reservoir Watershed from November 2001 to April 2002, which was the most minimal amount during the last 35 years. Due to insufficient amount of rainfall in the Shihmen Reservoir Watershed, a water shortage warning was announced, and the related governmental organizations had meetings to review the irrigation water for the fist crops and to discuss the related responses needed to face water shortages. By March 2002, the water level of the Shihmen Reservoir descended to a serious limited drawdown of the rule curve (water level was 220 m). And in April 2002, the very small rainfall amounts were, in general, everywhere in Taiwan; meanwhile, water shortage warnings were announced, respectively, in Miaoli, Taichung, Matsu, and so on. The following year in May 2002, the water level of Shihmen Reservoir went down to 204 m, which was less than the serious limited drawdown. During the periods from May to June, the government carried out stricter water limitation measurements according to the water shortage situation. In addition, they dug wells and did pumping tests in response to the situation after the reservoir dried up from the overpumping of groundwater. Finally in July 2002, relief came when the medium-size typhoon Rammasun distantly passed in a northwesterly direction along eastern Taiwan and brought abundant rainfall. The drought in northern Taiwan finally came to an end.

Another water shortage occurred in Taoyuan in 2004 when the water treatment plants were shut down as a result of typhoon Aere (Fig. 1), which attacked northern Taiwan from August 23 to 26 in 2004 and brought abundant rainfall that averaged 973 mm, which was about 50% of the annual rainfall. The storm occurred mainly on August 24–25 with the rainfall intensity of 88 mm/h, peak flow of 86,000 cm, and a total flow of 700 million tons, which was about three times of the reservoir effective storage capacity. Such heavy rainfall that precipitated into the watershed caused soil erosion and landslides of 295 ha. In addition, because of the erosion of the river bank of Tahan Creek, about 20 million cubic meters of sand deposition were estimated, which is about 14 times of the reservoir annual deposition (1.4 million cubic meters) flowing into the reservoir. Therefore turbidity of the water in the reservoir quickly increased from 70,000 to 1.2 million NTUs, which was far from the treatable value (5,000 NTUs) set by the

Taiwan Water Corporation. As a result, the Taiwan Water Corporation had to shut down all water treatment plants that supplied water to Taoyuan on August 26 and stopped water supply from August 26 until September 8. This measure affected about 1.45 million families in Taoyuan and Taipei. A comprehensive water supply started up after September 8.

Rainfall-Induced (or Climate-Induced) Disasters

These kinds of disaster include debris flows, landslides, land subsidence, active faults, and so on, and the debris flow is the most frequent disaster resulting from typhoons or heavy rainfalls. According to the data of the Soil and Water Conservation Bureau, there are 1,503 potential streams of debris flows in Taiwan, and Taipei County and Nantou County possess the largest quantities. The heavy rainfall brought by typhoons is the major inducing factor causing debris flows. Based on the last rainfall characteristics of debris flows, most of them resulted from typhoons; except in some regions, the debris flows were caused by normal rainfall. The statistical data indicated that there were six typhoons on average that attacked Taiwan from 1991 to 2008. But the annual number of typhoons was greater than this value during 2003 and 2004 as a result of nine typhoons occurring for each year of 2003 and 2004, respectively. The number of typhoons that would result in a debris flow is 2.5; that is, about 40% of typhoons may induce debris flows annually.

More recently, typhoon Morakot (Fig. 1) attacked Taiwan on August 7, 2009 and brought severe damages to Taiwan, especially in southern and southeastern Taiwan with landslides and debris flows (Figs. 3 and 4). The Central Weather Bureau predicted rainfall amounts were about 2,500 mm in Pingtung, which were equal to the annual rainfall normally precipitated in three days. Such huge amounts of rainfall caused the roads and buildings to be destroyed in great numbers in the townships. Regarding the populace, death, and injuries were most severe in the mountainous villages of Kaohsiung. In addition, Tainan and Nantou received severe damages as well. The images of house destructions and traffic disruptions caused by landslides and debris flows were displayed through the media and became a high point of focus. But before typhoon Morakot attacked Taiwan, the government was contemplating on how to solve the problem of drought. No one imagined that the rainfall brought by Morakot would not only resolve the drought but also turn into a flood. Such extreme change warns the people about the power of nature. With climate change,

Fig. 3. Landslide Occurred in Chayi County During Typhoon Morakot (Photograph Taken by Tsai Mu-Fan).

we predict that the number of strong typhoons may increase in the future, and we have to take adequate disaster response measures to protect people's lives and property to prevent such regrets from occurring again.

ADAPTATION STRATEGIES AND DISASTER PREVENTION INITIATIVES

In order to protect the global environment and to avoid adverse impacts from international politics and trade, the Republic of China (Taiwan), though not a member of the United Nations, has always complied with international environmental treaties and acted as a responsible member of the global world. In May 1992, the Executive Yuan established the ministerial-level interdepartmental agency the Global Change Working Group to coordinate activities related to the United Nations Framework Convention on Climate Change (UNFCCC) and other global

Fig. 4. Debris Flow Occurred in Chayi County During Typhoon Morakot (Photograph Taken by Tsai Mu-Fan).

environmental issues. In August 1994, the working group was further reorganized and elevated as the Committee on Global Change Policy (CGCP), under which the UNFCCC Working Group was established. Due to the growing importance of sustainable development in the countries around the world, the government decided to integrate all domestic matters related to sustainable development. In August 1997, CGCP was expanded to form the National Council for Sustainable Development (hereinafter referred to as the "Council"), while the Atmospheric Protection and Energy Working Group under the Council was responsible for affairs related to the Montreal Protocol and the UNFCCC. In 1999, the Council was further elevated and chaired by the Vice Premier, thus showing our commitment in response to the UNFCCC (Environmental Protection Administration, Executive Yuan, R.O.C. (Taiwan), 2002).

To address the impacts of climate changes on Taiwan, in addition to related studies on greenhouse gases, adaptation strategies are being considered to lessen the levels of impact from climate change. Considering the results of related studies and policies, the relevant government agencies have devised five adaptation strategies in regard to sea level rise, water resources, industries, public health, and ecological systems. Two of those, sea level rise and water resources, include disaster prevention strategies. The essentials of these two strategies are as follows (Council for Economic Planning and Development, 2007; Environmental Protection Administration, Executive Yuan, R.O.C. (Taiwan), 2002).

Sea Level Rise

Like other islands, Taiwan will increasingly face serious coastal erosion. As precautionary measures, the following adaptation strategies will be taken:

(1) Complete an impact assessment of sea level rise of the entire Taiwan region: Delimit potentially flooded areas, affected population, and socioeconomic impacts according to the local terrain, nearby industries, population, and culture as useful references for preparing appropriate strategies and prevention measures.
(2) Protect coastal zone and establish a new tidal wave prevention and drainage system: Establish and draw up new design standards for coastal embankment, tide, and flood prevention as well as drainage systems; draft new prevention methods and carry out flood modeling analysis in order to respond to the future conditions of protecting life and property in the coastal regions.
(3) Reduce impacts from sea level rise: Establish land use restriction in coastal areas; control building in low-lying areas; establish industry transformation assistance, wetland protection, flood protection, and sand dune protection.
(4) Preserve coastal wetlands and ecosystems: Using transplantation, cofferdams, or other artificial means to protect rare coastal ecosystems.
(5) Control large-scale coastal developments: Reevaluate plans for coastal industrial parks and integrate sea level rise as an element of environmental assessment of large coastal projects.
(6) Obtain experiences from international cooperation: Actively participate in international cooperation projects.
(7) Establish transitional assistance and planning for nonrecoverable industries: For flooded areas not able to be recovered by preventive strategies, assist the residents in relocation and job changes in order to minimize socioeconomic impacts.

(8) Define flood plain areas and implement flood insurance systems: Establish different levels of flood areas, prevent reckless development, and implement a flood insurance system.
(9) Establish monitoring systems: Including a sea level monitoring system, a coast and coastal structure monitoring system, expand on the subsidence monitoring system, and enhance the establishment of the coastal database.

Water Resources

In order to reduce the level of impacts of climate change on water resources in Taiwan, suitable adaptation strategies are needed in three major areas: water resource conservation, development, use, and management; strengthening flood control and rescue; and drought prevention and emergency response. Below are listed the strategies for each.

(1) Water resources conservation, development, use, and management
   (a) Develop surface water as the primary objective by giving consideration to the base flow of the river ecology, and development of groundwater as a supplementary source with increased development of other water resources.
   (b) Actively develop and allocate water resources.
   (c) Establish management of water rights and collect water right tariffs, and develop a standard function for water measurement equipment in order to put into practice the user pays principle.
   (d) Strengthen water utilization management, raise water utilization efficiency, and allocate waster resources.
   (e) Reinforce the management and conservation of watersheds, conserve water sources, and maintain the natural ecological function of water systems.
   (f) Promote modernization of hydrological monitoring, develop a groundwater observation network, and establish the principles for groundwater use and protection.
   (g) Promote research and development of water resource technologies in order to meet the technological needs for water supply, water conservation, diversion dams, flood control, coastal protection, groundwater conservation, and subsidence prevention.
(2) Strengthen flood prevention and rescue
   (a) Construct flood prevention structures: Construct river and sea embankment and regional drainage projects.

(b) Strengthen maintenance management: Reinforce river management plans, establish a safety inspection system for water structures, and integrate protection plans for small and medium dams.
(c) Establish a disaster prevention and response mechanism.
(d) Perform research in disaster prevention technology: Focus research on flood prevention and rescue databases, typhoon disaster potential analysis, and a flood forecast network.
(e) Establish nonengineering flood control measures: Strengthen the review of land use planning for flood plains, promote a flood insurance system, and establish a flood warning system.
(f) Integrate flood control measures for reservoirs: Strengthen water conservation, increase surface permeation, decrease flood peak flow, and take into consideration ecological protection and the environmental landscape.

(3) Drought prevention and emergency response
(a) Establish drought prevention and response systems.
(b) Promote related work on prevention and response: Enhance reservoir protection and groundwater control and monitoring, research water conservation technologies for industries, develop alternative water resource technologies, develop new water resources, protect water quality, regularly publish water supply and demand conditions, and promote education on water resource protection and conservation. Special emergency response measures for droughts include: regularly publish drought information, stop fallow irrigation, and make artificial rain.
(c) Devise future development goals and response measures: Establish a central to local drought prevention and relief system, strengthen research on the drought warning system, promote education on water resources and watershed protection, and review coordination problems among the existing policies and regulations.

Nevertheless, disaster prevention adaptation strategies devised by the Taiwan government do address the impact from climate change and do focus on sea level rise and water resources. The conservation of watersheds included in the water resource strategy closely interrelates with debris flow prevention in Taiwan. Therefore, the adaptation strategies include all kinds of disaster issues. Recently, many works that follow adaptation strategies are still in the research stage. These two more efficient works are the main establishment of the disaster prevention mechanism and the Flood-Prone Area Management Plan (FPAMP) (National Fire Agency, Ministry of the Interior, 2008).

The capacity of the disaster prevention and response system in Taiwan met an unprecedented challenge when the Chi-Chi earthquake occurred in 1999. The central government (Executive Yuan) instructed the National Science Council (NSC) to assist the National Fire Agency (NFA) in drafting the Disaster Prevention and Protection Act (DPPA). This chapter is consistent with the conditions of real disasters and similar bills of the United States and Japan. The draft of DPPA was adopted by the Executive Yuan on December 25, 1999. And then the draft was considered by the Legislative Yuan and put into practice on July 19, 2000. DPPA is the first law dealing with the disaster prevention and response issues. It includes eight chapters entitled General Principles, Disaster Organization, Plan of Disaster Prevention and Protection, Disaster Prevention, Disaster Response Countermeasures, Recovery and Reconstruction after Disaster, Penal Provisions, and Supplementary Provisions, respectively. DPPA explicitly defines the responsibility of the government, the populace, the community, the civil defense, and the army when disaster occurs. DPPA is the superior law regarding all disaster prevention issues in Taiwan. In recent years, the regional plans of disaster prevention and protection for each county in Taiwan have been subsequently devised. The mechanism of disaster prevention and response is currently more intact (National Fire Agency, Ministry of the Interior, 2008). Fig. 5 shows the framework of the current disaster management.

In order to effectively resolve inundation problems in land subsidence areas, low-lying terrain, and urban planning areas, the central government announced the "Special Act for Flood Management (SAFM)" on January 27, 2006. SAFM is the superior law pertaining to flood prevention action. It explicitly indicates the participatory organizations, needed budget, and necessary measures to prevent flood disasters. According to the purpose of SAFM, the Ministry of Economy Affairs (MOEA) proposed the FPAMP, which was adopted by the Executive Yuan on March 3, 2006. FPAMP has an NT$116 billion budget to be dispensed over a period of eight years. The budget aims to systematically harness 35 county and city rivers, 238 regional drainage systems, 4 enterprise-related sea dikes, and 350,000 ha of hillsides. Almost 500 km^2 of flood-prone areas will be improved. FPAMP is the largest flood protection action plan. The main four objectives of FPAMP are to decrease the 12 billion NT dollars financially lost by floods every year, to protect 250 million people by avoiding floods every year, to reduce the influence of floods on 500 km^2 of flood-prone areas, and to raise the complete ratio of the flood protection structure from 30% to 60% (Ministry of Economic Affairs, 2006).

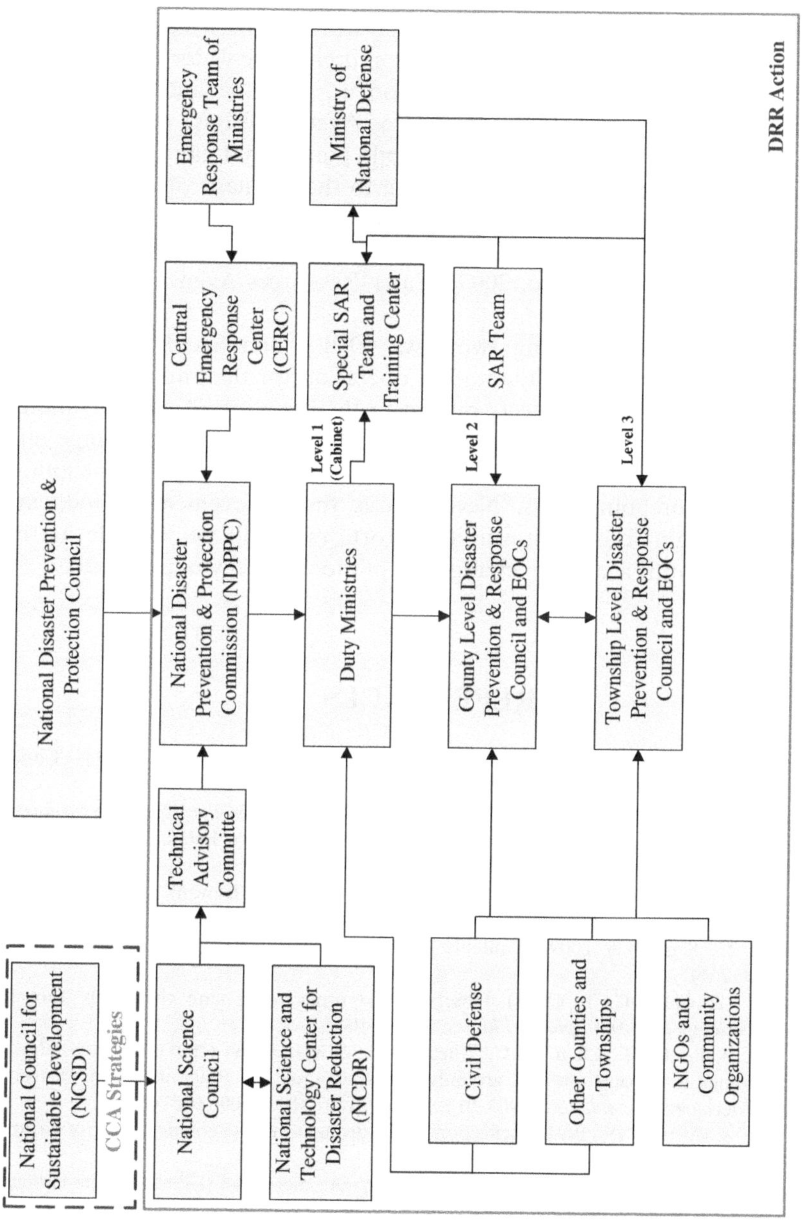

Fig. 5. Framework of Current Disaster Management Organizations in Taiwan (Diagram Linking CCA and DRR).

Recently, all works of the first stage of FPAMP are completed. The government finished 634 dredge channel works with a total length of 1868 km, 724 emergency works, and 123 comprehensive water control plans. Furthermore, in order to delay the flood arrival time and increase the mobility of drainage, the Water Resources Agency (WRA) purchased 150 mobile pumps. Today, there are 642 mobile pumps that can be dispatched immediately to drainage areas. According to the drainage of narrow areas where mobile pumps cannot be used, WRA purchased 661 small pumps that would protect almost 100,000 people, 258 villages, and 30 schools by avoiding water disasters (Lee, 2009; Water Resources Agency, Ministry of Economic Affairs, 2008).

Even though the foregoing two laws (DPPA and SAFM) have been announced, there still is no adaptation operation for the impact of climate change. Recently, many subjects related to the adaption of climate change have been studied. The uncertainty of the impact of climate change still cannot be quantified. Hence, the adaption of climate change is not joined with the two foregoing laws. Nevertheless, the achievement of reducing disasters through disaster prevention efforts may still be possible at the current extent until a climate change adaptation strategy is proposed.

REFERENCES

Central Weather Bureau. (2004). *100 Questions about typhoons* (in Chinese). (p. 33). Taipei: Central Weather Bureau.

Council for Economic Planning and Development. (2007). *Study on the adaptation strategies for global warming in Taiwan* (in Chinese). Taipei: Council for Economic Planning and Development.

Environmental Protection Administration, Executive Yuan, R.O.C. (Taiwan). (2002). UNFCCC national communication of the Republic of China (Taiwan) (in Chinese), July 2002, Web 21, September 2009. Available at http://epa.gov.tw/FileLink/FileHandler.ashx?file=12434

Hsu, H. H., & Chen, C. T. (2002). Observed and projected climate change in Taiwan. *Meteorology and Atmospheric Physics*, *79*, 87–104.

Hsu, K. C., Wang, C. H., Chen, K. C., Chen, C. T., & Ma, K. W. (2007). Climate-induced hydrological impacts on the groundwater system of the Pingtung Plain, Taiwan. *Hydrogeology Journal*, *15*(5), 903–913. DOI 10.1007/s10040-006-0137-x.

Juang, B. J., & Chen, S. H. (1994). Reflection of the drought in 1993-counterattack of nature. *Water resources conference* (in Chinese).

Lee, Y. P. (2009). A review of the flood-prone area management plan (FPAMP). *International conference on the disaster prevention technology mitigation education*, Yunlin, Republic of China (Taiwan).

National Fire Agency, Ministry of the Interior. (2008). *Disaster prevention and protection act, 2008* (in Chinese). Taipei, Republic of China (Taiwan): National Disaster Prevention and Protection Commission, R.O.C.

National Science Council. (2005a). *The impact evaluation and response measures of climate change on hydrological environment and water resources management* (in Chinese). Taipei: National Science Council.

National Science Council. (2005b). *The impacts and adaptations of climate change on Taiwan-about public health, ecology, water resources, and disaster prevention* (in Chinese). Taipei: National Science Council.

National Science Council. (2008). *Sub-project 4: Hydrological environment research in Taiwan (β). Research of hydrological cycle and hydrological environment in Taiwan* (in Chinese). Taipei: National Science Council.

Ministry of Economic Affairs. (2006). *Special act for flood management*. Laws and Regulations database of the Republic of China, 2006. Available at http://law.moj.gov.tw/eng/LawClass/LawContent.aspx?PCODE = J0110072

Tung, C. B., & Li, T. Y. (2001). The impact evaluation of climate change on Chi-Jia-Wan Creek. *Journal of Chinese Agricultural Engineering*, *47*(1), 65–74. (in Chinese).

Tung, C. P., & Haith, D. A. (1995). Global warming effects on New York streamflows. *Journal of Water Resources Planning and Management*, *121*(2), 216–225.

Water Resources Agency, Ministry of Economic Affairs. (2008). *Regulation projection of flood-prone areas*. Update: November 6, 2008. Available at http://fcp.wra.gov.tw/. Accessed on September 21, 2009.

CHAPTER 17

WATER INSECURITIES AND CLIMATE CHANGE ADAPTATION IN THAILAND

Louis Lebel, Phimphakan Lebel and Rajesh Daniel

INTRODUCTION

In Thailand, as in many other developing countries, a significant and coherent policy response to the challenges posed by climate change is just beginning to emerge. The initial emphasis was on meeting international reporting obligations and building a better understanding of the issues (OEPP, 2000). Most climate policy attention has focused on mitigation, in particular of the difficulties, and occasionally taking advantage of the opportunities, in decoupling growth in greenhouse gas emissions from social and economic development. While early concerns were expressed about impacts on water resources and agriculture, not much attention has been given to implementing specific adaptation measures.

More than a decade after the Initial Communication in 2000 by Thailand (OEPP, 2000) was made to the United Nations Framework Convention on Climate Change (UNFCCC), subsequent draft strategies and statements by politicians and leading bureaucrats are finally showing the emerging outlines and threads of an adaptation policy (Lebel, Lebel, & Daniel, 2009c). Given the importance of the Asian monsoon to Thailand's agriculture output and

Climate Change Adaptation and Disaster Risk Reduction: An Asian Perspective
Community, Environment and Disaster Risk Management, Volume 5, 349–372

ISSN: 2040-7262/doi:10.1108/S2040-7262(2010)0000005023

flood risks in built-up areas, water-related insecurities are prominent in many of these policies.

In this chapter, we explore some of the institutional and political barriers to inclusion or integration of emerging policies on adaptation to climate change, looking in particular at policies on water resources and disaster management in Thailand. This chapter is organized into three main sections. First, we briefly view scientific understanding about climate trends and associated water insecurities. Second, we present a summary of recent disaster risk reduction and climate change adaptation policies and how they are linked or not to water management policies. Third, we propose some ways in which deficiencies in current policy process may be addressed and better adaptation policies developed.

CLIMATE AND WATER INSECURITIES

Across monsoon Asia, there is overwhelming evidence of warming: annual mean temperatures have increased in many locations, there are more hot days and warm nights, glaciers are retreating, and snow cover is decreasing (IPCC, 2007; Hsu & Chen, 2009). Observed trends in precipitation are more complex and uncertain (IPCC, 2007). Some evidence suggests weakening of the Asian monsoon between 1948 and 2003 (Hsu & Chen, 2009). Two recent studies focused on urban and coastal Thailand indicate trends toward more intense daily rainfall (Atsamon, Sangchan, & Thavivongse, 2009; Limjirakan, Limsakul, & Sriburi, 2009).

Early studies by the CSIRO using the CCAM model focused on Thailand and the surrounding Southeast Asia region projected a longer hot season – defined as days with maximum temperature over 33°C (McGregor, Katzfey, & Nguyen, 1998). Similar findings were observed using the PRECIS model and the GCM ECHAM4 dataset with outputs rescaled to fit observed weather in the region more closely. Increases in temperatures are expected over all of Thailand (Chinvanno, Luang-aram, & Thanakitmetavut, 2009). Precipitation under future climate in Thailand is more uncertain. Several global and regional analyses suggest more intense precipitation events; increased tropical cyclone intensities; increased droughts and floods; and increased precipitation variability (IPCC, 2007). Projections for precipitation with the PRECIS model suggest increases are strongest after the middle of this century especially in east and south Thailand (Chinvanno, 2009; Chinvanno et al., 2009).

Exactly how global warming will interact with regional climate processes, however, is still the subject of research and debate (Hsu & Chen, 2009). Since 1980, for example, the relationship between summer monsoon rainfall over Thailand and the El Nino-Southern Oscillation (ENSO) has strengthened (Singhrattna, Rajagopalan, Kumar, & Clark, 2005). Even without more detailed understanding of trends in all climate variables, useful assessments of water insecurities may be possible by drawing on local understanding of how different sectors, ecosystems, or human activities interact with, and are otherwise sensitive to, current climate variability.

Water, Agriculture, and Rural Livelihoods

Changes in rainfall by altering flood regimes could easily exacerbate many existing water resource management challenges (Boonprakrob & Hattirat, 2006; Lebel, Foran, Garden, & Manuta, 2009a). Given a predictable monsoonal climate farming system, livelihoods and lifestyles in lowland areas were strongly adapted to seasonal changes in water levels. The availability of pumps, irrigation infrastructure, and protective barriers has in many areas reduced seasonal constraints, for example, by extending cropping seasons into what were dry periods and fallow or protecting fields from early- or late-season flood peaks. The effects on resilience of technologies and infrastructure are complex as they also lead to changes in risk-taking behavior. Nevertheless, experiences in dealing with current flood regimes impacted by land uses and interventions in river courses are relevant to dealing with additional compounding influence of climate change (Lebel et al., 2009a).

The challenges of allocating water among different users are likely to be more difficult under a changed climate and will interact with ongoing changes in both land- and water-use as well as water-related infrastructure (Costa-Cabral et al., 2007). Eastham and colleagues (2008) projected that dry season rainfall in the Mekong river basin within Thailand would increase in northern and decrease in eastern Thailand even though annual precipitation and runoff was expected to increase. With higher temperatures, evaporation would increase leading to either falls in irrigated rice yields or increased demand for water – as much as 8% higher in more extreme climate scenarios.

Most agriculture in Thailand, however, is rain-fed. At the regional level, the period for sowing and transplanting is wide due to variable water, soil conditions, and topographic sequence; in contrast, the period for heading and harvesting is narrow due to widespread use of similar photosensitive varieties (Sawano et al., 2008). Chinvanno, Souvannalath, Lersupavithnapa, Kerdsuk, and Thuan (2008a) studied impacts on rice yields in 18 villages in the Ubon Ratchathani province and found mild positive impacts on average climate due to elevated CO_2, increased rainfall, and modest projected temperature increases. Taking into account flooding and timing of rainfall, risks that farmers are concerned about suggest much more mixed outcomes (Chinvanno et al., 2008a). The two most important climate risks, according to farmers, are prolonged midseason dry spells after sowing or transplanting seedlings and flooding near the harvest time (Chinvanno, Souvannalath, Lersupavithnapa, Kerdsuk, & Thuan, 2008b). As dependence on rice production in this area is high, and many households are poor, coping capacities were often low. Rice farmers used standard rather than local varieties and had relatively limited scope to adjust crop calendars. Off-farm income, in particular, through seasonal or more permanent migration with remittances back, was the primary household measure to deal with climate risks. This experience to manage climate risks provides a foundation on which to consider options for adapting to climate change, but by itself is unlikely to be sufficient (Chinvanno et al., 2008b).

Pannangpetch and colleagues in a recent major study funded by the Thailand Research Fund assessed the impacts of climate change on four key annual crops – rice, sugarcane, cassava, and maize (Pannangpetch et al., 2009). They used datasets from the ECHAM4 A2 GCM model downscaled with the PRECIS regional climate model. Simulated yields for 2090–99 of cassava and maize fell by 43% and 15%, respectively, whereas those of sugar increased by 6% (Boonpradub et al., 2009). As a national level assessment this is a very significant step forward from past history of site-specific studies as it allows consideration of vulnerable areas through "hot-spot" analyses (Boonpradub et al., 2009). Yield projects were highly variable across the country reflecting interactions between climate and soil. The report also makes suggestions about how to adjust crop production systems to adapt to climate change (Pannangpetch et al., 2009).

Flood and Water Quality in Human Settlements

Over the past 30 years the number and impacts of flood disasters have increased (Dutta & Herath, 2004; ABI, 2005). This has occurred despite vastly improved abilities to monitor, warn, and describe floods. In Thailand this in part reflects growth in absolute numbers of people living in flood-prone areas and higher values of infrastructure at risk (Nicholls et al., 2007). Around Bangkok and many other urbanizing regions, new flood-sensitive settlements and land uses have expanded into low-lying wetlands and rice paddy landscapes.

Rapid urbanization has been a key factor in the economic development of Thailand. Bangkok illustrates many of the flood and water management challenges as human settlements expand into flood-prone areas and place valuable property and disadvantaged communities at risk. Parts of Bangkok in the Chao Phraya delta are flooded almost every year, and from time to time, much wider areas are flooded for extended periods (Dutta & Tingsanchali, 2003; Dutta, Babel, & Gupta, 2005). The network of irrigation and transport canals that dissects the delta and wetland areas has been substantially transformed by agrarian development and urbanization (Ross, Poungsomlee, Punpuing, & Archavanitkul, 2000; Molle, 2004). At the same time daily rainfall intensity over the Bangkok metropolitan area has increased over the last four decades (Limjirakan et al., 2009) with runoff challenges now compounded by infrastructure. Land subsidence, due to deep well groundwater extraction, further exacerbates current (Babel, Gupta, & Domingo, 2006) and future risks from floods. Moreover, infrastructure to protect certain areas from floods both within and upstream from the city often has significant impacts on others, often poorer, already more vulnerable groups (Lebel et al., 2009a). Both shortages and floods could exacerbate already serious problems with river and canal water quality in Bangkok (BMA, GLF, & UNEP, 2009). Small and intermediate-sized cities in Thailand face a different mixture of adaptation challenges. As they are unlikely to receive significant resources to build or maintain flood protection infrastructure, unless a major event has already struck, they depend more on early-warning systems to reduce either damages to perishable property or loss of life (Lebel et al., 2009d).

The tension between agricultural and urban interests is central to the politics of climate change adaptation in Thailand. For example, a recurrent annual challenge in operating water infrastructure upstream from Chiang Mai city is to balance flood control (maintaining sufficient reserve to accommodate possible late wet season rainfall) with the objective of maximizing storage for irrigation and domestic consumption during the dry season (Lebel et al., 2009a). Late depressions or cyclones can pose major risks to urban flooding at a time when river levels are normally already high. Global climate change effects on regional cyclone activity or other features of the Asian monsoon could easily have major implications for capacities to cope and respond to flood disasters. Other recent studies of risk management in the water sector also noted that for Northeast Thailand, risks associated with drought and floods need to be considered jointly (Chitradon, Boonya-aroonnet, & Thanapakpawin, 2009).

Watershed Services, Wetlands, and Fisheries

Climate drives the regional hydrological cycle and land-surface hydrology (Chen & Chappell, 2009). Interactions between storm and longer term climate with land uses influence flows of water and sediments to river (Xi & Higgitt, 2009) and, in turn, are impacted by dams and diversions (Kummu, Varis, & Sarkkula, 2009). Changes in land surface characteristics, such as loss of forest cover, may have regional-scale consequences for recycling of water into the boundary layer and thus on precipitation and water resources (Bates, Kundzewicz, Wu, & Palutikof, 2008). Declining regional precipitation over Northeast Thailand due to deforestation is predicted by many simulation models, but only observable in September when strong monsoon winds, which bring substantial moisture from the west and over the Indochina Peninsula, peter out (Kanae, Oki, & Musiake, 2001). Orographic effects – the cooling of air as it is pushed to higher elevations raises relative humidity and likelihood of precipitation – are also important and easily confound lay observations about associations between rainfall and vegetation (Dairaku, Emori, & Oki, 2004). Contrary to common perceptions about relationships between upstream and downstream water use, a modeling study in Mae Chaem found that future scenarios with plausible expansion of crops at the expense of forests in the uplands led to slightly higher annual and wet-season water yields compared to similar expansion in the lowland–midland zone (Thanapakpawin et al., 2006).

A diverse range of ecosystem services are provided by upper tributary watersheds apart from flows of water to downstream crops or immediate production of food. Services include a variety of nontimber products, and services, such as water purification, pollination, and sites for cultural activities, important to the well-being and livelihoods of upland farmers (Lebel, Daniel, Badenoch, & Garden, 2008; Lebel & Daniel, 2009). For lowland areas, flood protection services often receive significant attention from watershed organizations and other policy bodies. Not much is known about how climate change could effect provision of watershed services in Thailand.

Further downstream the hydrological regimes of wetlands are also sensitive to changes in climate with impacts on individual species and ecosystem productivity. Seasonality produces changes not only in water volumes, but also in water quality in wetland ecosystems that are important for ecological structure and function. Likely effects of the mean global temperature increase on freshwater systems include higher water temperatures, lower dissolved oxygen levels, and higher toxicity of pollutants. Poor water quality is already a major problem in heavily industrialized areas in Thailand.

In the Songkram wetlands, pH, dissolved oxygen, and conductivity increase with falling water levels in dry (and cooler) seasons (Satrawaha, Prathepha, Andrews, & Petney, 2009). In this area, the majority of residents still view most seasonal flood events as positive as they can catch more fish and incomes are higher (Friend, 2007). More extreme floods, however, do damage paddy rice crops and affect drinking water supplies. An assessment of possible climate change impacts concluded that, overall, droughts would have more adverse impacts than floods, but even so there was substantial resilience to changing climate in society with the ongoing expansion of livelihood options on- and off-site (MWBP, 2005).

Freshwater fish life histories have evolved to particular flow regimes and temperatures and thus likely to be impacted (Ficke, Myrick, & Hansen, 2007). Typical adaptation measures, for instance, water storage and diversion, could make the challenge of shifting ranges and otherwise coping even more difficult (Palmer et al., 2008; Baran & Myschowoda, 2009). The overall impacts of climate change on fisheries and aquaculture are likely to be significant with both positive and negative impacts (FAO, 2008). Aquaculture is sensitive to severe weather and storm surges. Other possible impact pathways are through broodstock supplies, temperature regimes in inland waters, and salt water intrusion (NACA, 2007). Impacts on fisheries, because of importance in feeds, would also impact aquaculture.

EMERGING POLICIES

Climate change has potentially important implications for multiple dimensions of water insecurity – from having sufficient water to grow food, through risks of floods damaging crops or contaminating drinking water supplies, to loss of watershed and wetland ecosystem services. Taken together this suggests that the challenge for policy is, in part, a problem of integrated disaster risk management. In this section, we briefly describe key climate change adaptation and disaster risk reduction policies and identify some of the institutional and political barriers to more inclusive and integrated policy. As adaptation to climate change is a relatively new issue for Thailand, most government policy statements remain vague and arguments are still at the level of framing the policy problem (Lebel et al., 2009c).

Climate Change Adaptation

The Office of Natural Resources and Environmental Policy and Planning (ONEP, formerly OEPP) within the Ministry of Natural Resources and the Environment (MONRE) is the formal contact and coordination point for climate change policy in Thailand. Most of the initial responses within the bureaucracy were to push climate change toward the environment ministry; as the implications of international negotiations and reporting on emissions as well as departmental budget allocations have become clearer, industrial and energy bureaucracies have become more strongly involved.

From the beginning of climate change as a policy issue, lack of knowledge and uncertainties have been highlighted by Thailand's experts and policy makers as crucial barriers. In Thailand's Initial Communication to the UNFCCC, uncertainties were identified as hindering effective responses: *"The high level of uncertainty in vulnerability studies hinders the advancement of adaptation analysis that can lead to more meaningful policy recommendations"* (OEPP, 2000, p. 74). The proposed solution was technical capacity building to improve the quality of vulnerability assessments through better use of global climate models and development of more relevant regional models (OEPP, 2000).

The Initial Communication by Thailand to the United Nations Framework Convention on Climate Change (UNFCCC) in 2000 offered the following for the agricultural sector: conserving and developing drought-resistant crop varieties, promoting crop diversification, and water conservation (OEPP, 2000). Other proposed measures included water pricing, integrated water management, and water conservation in agriculture (OEPP, 2000).

The Office of Climate Change Coordination was initially set up under a Cabinet Resolution on August 15, 2006, which also called for setting up a National Board on Climate Change Policy and a Thailand Greenhouse Gas Organisation. The Office sits in the ONEP and is responsible for providing education, training, and public awareness; promoting development and technology transfer; initiating clean development mechanisms; capacity building; conducting research and observation; preparing national communication report; undertaking measures for the mitigation of greenhouse gases; and undertaking measures to adapt to climate change. The Office is supposed to coordinate the activities of many other government organizations (ONEP, 2009b).

The draft Five-Year Strategy on Climate Change (2008–12) ordered relevant Ministries to form subcommittees, follow progress, and coordinate activities (ONEP, 2007). The strategy has six components: (1) building capacity to adapt and reduce vulnerabilities to climate change impacts, (2) supporting reduction in greenhouse gas emissions, (3) supporting research and development to better understand climate change impacts, and adaptation and mitigation options, (4) raising awareness and promote public participation, (5) building capacity of relevant personnel and institutions and establishing a framework of coordination and integration, and (6) supporting international cooperation to achieve the common goal of climate change mitigation and sustainable development. The emphasis is on capacity building and coordination of government stakeholders that have historically compartmentalized climate change analyses and reactions. Adaptation and mitigation issues are strongly intertwined in the strategy document. There is very little substantive detail and no significant connections in the strategy are made to existing flood or disaster management initiatives.

A follow-up 10-year Strategic Plan on Climate Change and three-year Action Plan (2010–2012) were under review at the time of writing. Documents provided at consultation meetings to draft the plan (ONEP, 2009a) highlight building adaptive capacity as one of the two primary goals. The strategies proposed are to improve capacities to: (1) sustain food security and agrobiodiversity, (2) handle natural disasters and coastal degradation, (3) prepare the tourism sector for impacts, (4) handle health problems, (5) conserve and improve forests, and (6) adapt or establish public utilities. Although a lot of hope has been pinned on the strategic and action plans, it is not clear that a strategy based on a single overarching policy for climate change adaptation is the most effective way forward. Nor is it clear that it will change the fragmented and sectoral situation that currently exists. A lot still depends on actions by individual ministries who in some instances are already working on their own strategies that fit with their areas of daily work.

Efforts to mainstream climate change adaptation policies in disaster management, water resources management, or development planning have been slow to emerge and remain vague. This is illustrated by another policy document that came out soon after the five-year climate change strategy.

The 2008 National Sustainable Development Strategy of Thailand developed in response to the World Summit on Sustainable Development 2002 mentions "climate change" only twice in 96 pages, both times in reference to existing policy commitments under the UNFCCC (NESDB, UNEP, & TEI, 2008).

Recent political instability in Thailand has affected efforts to initiate a coherent set of responses to climate change (and many other issues). In early 2008, for example, Dr. Yongyut Yuthawong, the head of Ministry of Science and Technology, said he would not work on strategies for responding to global warming, but leave the details for the next government to work out (The Manager, 2008). In October 2008, the interim Prime Minister, Somchai Woongsawat, listed 16 points they would follow in the coming year to tackle global warming; the 15th was to "*rush measures and projects to reduce impacts, including adaptations to, critical threats from global warming*" (Suthichaiyoon, 2008). In a few weeks, another government replaced his administration and his policy measures were forgotten.

Disaster Risk Reduction

At more local levels, there has been more significant progress. The Bangkok Metropolitan Administration (BMA) in its recent assessment on climate change considered several explicit changes to flood disaster management, including changing flood plain mapping to adjust to new realities in which a 100-year flood may now be a 50-year flood; land-use restrictions for flood plains; infrastructure adjustments, such as flood channels, dams, and dykes; new programs for disaster recovery and support at district level; and modifying disaster response protocols to take into account possibly higher flooding frequency (BMA et al., 2009, p. 72). Another specific focus area is on buildings and infrastructure with emphasis on climate-sensitive designs for new buildings and monitoring old structures so that they can be modified in time. The assessment is an impressive document; what remains to be seen is whether it will have much influence on spatial planning and development.

The Royal Irrigation Department (RID), Ministry of Agriculture and Cooperatives, with responsibilities for construction of dams, dykes, and pumping systems as well as day-to-day operations has had a very significant role in management of disaster risks related to floods and water shortages. The RID has operation centers at national and regional levels for disaster preparedness that monitor and report on water storage and flow conditions. The DDPM and RID are largely independent operations with the RID relatively more active in river basin management organizations (Lebel, Garden, Subsin, & Nan, 2009b), which are promoted and supported by the Department of Water Resources (DWR) in the Ministry of Natural Resources and Environment. Views within the RID about appropriate ways to manage floods are evolving. As the director general of RID, Pramote Maiklad notes that "*floods are a natural phenomenon and in trying to overcome the adverse impact of floods by numerous measures, one tends to forget most of the time that water needs space too*" (Maiklad, 1999, p. 65). Moreover, infrastructure is widely perceived as the best solution to reducing risks of exposure from floods, but can lead to false sense of security: "*once a dam is built, most of the people living downstream of the dam tend to believe that they will be completely safe from floods*" (Maiklad, 1999, p. 107).

In most public discussions, the emphasis on infrastructure in traditional flood management has been transplanted into responses to climate change (Lebel et al., 2009e). A group of politicians and experts proposed to protect Bangkok from sea-level rise by building an 80-km-long wall 300 m offshore and 3 m higher than moderate sea level (Wipatayotin, 2007). Calls for an even higher offshore sea wall to protect Bangkok and neighboring provinces later resurfaced (Bangkok Post, 2008b) after predictions by the director of the NDWC of storm sea surges (Bangkok Post, 2008c). The conventional focus on physical infrastructure should be understood as arising from planning norms and organizational interests of the traditional agencies involved in flood-related disaster management and irrigation. Climate change can be used as a tag and rationale for infrastructure investment (Lebel, 2007) and exploitation of water resources. Arjun Thapun, director general for Asian Development Bank's Southeast Asia Department, for example, wrote: "*There are disproportionate risks from climate change, including falling rice and agricultural production, higher intensity storms, typhoons and natural disasters, and diminishing clean water supplies. So, in addition to climate change adaptation measures, the countries need to consider increased investments in a new generation of carbon-neutral subregional infrastructure projects, such as railways, inland water transport, and clean energy – projects that will create jobs and stimulate growth, while protecting the environment*" (Thapun, 2009). An effective response to climate change, however, may require approaches that also envisage modifications to, or removal of, existing infrastructure (Palmer et al., 2008).

"Droughts" in Thailand are a recurrent "crisis" – created by the expansion of irrigation and subsequently dry season cropping into areas, where in the past, only wet season agriculture was practiced, given the high seasonality of rainfall (Lebel et al., 2009b). Politicians and bureaucrats frequently use mass media outlets to "talk up" the approaching dry season as a "drought" almost every year (e.g. Samabuddhi, 2004) to gain support for their supply augmentation projects and appeal to large rural electorates (Molle & Floch, 2008). Where water is shared with other large users, the seasonal conflicts can be even more intense. In the central plans and eastern region of Thailand, water shortages in 2004–2005 resulted in low rice yields. Major industries in the eastern seaboard region of Map Ta Phut faced severe water shortages. In response, the government has promised various storage and diversions projects for the eastern seaboard that were contested by farmers who would lose out in the reallocations. Authors of a report for Greenpeace discussing this drought said "*one of the main causes of the current prolonged drought in Thailand can be attributed to global warming*" (Boonprakrob & Hattirat, 2006, p. 11). Attributing individual events to climate change is a common tactic, but in the case of water resources management may be counterproductive as it shifts policy attention away from serious internal challenges of growing demand and allocation trade-offs to focusing on "supply-side" issues.

Many agencies claim or otherwise shoulder responsibilities for responding to, or preventing, disasters and reducing water-related insecurities. A significant institutional reform in the last decade was the establishment of the Department of Disaster Prevention and Mitigation (DDPM) in October 2002. The DDPM started with a large mandate, but modest resources (Manuta, Khrutmuang, Huaisai, & Lebel, 2006). DDPM's primary role was coordination, for example, by preparing master plans from the individual plans of various agencies, but its authority and boundaries on responsibilities were always clear or acceptable to pre-existing organizations. Significant disconnects among agencies responsible for preventing and preparing for disasters, on the one hand, and those more oriented toward relief and emergency measures, on the other, persisted despite the reforms (Manuta et al., 2006). For example, after the Indian Ocean Tsunami, the National Disaster Warning Centre (NDWC) was set up under the Ministry of Information and Communication Technology (MICT) rather than the Ministry of Interior (MOI), which houses the DDPM, renewing bureaucratic rivalries. Later the NDWC was placed under the Meteorological Department leading to its founding Director resigning in May 2008 to protest the loss of independence of the NDWC (Bangkok Post, 2008a). The Meteorological Department (in MICT) primarily deals with weather monitoring and forecasting and thus has a crucial role to play in early-warning systems. The disaster management reforms included the replacement of the 1979 Civil Defense Act by the 2007 Disaster Prevention and

Mitigation Act, strengthening the legal status of the DDPM (Government of Thailand, 2007). The new act was expected to help improve interagency collaboration. The latest director of the NDWC pointed out to us the challenges their agency faces in dealing with 40 agencies in their work, much of it related to news dissemination and communication: *"Within Thailand, the strength is that each agency has its own information; but they keep it[to themselves] when what we need is information-sharing and coordination"* (NDWC, 2009).

Fragmentation and inter-Ministerial competition has hindered policy coordination and the mainstreaming of adaptation measures to climate change. Disaster management planning has until very recently paid little attention to issues of climate change, whereas climate change strategies frequently refer to risks of flood and drought-related disasters. Under modified flood regimes, new areas will be exposed requiring expansion of early-warning systems and disaster preparedness activities. In locations with experience coping with floods, impacts of floods may become more severe, exceeding current coping capacities. Responding to more extreme rainfall events under a changed climate will thus require better cross-level coordination of emergency relief, financing of rehabilitation, and mobilization of human resources. Establishing new community-based capacity, and coordinating these with existing administrative hierarchies, will also require attention to ways in which stakeholders are engaged. Some recent initiatives under the DDPM suggest efforts are being made at recognizing importance of greater local government or community involvement – such as through the OTOS or "One Tambon One Search and Rescue Team" and the "Mr. and Mrs. Early Warning" programs (2009). At the same time much national-level programming, for instance, the Strategic National Action Plan (SNAP) for 2007–2016, continues to treat public participation as involvement in trainings and evacuation exercises and not in planning.

TOWARD BETTER ADAPTATION POLICIES

From the perspective of water-related insecurities – whether the problem is too much or too little water – there are clearly still major gaps between various policy initiatives. On the one hand, some differences in framing are to be expected, given differing organizational mandates, responsibilities, and capacities (Lebel et al., 2009c). On the other hand, the potential for climate change, as another cross-cutting issue, to help strengthen integration of longer term strategic planning has not yet been realized.

Challenges

Thailand is still struggling to develop an effective and sustained policy response to the challenges posed by climate change to water resources management. The difficulty arises from several sources.

First, climate change is but another consideration on top of an already significant litany of conflicts in the allocation of dry season water flows to agriculture and other uses (Molle, 2004), and concerns over the costs and side-effects of flood protection in the wet season (Lebel et al., 2009a). Although widely promoted, the integrated water resource management paradigm has had only modest impacts on water management practices and little on decision-making about much policies or infrastructure (Molle, 2008).

Second, adaptation has been seen, like water allocation and flood disaster management before it, as primarily a technical planning exercise (Lebel, 2007; Lebel & Sinh, 2007). Stakeholders outside government bureaucracies have not been important to assessment or strategic planning; they remain primarily involved as subjects that need to be made aware of risks and governmental plans (Manuta et al., 2006).

Third, climate change was initially classified by the government as an environmental issue and as a consequence has remained of only marginal relevance to water management, disaster risk reduction, and larger social development policy-making. Increasingly, however, it has become apparent that much better coordination among stakeholders, even just considering those in the government, is needed for an effective response to emerge. The DDPM, for example, recently provided the Ministry of Natural Resources and Environment with inputs on their draft master plan (DDPM, 2009) and is also working closely with the National Economic and Social Development Board (NESDB) in including climate change as part of the next national strategy (MONRE, 2009). These new coordinated relationships need to be encouraged.

Fourth, the potential impacts on quantities and qualities of water resources and flood risks are not yet well understood or appreciated. Genuine misunderstandings, heterogeneous risks, uncertainties, and lack of basic information are all contributing factors. The need for capacity building in government agencies and the wider community to enable appropriate assessment of risks and exploration of adaptation responses is increasingly recognized.

These difficulties reflect some broader limitations in policy-making styles in Thailand. Despite bureaucracies with significant technical expertise, key decisions on projects – whether to augment supply or protect the city – often seem to have their origins in the promises of politicians rather than as outcomes of assessment or longer term planning strategies (Molle & Floch, 2008; Lebel et al., 2009e). This significantly constrains the modes by which other stakeholders, including people likely to be affected by actions taken in the name of reducing water insecurities, can meaningfully influence decisions. It has also reduced the effective implementation of policies and projects as there is often inadequate follow-up, and opportunities to learn from doing are lost.

Ways Forward

Climate change is a confounder; it adds to existing stresses on sustainable development (Cruz et al., 2007). In Thailand, for the most part, climate change has been seen as a discrete issue rather than a set of risks that could be managed as part of development. When exploring policy responses, it may be best to begin with existing alternative policies many of which were not formulated with adaptation to climate change in mind (Dovers, 2009). The Bangkok Metropolitan Authority (BMA) rightly argues that *"many measures that should be considered adaptation actions are already in place within some proactive programs ... an adaptation action may be simply to increase the frequency or magnitude of existing programme implementation and monitoring"* (BMA et al., 2009, p. 69). This does not, however, imply that past flood and drought disturbance regimes are by themselves a sufficient guide to how policies and practices have to be adjusted.

Current water infrastructure and practices have been designed based on experiences of past climate and water flows. Many projects and investments in planning have not taken into account climate change. Under an uncertain climate future, such strategies will need to be revised (Bates et al., 2008). Organizations like the RID and BMA may need to incorporate higher safety margins, for example, into flood protection infrastructure. Side-effects of infrastructure will also need to be considered not just the "benefits" (Lebel & Sinh, 2009).

The issue is not just that the past may not be a good guide to future conditions, but also that the amount of change is hard to anticipate and climate change is such that uncertainties typically expand as longer time horizons are considered. Increasing uncertainties about future climate implies that effective decision-making must allow for a wider range of future risks and adjustments. Flexible and reversible options will need to be increasingly favored (Hallegatte, 2009): for example, flood protection walls that are inflatable and stored away when not in use or which may have value for water storage during unusually dry seasons.

Existing local capacities and strategies should not be overlooked. In many flood-prone lowland areas, residents adopt practices that contribute to resilience to both seasonal and interannual climate variability. These include styles of architecture that can cope with moderate flooding, such as two-storey homes or houses on stilts and use of boats for more prolonged periods of inundation. Local knowledge about managing climate risks is highly relevant to building adaptive capacities and larger-scale adaptation planning, but often overlooked (Chinvanno et al., 2008b; Resurreccion, Sajor, & Fajber, 2008). Government policies therefore need to take into account various aspects and actors with respect to local adaptation to climate change (Fig. 1).

Maintaining and restoring the resilience of ecosystems, and securing access to those services, is crucial to many vulnerable groups in rural areas.

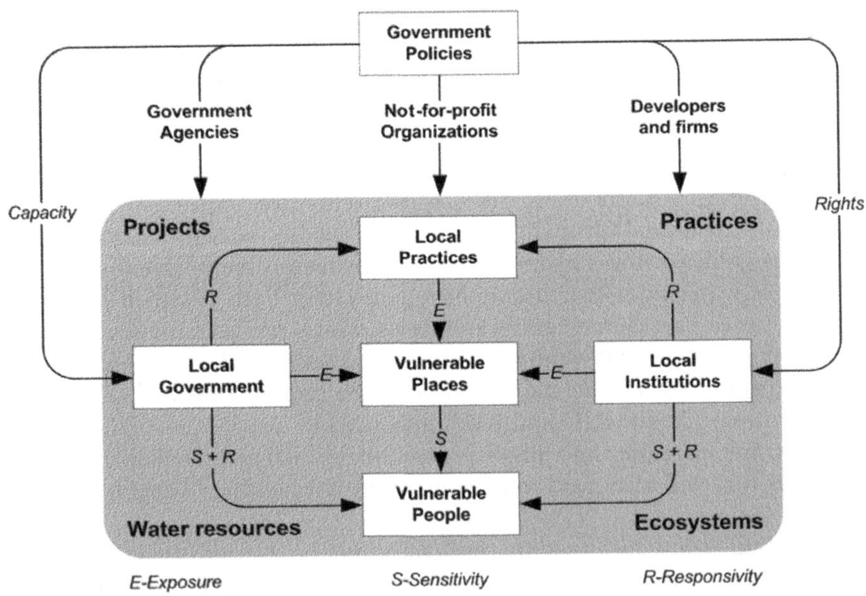

Fig. 1. Government Policies Can Influence Different Aspects and Actors with Respect to Local Adaptation to Climate Change and through Several Pathways with Corresponding Impacts on Different Components of Vulnerability – Exposure (E), Sensitivity (S), and Responsivity (R).

In urban areas, much more attention is needed to improve social resilience of vulnerable communities, for instance, through providing space and funds for self-organization in low-income informal settlements. Climate change impacts on flooding may require more than technical quick-fixes; managing for resilient social–ecological systems, such as restoring flood plains so there is space for water.

Local initiatives to improve soil and water management, including water harvesting and storage for climate-vulnerable groups, should also be considered (Lebel, 2008). Private investments in drought- and flood-tolerant crop varieties and water-saving technologies should be encouraged, where appropriate, in partnership with public research and extension agencies. Better integration and implementation of land- and water-use planning is clearly crucial to reducing vulnerabilities to changing flow regimes, including seasonal shortages, especially in upland watersheds. River sub-basin or watershed organizations provide an opportunity for building public awareness about changing risks and deliberating alternatives (Thomas, 2006). For the most part, however, they have stayed clear of flood management issues leaving their management to municipalities or the Royal Irrigation Department to handle. In principle, watershed organizations should also provide opportunities for various line agencies to work together across Ministerial and jurisdictional lines in sight of a mindful public. As such they could be important in developing stakeholder partnerships. In practice, there are significant barriers with the move from small relatively homogeneous water user associations through attempts to create genuine multistakeholder platforms in complex periurban settings. Under a changed climate, integrated approaches to water resources management will also need to become much more learning oriented, if it is to be able to adapt to shifts in flow and disturbance regimes (Pahl-Wostl, 2007).

Expanding meaningful public participation in managing climate risks is crucial to gaining acceptance for difficult decisions about zoning or new flood protection infrastructure. The importance of community-based disaster management is increasingly recognized by governments (Maiklad, 1999; ADPC, 2000); however, coordinating their activities with those of local authorities and higher levels remains a challenge. Disaster management agencies still view participation as "training" and "responsibility sharing" rather than ways to share knowledge and needs. Many state-led river-basin or watershed organization efforts suffer from similar limitations. Assessing risks of exposure and planning on how to reduce them for affected and vulnerable groups needs diverse inputs, not just those of line agencies.

Stakeholder participation is important for observing and monitoring changes in climate, flood regimes, and vulnerabilities. Vulnerable groups need to be fully engaged in exploring alternatives response options. Emergency response efforts need local knowledge, but also trust of authorities. Negotiation may be needed to balance benefits for protection and compensation schemes for those whose property cannot be so protected. Deliberative approaches to flood and disaster management are often needed under current flood regimes; uncertainties and surprises from climate change make social learning imperative (Lebel, Grothman, & Siebenhüner, 2010). Interventions as projects or policies will require careful monitoring to evaluate whether they meet their climate adaptation objectives and also contribute to sustainable development.

As climate change is a relatively new issue for many decision-making and management bodies in Thailand, strengthening links with technical experts is valuable in several ways: to improve understanding of credible and current scientific knowledge; explore and extract more salient findings; shape research agendas to fit decision-maker needs but with scientific plausibility taken into account; and develop joint assessment activities that could lead to new actionable knowledge. Attention should be placed on capacities to use climate information, with medium term climate or seasonal forecasting an area that not only promises benefits to some groups in the shorter term, but also develops the sorts of skills, which could be useful in dealing with longer term trends in climate. A promising initiative along these lines was the creation in late 2009 of the Climate Change Knowledge Management Center under the National Science and Technology Development Agency in the Ministry of Science and Technology. The Center has a mission to collect, synthesize, and disseminate knowledge on climate change to support strategic planning of government agencies, the private sector, as well as local communities.

CONCLUSION

Climate changes pose significant, but uncertain, risks to water and water-related resources, livelihoods, and infrastructure. In Thailand, disaster risk reduction initiatives and the emerging policies for adaptation to climate change have had parallel histories with little constructive interaction. Reducing the risks of disaster is important to sustainable development and is only likely with long-term adaptive planning. Given large uncertainties about key aspects of future climate and the effectiveness of different policies and management interventions, it is clear that learning processes and public engagement should be an integral part of adaptation governance. Developing effective institutions and capacities requires broad engagement with stakeholders and a paradigm shift from reactive to adaptation and proactive risk management.

ACKNOWLEDGMENTS

This work by the authors was funded by grants from Echel Eau and the International Fund for Agricultural Development under a grant from the Challenge Program on Water and Food (PN50), Asia-Pacific Network for Global Environmental Change Research (Grant ARCP2009-03CMY-Nikitina), and Stockholm Environment Institute (Adaptation Knowledge Platform). Thanks to Elena Nikitina and an anonymous reviewer for their feedback on earlier drafts.

REFERENCES

ABI. (2005). *Financial risks of climate change. Summary report*. Association of British Insurers (ABI), London.

ADPC. (2000). *Community based disaster management*. Trainer's Guide. Bangkok: Asian Disaster Preparedness Centre.

Atsamon, L., Sangchan, L., & Thavivongse, S. (2009). Assessment of extreme weather events along the coastal areas of Thailand. In: *89th American Meteorological Society annual meeting* (pp. J21–22). Phoenix, Arizona.

Babel, M. S., Gupta, A. D., & Domingo, N. D. S. (2006). Land subsidence: A consequence of groundwater over-exploitation in Bangkok, Thailand. *International Review for Environmental Strategies*, *6*, 307–327.

Bangkok Post. (2008a). A disaster in the making. Editorial. *Bangkok Post*, May 17, 2008, Bangkok.

Bangkok Post. (2008b). Gulf communities 'face devastation'. *Bangkok Post*, August 15, 2008, Bangkok.

Bangkok Post. (2008c). Storm surge fears. Is the threat real? *Bangkok Post*, October 5, 2008, Bangkok.

Baran, E., & Myschowoda, C. (2009). Dams and fisheries in the Mekong basin. *Aquatic Ecosystem Health & Management*, *12*, 227–234.

Bates, B., Kundzewicz, Z. W., Wu, S., & Palutikof, J. (Eds). (2008). *Climate change and water. Technical Paper VI of the intergovernmental panel on climate change*. IPCC Secretariat, Geneva.

BMA, GLF, & UNEP. (2009). *Bangkok assessment report on climate change 2009*. Bangkok: Bangkok Metropolitan Administration, Green Leaf Foundation, United Nations Environment Programme.

Boonpradub, S., Ratanasriwong, S., Sarawat, V., Kapetch, P., Ek-un, K., Damrhikemtrakul, W., Buddhasimma, I., & Pannangpetch, K. (2009). Impact of global warming on three major field crops production in Thailand (in Thai). *KKU Research Journal, 14*, 626–649.

Boonprakrob, K., & Hattirat, S. (2006). *Crisis or opportunity: Climate change impacts and Thailand*. Bangkok: Greenpeace Southeast Asia.

Chen, Y. D., & Chappell, N. (2009). Climate regulation of Southeast Asian hydrology. In: L. Lebel, A. Snidvongs, C. Chen & R. Daniel (Eds), *Critical states: Environmental challenges to development in monsoon Southeast Asia* (pp. 205–220). Selangor: SIRD/Gerakbudaya.

Chinvanno, S. (2009). Future climate projection for Thailand and surrounding countries: Climate change scenario of 21st century. In: F. Murray (Ed.), *Climate change and regional air pollution in Thailand, Vietnam and Guangxi and Yunnan provinces of PRC*. Manila: Asian Development Bank.

Chinvanno, S., Luang-aram, V., & Thanakitmetavut, J. (2009). Global warming and climate change in Chi-Mun River basin. *KKU Research Journal, 14*, 666–682.

Chinvanno, S., Souvannalath, S., Lersupavithnapa, B., Kerdsuk, V., & Thuan, N. T. H. (2008a). Climate risks and rice farming in the lower Mekong River countries. In: N. Leary, J. Adejuwon, V. Barros, I. Burton, J. Kulkarni & R. Lasco (Eds), *Climate change and vulnerability* (pp. 333–350). London: Earthscan.

Chinvanno, S., Souvannalath, S., Lersupavithnapa, B., Kerdsuk, V., & Thuan, N. T. H. (2008b). Strategies for managing climate risks in the lower Mekong River basin: A place-based approach. In: N. Leary, J. Adejuwon, V. Barros, I. Burton, J. Kulkarni & R. Lasco (Eds), *Climate change and adaptation* (pp. 228–246). London: Earthscan.

Chitradon, R., Boonya-aroonnet, S., & Thanapakpawin, P. (2009). Risk management of water resources in Thailand in the face of climate change. *Sasin Journal of Management*, 64–73.

Costa-Cabral, M., Richey, J., Goteti, G., Lettenmaier, D., Feldkotter, C., & Snidvongs, A. (2007). Landscape structure and use, climate, and water movement in the Mekong River basin. *Hydrological Processes, 22*, 1731–1746.

Cruz, R. V., Harasawa, H., Lal, M., Wu, S., Anokhin, Y., Punsalmaa, B., Honda, Y., Jafari, M., Li, C., & Ninh, N. H. (2007). Asia. In: M. L. Parry, O. F. Canziani, J. P. Palutikof, P. van der Linden & C. E. Hanson (Eds), *Climate change 2007: Impacts, adaptation and vulnerability. (Contribution of Working Group II to the Fourth Assessment Report of the Intergovernmental Panel on Climate Change)* (pp. 469–506). Cambridge, UK: Cambridge University Press.

Dairaku, K., Emori, S., & Oki, T. (2004). Rainfall amount, intensity, duration, and frequency relationships in the Mae Chaem watershed in Southeast Asia. *Journal of Hydrometeorology, 5*, 458–470.

DDPM. (2009). *Interview with Montree Chanachaiviboonvat, 2 November 2009*. Bangkok: Bureau of Disaster Management Policy, Department of Disaster Prevention and Mitigation.

Dovers, S. R. (2009). Normalizing adaptation. *Global Environmental Change, 19*, 4–6.

Dutta, D., Babel, M. S., & Gupta, A. D. (2005). *An assessment of the socio-economic impacts of floods in large coastal areas*. Bangkok: Asian Institute of Technology.

Dutta, D., & Herath, S. (2004). Trend of floods in Asia and flood risk management with integrated river basin approach. In: *2nd Asian Pacific Association of Hydrology and Water Resources conference*. Singapore. Kyoto: Water Resources Research Centre, Kyoto University.

Dutta, D., & Tingsanchali, T. (2003). Development of loss functions for urban flood risk analysis in Bangkok. In: *International symposium on new technologies for urban safety of mega cities in Asia.* International Center for Urban Safety Engineering, University of Tokyo, Tokyo, Japan.

Eastham, J., Mpelaskoka, F., Mainuddin, M., Ticehurst, C., Dyce, P., Hodgson, G., Ali, R., & Kirby, M. (2008). *Mekong River basin water resources assessment: Impacts of climate change.* Canberra, Australia: CSIRO: Water for a Healthy Country National Research Flagship.

FAO. (2008). *Climate change for fisheries and aquaculture. Technical background document from the expert consultation held on 7–9 April 2008, Rome.* Rome: Food and Agricultural Organization of the United Nations.

Ficke, A., Myrick, C., & Hansen, L. (2007). Potential impacts of global climate change on freshwater fisheries. *Reviews in Fish Biology and Fisheries, 17*, 581–613.

Friend, R. (2007). *Securing sustainable livelihoods through wise use of wetland resources: Reflections on the experience of the Mekong Wetlands Biodiversity Conservation and Sustainable Use Programme (MWBP).* Vientianne, Lao PDR: Mekong Wetlands Biodiversity Conservation and Sustainable Use Programme.

Government of Thailand. (2007). Royal Gazette Vol. 124, Part 52 A 7 September 2007 (B.E. 2550). Thailand: Government of Thailand.

Hallegatte, S. (2009). Strategies to adapt to an uncertain climate change. *Global Environmental Change, 19*, 240–247.

Hsu, H.-H., & Chen, C.-T. (2009). Stability and change in the monsoon climate-atmosphere system. In: L. Lebel, A. Snidvongs, C. Chen & R. Daniel (Eds), *Critical states: Environmental challenges to development in monsoon Southeast Asia* (pp. 97–111). Selangor: SIRD/Gerakbudaya.

IPCC. (2007). *Climate change 2007: The physical science basis.* Contribution of Working Group I to the Fourth Assessment Report of the IPCC. Cambridge, UK: Cambridge University Press.

Kanae, S., Oki, T., & Musiake, K. (2001). Impact of deforestation on regional precipitation over the Indochina peninsula. *Journal of Hydrometeorology, 2*, 51–70.

Kummu, M., Varis, O., & Sarkkula, J. (2009). Impacts of land surface changes on regional hydrology in mainland Southeast Asia. In: L. Lebel, A. Snidvongs, C. Chen & R. Daniel (Eds), *Critical states: Environmental challenges to development in monsoon Southeast Asia* (pp. 221–238). Selangor: SIRD/Gerakbudaya.

Lebel, L. (2007). Adapting to climate change. *Global Asia, 2*, 15–21.

Lebel, L. (2008). Climate change, water insecurities and food systems in Monsoon Asia. In: C. Loh, A. Stevenson & S. Tay (Eds), *Climate change negotiations: Can Asia change the game?* (pp. 88–99). Hong Kong: Civic Exchange.

Lebel, L., & Daniel, R. (2009). The governance of ecosystem services from tropical upland watersheds. *Current Opinion in Environmental Sustainability, 1*, 61–68.

Lebel, L., Daniel, R., Badenoch, N., & Garden, P. (2008). A multi-level perspective on conserving with communities: Experiences from upper tributary watersheds in Montane mainland Southeast Asia. *International Journal of the Commons, 1*, 127–154.

Lebel, L., Foran, T., Garden, P., & Manuta, B. J. (2009a). Adaptation to climate change and social justice: Challenges for flood and disaster management in Thailand. In: F. Ludwig, P. Kabat, H. van Schaik & M. van der Valk (Eds), *Climate change adaptation in the water sector* (pp. 125–141). London: Earthscan.

Lebel, L., Garden, P., Subsin, N., & Nan, S. N. (2009b). Averted crises, contested transitions: Water management in the upper Ping River basin, Northern Thailand. In: D. Huitema & S. Meijerink (Eds), *Water policy entrepreneurs. A research companion to water transitions around the globe* (pp. 137–157). Cheltenham, UK: Edward Elgar.

Lebel, L., Grothman, T., & Siebenhüner, B. (2010). The role of social learning in adaptiveness: Insights from water management. International Environmental Agreements: Politics, Law and Economics, DOI: 10.1007/s10784-010-9142-6.

Lebel, L., Lebel, P., & Daniel, R. (2009c). *Emerging policy frames on climate change adaptation in Thailand. USER Working Paper WP-2009-10*. Unit for Social and Environmental Research, Chiang Mai University, Chiang Mai.

Lebel, L., Perez, R. T., Sukhapunnaphan, T., Hien, B. V., Vinh, N., & Garden, P. (2009d). Reducing vulnerability of urban communities to flooding. In: L. Lebel, A. Snidvongs, C.-T. A. Chen & R. Daniel (Eds), *Critical states: Environmental challenges to development in Monsoon Asia* (pp. 381–399). Selangor, Malaysia: Strategic Information and Research Development Centre.

Lebel, L., & Sinh, B. T. (2007). Politics of floods and disasters. In: L. Lebel, J. Dore, R. Daniel & Y. S. Koma (Eds), *Democratizing water governance in the Mekong region* (pp. 37–54). Chiang Mai: Mekong Press.

Lebel, L., & Sinh, B. T. (2009). Risk reduction or redistribution? Flood management in the Mekong region. *Asian Journal of Environment and Disaster Management, 1*, 23–39.

Lebel, L., Sinh, B. T., Garden, P., Seng, S., Tuan, L. A., & Truc, D. V. (2009e). The promise of flood protection: Dykes and dams, drains and diversions. In: F. Molle, T. Foran & J. Kakonen (Eds), *Contested waterscapes in the Mekong region* (pp. 283–306). London: Earthscan.

Limjirakan, S., Limsakul, A., & Sriburi, T. (2009). Increasing trend of extreme rain events over Bangkok metropolitan area. In: *89th American Meteorological Society annual meeting* (p. 13B.12). Phoenix, Arizona.

Maiklad, P. (1999). Development and achievements in flood control and management in Thailand. In: UN-ESCAP (Ed.), *Regional cooperation in the 21st century in flood control and management in Asia and the Pacific* (pp. 59–111). New York: United Nations.

Manuta, J., Khrutmuang, S., Huaisai, D., & Lebel, L. (2006). Institutionalized incapacities and practice in flood disaster management in Thailand. *Science and Culture, 72*, 10–22.

McGregor, J. L., Katzfey, J., & Nguyen, K. (1998). *Fine resolution simulations of climate change for Southeast Asia. Final report for a research project commissioned by the Southeast Asian Regional Committee for START (SARCS)*. CSIRO Division of Atmospheric Research, Aspendale, Australia.

Molle, F. (2004). Technical and institutional responses to basin closure in the Chao Phraya River basin, Thailand. *Water International, 29*, 70–80.

Molle, F. (2008). Nirvana concepts, narratives and policy models: Insights from the water sector. *Water Alternatives, 1*, 23–40.

Molle, F., & Floch, P. (2008). Megaprojects and social and environmental changes: The case of the Thai "water grid". *Ambio, 37*, 199–204.

MONRE. (2009). *Report on advice for preparedness to climate change by the National Economic and Social Advisory Council (in Thai)*. Bangkok: Ministry of Natural Resources and the Environment.

MWBP. (2005). *Vulnerability assessment of climate risks in the lower Songkhram River basin, Thailand.* Vientiane, Lao PDR: Mekong Wetlands Biodiversity Conservation and Sustainable Use Programme.

NACA. (2007). *Research needs in sustaining aquaculture sector in Asia-Pacific to Year 2025 and beyond. Workshop report, 4–7 June 2007*, Rayong, Thailand. Bangkok: Network of Aquaculture Centres in Asia-Pacific.

NDWC. (2009). *Interview with director, Wiriya Mongkolweerapan, 26 November 2009.* Bangkok: National Disaster Warning Centre.

NESDB, UNEP, & TEI. (2008). *National sustainable development strategy for Thailand: A guidance manual.* Bangkok: Office of National Economic and Social Development Board, United Nations Environment Programme, Thailand Environment Institute.

Nicholls, R. J., Hanson, S., Herweijer, C., Patmore, N., Hallegatte, S., Corfee-Morlot, J., Chateau, J., & Muir-Wood, R. (2007). *Ranking of the world's cities most exposed to coastal flooding today and in the future.* Paris: Organisation for Economic Co-operation and Development (OECD).

OEPP. (2000). *Thailand's initial national communication under the United Nations framework convention on climate change.* Bangkok: Office of Environmental Policy and Planning, Ministry of Science, Technology and Environment.

ONEP. (2007). *Five-year strategy on climate change (2008–12).* Bangkok: Office of Natural Resources and Environmental Policy and Planning, Ministry of Natural Resources and the Environment.

ONEP. (2009a). *Consultation meeting document for draft master strategic plan on climate change (2010–2019).* [in Thai]. Bangkok: Office of Environmental Policy and Planning, Ministry of Environment and Natural Resources.

ONEP. (2009b). *Interview with ONEP staff, 20 October 2009.* Bangkok: Office of Natural Resources and Environmental Policy and Planning.

Pahl-Wostl, C. (2007). Transitions towards adaptive management of water facing climate and global change. *Water Resources Management, 21*, 49–62.

Palmer, M. A., Liermann, C., Nilsson, C., Florke, M., Alcamo, J., Lake, P. S., & Bond, N. (2008). Climate change and the world's river basins: Anticipating management options. *Frontiers in Ecology and Environment, 6*, 81–89.

Pannangpetch, K., Sorawat, V., Boonpradub, S., Ratanasriwong, S., Kongton, S., Nilaphan, S., Putthisima, I., Gapetch, P., Ekoun, C., Damrikematrakul, W., Buddhaboon, C., & Khunket, K. (2009). *Impacts of global warming on rice, sugarcane, cassava, and maize production in Thailand. Final Technical Report. 31 October 2009.* Bangkok: Thailand Research Fund.

Resurreccion, B. P., Sajor, E. E., & Fajber, E. (2008). *Climate adaptation in Asia: Knowledge gaps and research issues in South East Asia.* Kathmandu: ISET-International.

Ross, H., Poungsomlee, A., Punpuing, S., & Archavanitkul, K. (2000). Integrative analysis of city systems: Bangkok "Man and the Biosphere" programme study. *Environment & Urbanization, 12*, 151–161.

Samabuddhi, K. (2004). *Water crisis looms, says grid study. Bangkok Post*, June 12.

Satrawaha, R., Prathepha, P., Andrews, R. N. L., & Petney, T. (2009). Fundamental hydrochemical parameters of the Songkhram River in Northeast Thailand: Foundation data for the study of an endangered tropical wetland ecosystem. *Limnology, 10*, 7–15.

Sawano, S., Hasegawa, T., Goto, S., Konghakote, P., Polthanee, A., Ishigooka, Y., Kuwagata, T., & Toritani, H. (2008). Modeling the dependence of the crop calendar for rain-fed rice on precipitation in Northeast Thailand. *Paddy and Water Environment, 6*, 83–90.

Singhrattna, N., Rajagopalan, B., Kumar, K. K., & Clark, M. (2005). Interannual and interdecadal variability of Thailand summer monsoon season. *Journal of Climate, 18*, 1697–1708.

Suthichaiyoon. (2008). Explaining the policy of Somchai's government: Responding to critical threats to the economy through deliberation and fixing the constitution [in Thai] *Suthichaiyoon*, October 2, 2008, Suthichaiyoon.com, Bangkok.

Thanapakpawin, P., Richey, J., Thomas, D. E., Rodda, S., Campbell, B., & Logsdon, M. (2006). Effects of landuse change on the hydrologic regime of the Mae Chaem River basin, NW Thailand. *Journal of Hydrology, 334*, 215–230.

Thapun, A. (2009). *A 2020 vision for the Mekong's subregion. The Nation*, June 16, 2009, Bangkok.

The Manager. (2008). *Out of fear of losing good manners "strategies for global warming" left to the next government to work out the details.* (in Thai). *The Manager*, January 28, 2008, Bangkok.

Thomas, D. E. (2006). *Participatory watershed management in Ping watershed: Final report.* Thailand, Bangkok: Office of Natural Resources and Environmental Policy and Planning, Ministry of Natural Resources and Environment.

Wipatayotin, A. (2007). *Group floats plan to keep rising sea out.* Bangkok: The Bangkok Post.

Xi, L. X., & Higgitt, D. (2009). Soil erosion, landslides and sediment yields. In: L. Lebel, A. Snidvongs, C. Chen & R. Daniel (Eds), *Critical states: Environmental challenges to development in monsoon Southeast Asia* (pp. 253–267). Selangor: SIRD/Gerakbudaya.

CHAPTER 18

CLIMATE CHANGE ADAPTATION AND DISASTER RISK REDUCTION IN VIETNAM

Huy Nguyen and Rajib Shaw

INTRODUCTION AND CONTEXT

Background

Vietnam is a stretch of land strengthening along Indochinese peninsula, which is located in Southeast Asia. Vietnam's mainland stretches from 23023′ to 08002′ north latitude and widens from 102008′ to 109028′ east longitude. Length counted in straight line from north to south stays at about 1,650 km, width from west to east maximizes at 600 km and minimizes at 50 km. Entire territory of Vietnam includes 331,112 km^2 of mainland and 1 million square kilometers of territorial sea. Currently, Vietnam enjoys 63 provinces, central cities with 622 administrative units of districts and cities and 10,511 administrative units of communes and wards.

Vietnam has inland border of about 3,730 km, shares its border with People's Republic of China in the north with border length of 1,150 km, People's Democratic Republic of Laos with length of 1,650 km, and Cambodia with length of 930 km in the west. Vietnam has sea border in the east, south, and southeast, over the China Sea and Thailand gulf, the Republic of Philippines, Indonesia, Singapore, Brunei, and Malaysia.

Climate Change Adaptation and Disaster Risk Reduction: An Asian Perspective
Community, Environment and Disaster Risk Management, Volume 5, 373–391

ISSN: 2040-7262/doi:10.1108/S2040-7262(2010)0000005024

Vietnam covers relatively complicated terrain: countless mountains, numerous rivers, and stretching and meandering coastline; the percentage relation between mountains and plains in mainland area is indifferent among regions.

With an area of about 331,112 km^2, Vietnam has a population of 85,789,573 people, according to the Census data released on April 1, 2009, and has a rapid population growth, with an average annual growth rate of 2.23%. Vietnam is in a process of rapid social and economic changes. It is foreseen that it will become a middle income country before the year 2010. However, challenges to an equitable and just development still remain. The gap between the very rich and the very poor is increasing, especially between rural and urban populations. To counteract this, the Vietnamese government has stressed the importance of an inclusive agenda in its socio-economic development plan.

Disaster Risk and Disaster Impacts

The location and topography of Vietnam make it one of the most disaster-prone countries in the world, suffering from typhoons, tropical storms, floods, drought, seawater intrusions, landslides, and forest fires (Jelligos, Lunde, Kawate, & Dzung, 2005). Of these, the most damaging and frequent are typhoons, tropical storms, floods, and drought. The northern and central regions are affected on average by six to eight typhoons or tropical storms from the China Sea per year; they also occur in the southern areas, but less frequently. Because of its topography, Vietnam is susceptible to several types of natural disasters, as shown in Table 1.

Typhoons and floods are the most frequent and most devastating of these hazards. The storm season is from May to December. In most years, storms

Table 1. Vietnam's Relative Disaster Frequency.

High	Medium	Low
Flood	Tornado	Earthquake
Typhoon	Drought	Accident (technology)
Inundation	Landslide	Frost
	Flash flood	Damaging cold
	Fire	Deforestation

Source: Ministry of Agricultural and Rural Development (MARD) (2009).

hit the northern part of Vietnam from May through June, then move gradually south from July to December. Typhoons and floods often cause a heavy loss of life and damage to infrastructure and economic activity because of the concentration of population along the coastline and in low-lying deltas, particularly of the Mekong and Red Rivers. In recent years, disasters have continually occurred all over the country, causing vast losses in human life, property, socioeconomic and cultural infrastructure as well as environmental degradation. In the recent decade (1997–2006), natural disasters such as typhoons, floods, and droughts have caused significant losses, including 7,500 missing and dead people and asset damage equivalent to 1.5% of GDP. Natural disasters in Vietnam have been increasingly severe in terms of magnitude, frequency, and volatility. Table 2 provides some evident data on that.

Climate change will only worsen the frequency, severity, and intensity of the hydrometeorological disasters that Vietnam faces. According to the Intergovernmental Panel on Climate Change (Intergovernmental Panel on Climate Change [IPCC], 2007), over the coming decades many parts of the world, including Vietnam, are likely to experience significant changes in sea level, temperature, precipitation, and the frequency and severity of extreme weather events.

Table 2. Recent Damages of Disasters in Vietnam, 1999–2008.

Year	Event	No. of People Dead	No. of People Injured	No. of People Missing	Economic Loss (VND Billion)	Area Effected
2008	Storm Kammuri	133	91	34	1,939.733	9 north and central provinces
2007	Storm Lekima	88	180	8	3,215.508	17 north and central provinces
2006	Storm Xangsane	72	532	4	10,401.624	15 central and southern provinces
2005	Storm no. 7	68	28		3,509.150	12 north central of Vietnam
2004	Storm no. 2	23	22		298.199	5 central provinces
2003	Floods	65	33		432,471	9 central provinces
2002	Floods	171			456.831	The Mekong River delta
2001	Floods	393			1,535.910	
2000	Flash floods	28	27	2	43.917	5 northern province
1999	Floods	595	275	29	3,773.799	10 central provinces

Source: MARD, 2009

Dangers to the population have increased with the spread of economic activity and development into marginally suitable areas such as floodplains, coastal swamps, drainage channels, or other natural buffers. The result is that most of Vietnam's people (about18 million) now live in low-lying river basins and coastal areas, so they are exposed to risks from multiple natural hazards.

According to the Central Committee for Flood and Storm Control, an estimated 80–90% of the total population is affected by typhoons,. Vietnam experiences an average of six to eight typhoons or tropical storms of varying intensity each year, with the northern and central coastal regions being hardest hit in the early months of the storm season. Communities along the coast are directly affected, but so too are communities in upland areas, which can experience flash floods resulting from the heavy rains of typhoons. River plain flooding is extensive and prolonged throughout the wet season in the large deltas. Most of Vietnam's 2,360 rivers are short and steep, so that heavy rainfall in their basins produces intense, short duration floods.

Sizeable portions of the country – especially the Central Highland and Central Coast – are subject to heavy rainfall. Most recently, the floods of 2008 resulted in significant loss of life and property. Along with floods and typhoons, most regions in Vietnam, especially regions in the central to the south, were significantly affected by severe drought causing adverse impacts on livelihood and the national economy. According to a drought assessment of the Ministry of Agriculture and Rural Development (MARD), in 1997–1998 about 3 million people were affected and the total losses in terms of agricultural production were estimated to about 400 million US dollars. In addition, diseases related to the lack of food, water sanitation, and hot weather were observed. Some diseases became epidemics during the drought event; about 250,000 people were infected by dengue fever in Vietnam (Lien, 1998). There were no statistics on other sectors such as fishing and ecosystem losses, recreation and tourism, etc., but all of them were considerably affected (Lien, 1999).

Climate Change Impacts

In 2007, an assessment by the World Bank listed Vietnam as one of the five countries in the world potentially most affected by climate change because so much of its population, infrastructure, and economic production are located in costal lowlands and deltas. According to one estimate, a 1 m rise in the sea level would affect 39 of the 63 provinces in six of Vietnam's eight economic regions. About 20% of the communes could be wholly or partially inundated, with the Mekong River Delta being the most seriously affected area. Also according to this estimate, a 1-m rise in sea level would affect approximately 5% of Vietnam's land area, 11% of the population, 10% of total GDP, and 7% of the agricultural input (Dasgupta, Laplante, Meisner, Wheeler, & Yan, 2007).

Many of Vietnam's cities will be increasingly affected by natural disasters. This will pose a danger to the country not only because of the large concentrations of people in those urban areas but also because cities are a critical element of Vietnam's spectacular economic growth and poverty reduction story. Vietnam's GDP per capita has grown 80% from US$402 in 2000 to US$722 in 2006; economic growth reduced poverty from around 58% in 1993 to 15.5% in 2006, according to current government of Vietnam criteria. Even at today's relatively low level of urbanization (30% as per the 2009 Census), cities are the major contributors to GDP (70% as per a 2006 World Bank study). Further, the World Bank estimates suggest that 1 million people every year will be added to the urban populations until 2020. From 1990 to 1995, the number of rural–urban migrants was 1,037,800; this increased to 2,291,000 in the 1995–2000 period and is estimated to be 3,000,000 in the 2000–2005 period (GSO, 2005). As a result, the urban population, which is around 30% of the total, according to the latest Census, is estimated to increase to about 33% of the population by 2010 and around 35% (UN estimates) to 45% (according to the Vietnamese Ministry of Construction) of the total population, which is more than 45 million people by 2020.

CONSEQUENCES OF NATURAL DISASTERS TO SOCIOECONOMIC DEVELOPMENT

Socioeconomic Consequences

Natural disaster in Vietnam is the direct impediment to the economic development, sustainable development, and poverty reduction – the huge obstacle to the process of striving for the Millennium Development Goals. Vietnam has more than 80% of its population living at risk of direct impacts of natural disasters. Natural disaster has taken away many achievements of the national socioeconomic development. In the last 5 years (2002–2006),

Table 3. Summary of Natural Disasters in Vietnam from 1953 to 2007.

Hazard	No. of Events	Killed	Injured	Homeless	Total Affected	Damage US$ (000)
Typhoon		20,260	8,819	3,667,355	39,497,880	1,897,525
Average/event	70	289	126	52,391	564,255	27,108
Flood		4,843	1,165	375,650	24,119,355	1,385,600
Average/event	49	99	24	7,666	492,232	28,278
Landslide		96	39	0	39	0
Average/event	3	32	13	0	13	0
Drought		0	0	0	6,110,000	649,120
Average/event	5	0	0	0	1,222,000	129,824

Source: Ministry of Agriculture and Rural Development (MARD), 2008.

natural disaster has killed 1,700 people and caused losses of estimated VND 75,000 billion of assets. Natural disaster intensifies the division in residents' living standard, and hinders and lowers the hunger eradication and poverty alleviation, especially in areas frequently at risk of disaster. On average, millions of people are in need of assistance due to natural disasters every year. Many of them, who have just escaped from poverty, are impoverished due to the disasters.

Natural disaster affects educational development, destroys educational infrastructure, and interrupts school time, especially in mountainous areas and the Mekong River Delta. Natural disaster also causes negative impacts on vulnerable groups such as the old, the disabled, women, and children. Table 3 indicates the most affected disasters to the social economy of Vietnam in the last 50 years since 1953. Most disasters are water-related hazards. Asa result, we can see that the worst disaster in Vietnam is caused by typhoon, which results in great damage to human lives and properties; the flood and landslide are the second and third causes of disastrous effects in Vietnam after typhoons. Drought is ranked fourth, which affects a large area of Vietnam.

Environmental Consequences

- Natural disaster destroys, degrades, and pollutes environment and negatively affects production and community's life.
- Natural disaster results in water pollution and disease generation.

Consequences of Natural Disasters to National Defense and Public Security

- Destroy defense and security infrastructure
- Reduce the national reserve
- Cause social instability
- Cause chaos in social security and order

Over the last few decades, the damage due to disasters has increased dramatically, a trend that is likely to continue in the future as climate change is expected to alter the current storm system and rainfall patterns.

For a population of more than 85 million, there has been an impressive impact on poverty alleviation: per capita GDP has doubled from US$200 to US$400 during the 1990s. Recent household surveys highlight the success in reducing poverty, and from 1993 to 2004 the poverty rate, based on per capita consumption, was almost halved and fell from approximately 58% in 1993 to 19.5% in 2004, indicating a fall of almost 39% over 11 years. The rate of poverty in 2004 was only a third of that in 1993: in absolute terms, some 24 million people were lifted out of poverty over this period.

STATUS OF DISASTER RISK REDUCTION AND CLIMATE CHANGE ADAPTATION IN VIETNAM

Vietnam has a long history of preparedness for, and active response to, natural disasters. The extensive system of dykes and seawalls is evidence that citizens and leaders over the centuries have recognized the country's vulnerability to the consequences of typhoons and other tropical storms. The government of Vietnam is actively engaged in preparing for what could be even greater future challenges of climate change adaptation (CCA) and disaster risk reduction (DRR).

National Policy Framework

Many developing countries have developed (or are in the process of developing) National Action Plans for Adaptation to Climate Change (NAPA). At present, there is no overarching national strategy on CCA in Vietnam, but the increasing emphasis on adaptation is reflected by its inclusion, albeit in broad terms, in the 5-year socioeconomic development plan (SRV, 2009). Integration into provincial and local planning has so far not taken place to any appreciable extent.

The two overarching policy documents dealing with disaster risk management in the context of climate change in Vietnam are the National Strategy for Natural Disaster Prevention, Response and Mitigation to 2020, and the National Target Program (NTP) to respond to climate change. These national frameworks set the stage for other activities on climate change and are complemented by other dedicated ordinances and laws. The National Assembly has adopted numerous pieces of legislation related to natural disasters, notably the Law on Water Resources (1998), the Ordinance on Flood and Storm Control (1993), the Law on Dykes (2006), and the Environment Protection Law (1998). Even so, according to an analysis by the Global Facility for Disaster Reduction and Recovery: "Much of the legislation lacks clear institutional arrangements for enforcement, and the current organizational structures, mandates, annual budget earmarks and working agenda focus largely on disaster response rather than prevention. There is no professional and specialized cadre of staff who focus on disaster management. Instead, it is managed in an 'as-needed' basis, part-time, by staff of the agriculture and rural development sector, mainly under the irrigation and dyke management sub-sectors." These are the big challenges for the implementation of DRR.

National Institutional Framework

The Ministry of Natural Resource and the Environment (MONRE) has been designated as the leading agency for climate change coordination in Vietnam, while the MARD maintains overall responsibility for natural disaster mitigation and response. In addition, the Ministry of Construction maintains responsibility of drainage systems and major public works, the Ministry of Planning and Investment has purview of the issues of land use and master planning, and the Ministry of Science and Technology is involved in climate forecasts.

The Central Committee for Flood and Storm Control (CCFSC), chaired by the minister of MARD, provides a coordinating umbrella for disaster risk management in Vietnam. MONRE is represented as a member on the Committee. Other members include representatives from the ministries of planning and investment, finance, fisheries, transportation, science and technology, construction, health, industry, labor and social affairs, telecommunications, and foreign affairs; Vietnam Red Cross, Vietnam TV, Voice of Vietnam, and DDMFC are also represented.

Each sectoral ministry has its own Committee for Flood and Storm Control that cooperates with the national central committee and offices at provincial level. According to their own functions and duties, ministries, sectors, and local entities are responsible for effectively implementing relevant contents, objectives, duties, and solutions stated in the Strategy.

MARD: Leading the Dialogue on Disaster Risk Reduction

MARD is the lead agency for the National Strategy for Natural Disaster Prevention, Response and Mitigation to 2020, which was approved by the government in November 2007. It succeeded the Strategy and Action Plan for flood mitigation, which had been adopted in 1994. Vietnam does not have a law dealing with disaster risk management, although discussions about such a law have been under way. Although discussions have also been underway on preparing a strategic national action plan (SNAP) on DRR, but no such plan has been drafted yet. After the approval of the National Strategy for Natural Disaster Prevention, Response and Mitigation to 2020, all the ministries, provinces, and cities have to make their own strategic action plans, and based on these action plans the CCFSC is to make a SNAP. Recently, CCFSC has mobilized a consultancy team to work on this.

The National Strategy for Natural Disaster Prevention, Response and Mitigation to 2020 outlines Vietnam's strategy for disaster mitigation and management and focuses on floods, storms, and drought. In addition to setting policy for disaster response, the National Strategy for Natural Disaster Prevention, Response and Mitigation to 2020 also provides long-term strategic orientations. The strategy provides a very concrete and practical action plan including programs on improvement of legislation and policies, consolidation of organizational structures, community awareness-raising, forestation and protection of upstream forests, strengthening of disaster management capacities through science and technology, structural measures, and some programs on strengthening of warning and forecast capacities. All 63 provinces and cities of Vietnam are tasked with developing action plans to implement the National Strategy for Natural Disaster Prevention, Response and Mitigation to 2020. All the ministries and all the provinces and cities of Vietnam are tasked with developing action plans to implement the National Strategy for Natural Disaster Prevention, Response and Mitigation to 2020. As of March 2009, 10 ministries and 56 provinces and cities had approved their own action plans. The actual degree of implementation and funding of these plans varies widely, however.

MARD and CCFSC preside over the implementation of the National Strategy for Natural Disaster Prevention, Response and Mitigation to 2020. The implementation of the Strategy is urged by regulation in other ministries, sectors, and localities. CCFSC acts as the national focal point with international communities in the field. Specific implementation responsibilities have been assigned to ministries, sectors, and local entities as follows: Ministry of Planning and Investment takes lead and works in collaboration with Ministry of Finance, MARD, CCFSC, National Committee for Search and Rescue, and other relevant ministries and sectors to balance and arrange annual investment resources in accordance with the law on state budget and other funding resources to effectively implement the Strategy on DRR. MARD is in charge of inspection and assessment of the Strategy implementation by ministries, sectors, and local entities. It conducts review of the Strategy implementation every year and every 5 years draws out experience, and recommends to the prime minister suitable adjustments to the contents and solutions of the Strategy.

MONRE: Leading the Dialogue on Climate Change Adaptation

The prime minister tasked the MONRE with preparing an NTP on climate change, which was approved with Decision 158 in December 2008. The NTP is defining a set of actions from central level to adaptation efforts at provincial, district, and commune level following a three-phase approach: first phase (2009–2010), starting up; second phase (2011–2015), implementation; and third phase (after 2015), development. The targets to be achieved by 2010 are the following: complete climate change scenarios, especially about sea level rise; complete the assessment of level, trends, and nature of climate change parameters; and implement pilot projects to assess climate change impacts on "fields, sectors, and localities that are sensitive and vulnerable to climate change, i.e., water resources, irrigation, agriculture, health and livelihoods, deltas, and coastal areas."

The NTP aims to provide a framework and guideline for ministries, sectors, and provinces to develop their own action plans, primarily in response to rising sea levels. Namely, the NTP targeted the competent bodies to complete the development of methodologies and guidance for identifying measures to respond to climate change and ask for specific pilot project implementation in different sectors and areas, also calling for support and coordination among the ministries, sectors, and local entities and the NTP management and implementation apparatuses. More precisely, ministries, sectors, and local entities are required to "complete the development of action plans to respond to climate change" by 2010, and ministries, sectors, and local entities dealing with specific vulnerabilities or placed in vulnerable areas are required to "implement their action plans" by 2010.

It is a clear responsibility of MONRE to guide and assist ministries, sectors, and provinces in developing and implementing their action plans to respond to climate change. Ministry of Planning and Investment (MPI) should take the lead and coordinate with other ministries, sectors, and provinces to develop a standard framework procedure and guidelines for mainstreaming climate change issues into socioeconomic development strategies, programs, and planning. Finally, people's committees of provinces and major cities should themselves develop and implement action plans to respond to climate change in their provinces and cities (MONRE, 2008). This workbook is therefore quite timely, since it is intended to support the definition of methods and guidelines to develop action plans to respond to climate change particularly at city level.

At sector and provincial levels, climate change responses have not yet been addressed systematically. Thus far, only MARD has promoted on September 2008, an action plan for adaptation and mitigation to climate change, with specific initiatives beyond ongoing DRR measures, but this action plan does not refer to MONRE NTP (which was approved later, on December 2008). The coordination between MARD and MONRE is lacking on the activities of CCA (Fig. 1).

The synergies and opportunities to coordinate have not yet been established between the MARD's Action Plan and National Target Plan on climate change.

Links between CCA and DRR

In addition to understanding terminology, it is also important for participants to be able to grasp the relationships between CCA and DRR. The essential point for participants to understand is that climate change almost certainly will increase the frequency and intensity of the types of natural disasters that are most common in Vietnam, i.e., typhoons and other tropical storms, and could introduce new risks such as disease epidemics and physical risks to areas previously unexposed. Vietnam also is likely to

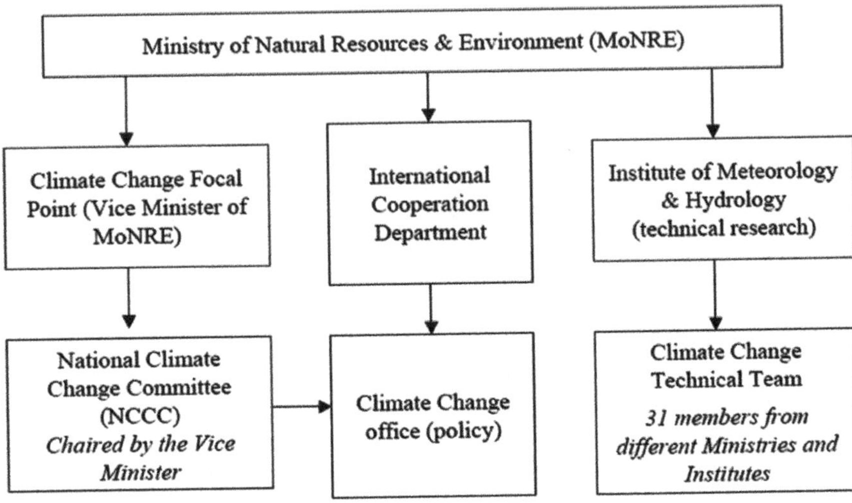

Fig. 1. Government Institutional Arrangement for Responding to Climate Change.

experience the consequences of rising sea levels linked to climate change; a rise of 1 m in mean sea level (now considered by many scientists to be probable by 2100) could displace 10% or more of Vietnam's current population. As a result, climate change will expose Vietnam to heightened risks of disasters, which must be managed for the sake of the people and their communities.

It is important to recognize that regular seasonal flooding, storms, or landslides may be disasters but may not reflect climate change per se. Climate change refers to changes in the predictable climate over time. For instance, annual mean temperature changes, precipitation changes, and changes to the levels of the sea at specific points over several decades can be the indication of climate change.

Disaster risk can be managed by taking steps that reduce the potential impacts of disasters, such as building dykes and preventing construction of housing and infrastructure in flood-prone areas. DRR can reduce the need for climate adaptation. However, not all climate change results in disasters; similarly, DRR can cover disasters other than those that are climate induced (e.g., seasonal flooding, seismic activity, etc.). Understanding these nuances is important to identifying the actors that could be involved in reducing vulnerabilities. Climate change and disaster concerns should not be isolated under the remit of any single ministry, but should be systematically integrated across all major development sectors. CCA and DRR policies need to be integrated into long-term planning for sustainable development and poverty alleviation policies. In particular, climate change needs to be incorporated into the next round of provincial Socio-Economic Development Plans (2011–2020). The "mainstreaming" of adaptation measures requires a comprehensive and integrated assessment of vulnerability, and how to address this through risk management.

NEED OF A NATIONAL PLATFORM FOR DRR IN CONTEXT OF CLIMATE CHANGE IN VIETNAM

Why a National Platform for DRR in Vietnam?

Vietnam will be increasingly impacted by natural disasters and climate change impacts. There is therefore an urgent need to invest in DRR and CCA activities and to mainstream DRR and CCA into sustainable development policies in order to reduce our society's vulnerability to these impacts.

Vietnam has already attained substantial achievements in building a multistakeholder approach to DRR and in integrating DRR into national development policies. Challenges, however, remain and no formal national platform for DRR has been established yet. The main challenges include (i) limited awareness and understanding of the DRR concept on climate change impacts, (ii) hindrance in community consultations because of lack of understanding of the legal development process, (iii) need to balance limited government budgets to cover both socioeconomic development needs and DRR investments in the context of increasing disaster impacts and related damages and losses, and (iv) need to identify roles and responsibilities of relevant stakeholders to be involved in the process and to gain their commitment

A national platform for DRR, which brings together Vietnam's multiple stakeholders to effectively advocate for DRR at different levels, would support

- Cooperation among key ministries and government bodies
- Gaining support from donors and nationally based UN agencies
- Increasing integration of DRR in international development and humanitarian assistance programs and strategies
- Infusing DRR into national development policies, strategies, legislation, or national plans

- Developing legislation related to DRR
- Adequately fitting elements of DRR to sustainable development issues
- Enhancing understanding of DRR and vulnerability at national and community level
- Integration of DRR into school curricula
- Encouraging community participation in DRR through awareness raising and training activities
- Involving media, private sector, and scientific community in DRR issues
- Establishing risk map and early warning systems and strengthen building codes
- Improving disaster preparedness and developing disaster contingency plans
- Strengthening the regional cooperation.

Proposed Membership of the National Platform

Membership should be open to all organizations and stakeholders involved in disaster risk management. Initially, the organizations listed in Table 4 below should be invited to have representatives on the national platform:

There will be several steps involved in the establishment of the national platform (Fig. 2). Initially it will be necessary to establish a small working group comprising nominated representatives from each relevant stakeholder group.
The primary focus will be on steps to advance the design and implementation of cooperative action and the elements required to maximize synergies through a multistakeholder approach toward DRR and CCA. A deeper appreciation of these enabling conditions will also provide timely insights to climate change negotiators.
Based on this review, the working group will draft an outline document describing the structure of a national platform, which will include the goal, objectives, prioritized principal activities, themes, the terms of agreement between partners, etc.

The national platform should be established based on the existing organizations that have been working in the field of DRR. Phase II of the NDM-P ended in June 2009 and a review team is currently examining the feasibility and finalizing recommendations for a new partnership agreement. The new partnership is likely to be both an agreement and an entity serving

Table 4. Proposed Membership of the National Platform.

Government	Academic Bodies/ Institutes	Donors	INGOs/ NGOs
MARD (Chair)	Water resources University	UN agencies	CARE
MONRE	Hanoi School of Public Health	World Bank	Oxfam
MoFA	Vietnam National University	AusAID	Save the Children
MoH	Institute of Geophysics	JICA	CECI
MoLISA	Thai Nguyen University	ADB	World Vision
MoC	Hue University	Norwegian Embassy	DWF
MoT	Da Nang University	Netherlands Embassy	ADRA
MPI	Ho Chi Minh University	EC/DIPECHO	PLAN
Ministries for Finance and Planning	Can Tho University	Representatives from business/private sector	ADPC
Vietnam Red Cross Society			IFRC
Women's Union			
Youth Union			
Fatherland Front			
PACCOM			
Representatives from six provinces (two from each region)			

the government, donors, NGOs, and other stakeholders involved in Disaster Risk Management (DRM) in Vietnam. The activities of a new partnership will likely include steps to fulfill the commitments to the Hyogo Framework for Action, including the establishment of a national platform. The platform, in the shape of a partnership, will have as members the government, donors, NGOs, academic institutions, regional/international organizations, and private sector and media representatives.

Climate change is a very real threat to Vietnam's continued socio-economic development. Increasingly erratic and variable rainfall, higher temperatures, more intense extreme weather events like typhoons, droughts, and heavy rainfall causing floods, and the rising seawater level will all have significant impacts across sectors, regions, and income groups, and particularly on livelihood security of the poorest rural people. Climate

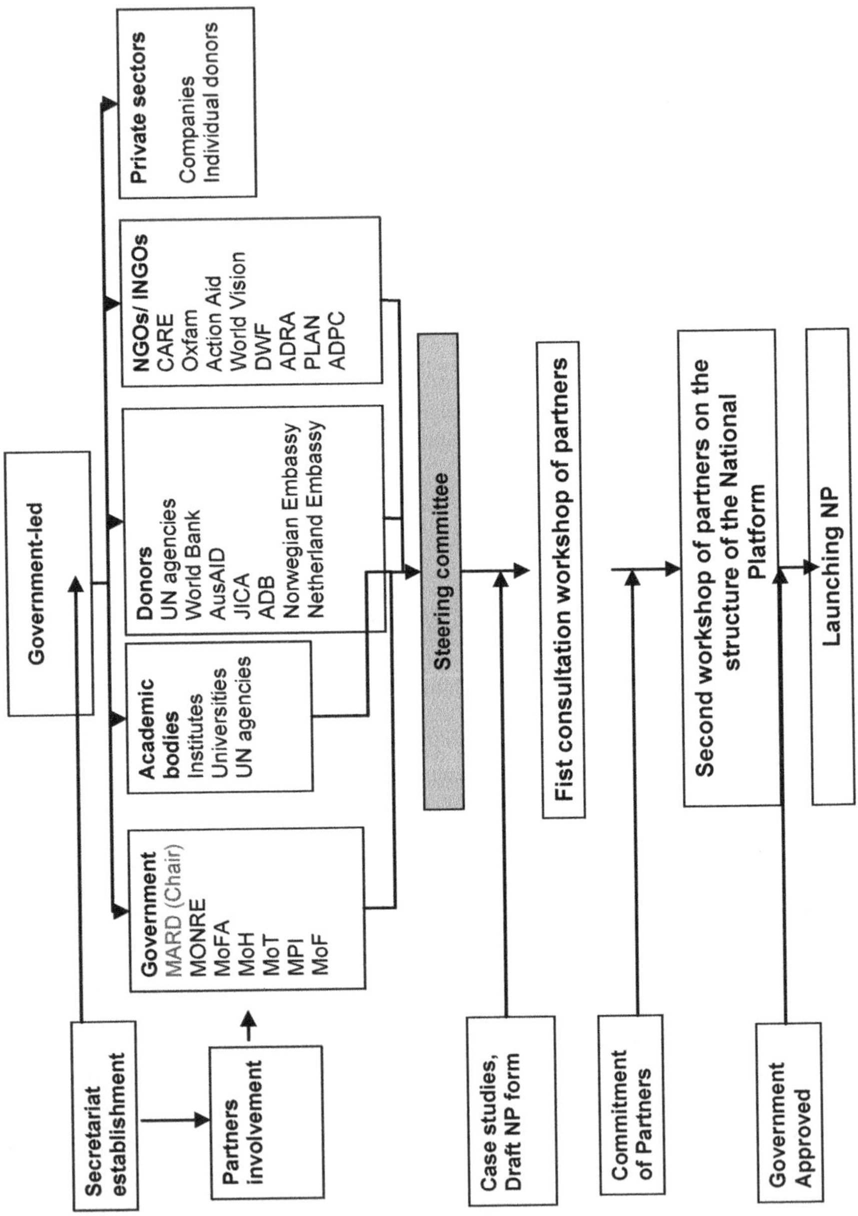

Fig. 2. A Roadmap for Establishing National Platform.

change brings closer together the three areas of the disaster management community, the environmental community, and the poverty reduction community – mainstreaming is the key to doing so, but it is important to stay focused. The threat of climate change is beginning to be acknowledged, but information and awareness remains at a low level. Vietnam does not yet have national or local CCA strategies, and national and local capacity building is urgently needed to ensure that policy responses are adequate and effective.

There is not so much focus on the gradual changes of climate change – reactive approach to large-scale visible impacts. Thus, it is needed to be proactive to the chronic changes of climate change that can and will lead to disasters of another sort. Many existing frameworks for sustainable development and livelihoods and for DRR are relevant tools that can be adapted to incorporate responses to known climate change impacts. However, there is a real need to connect the scientific community to the local scale where these changes and coping mechanisms are taking place.

Coordination between line ministries also needs to be urgently improved, and cooperation with international agencies and NGOs enhanced so that climate change can be addressed in an integrated way with long-term socioeconomic and poverty reduction efforts. Crucially, those most at risk from climate change, the rural poor living in provinces most affected by climate change have limited information or financial and technical support to adapt to their changing world, despite some localized successes. Their direct experience of climate change impacts should be incorporated into future responses, and solutions sought that build upon existing local adaptation practices, where appropriate. The local people who are directly affected by climate change and disasters should also be key participants in the planning and implementation of future CCA measures, particularly where these require relocation or significant dislocation of existing livelihood practices.

ACKNOWLEDGMENTS

The work published in this chapter was supported by the Ministry of Education, Culture, Sports, Science and Technology (MEXT) of Japan through doctoral fellowship to the primary author at Kyoto University, and Japanese Society for Promotion of Science (JSPS) project fund and Graduate School of Global Environmental Studies (GSGES) project fund through grand research to the secondary author. We also acknowledge

Ministry of Agriculture and Rural Development, Ministry of Natural Resource and Environment of Vietnam, UN International Strategy for Disaster Reduction, and World Bank in Vietnam for supporting us in this research.

REFERENCES

Dasgupta, S., Laplante, B., Meisner, C., Wheeler, D., & Yan, J. (2007). *The impact of sea level rise on developing countries: A comparative analysis*. Washington, DC: World Bank press.

Intergovernmental Panel on Climate Change (IPCC). (2007). *Climate change impacts, adaptation and vulnerability*. Contribution of Working Group II to the Fourth Assessment Report. Geneva, Switzerland: IPCC.

FURTHER READING

Chaudry, P., & Greet, R. (2007). *Climate change and human development in Vietnam, in occasional paper*. New York: United Nations Development Program.

Dasgupta, S., Laplante, B., Murray, S., & Wheeler, D. (2009). *Sea-level rise and storm surges: A comparative analysis of impacts in developing countries*. Policy Research Working Paper 4901. World Bank, Washington, DC.

General Statistics Office of Vietnam (GoV). (2005). Population and employment. Hanoi, Vietnam.

Global Facility for Disaster Reduction and Recovery (GFDRR). (2009). Disaster risk management programs for priority countries: East Asia and Pacific, Vietnam. Washington, DC: World Bank, 107.

Huy N., Jerry, V., & Praveen, P. (2009). A roadmap for establishing a national platform for DRR in Vietnam. *National Forum on DRR and CCA in Vietnam*, October 7, 2009.

Jelligos, S.R., Lunde, G., Kawate, H., & Dzung, T.V. (2005). Final evaluation: Vietnam Red Cross mangrove and disaster preparedness in the Red River Delta and Northern Coastal Vietnam (1994–2005).

Lien, T.V. (1998). Variation of dengue fever in Vietnam and its relationship to ENSO event. *Proceeding of the 7th scientific conference of IMH* (pp. 2, 50–58), (in Vietnamese).

Lien, T. V. (1999). *Relation between forest ecosystem and hydro-climate in the central highland of Vietnam*. Report of national program on forest inquire and planning of the period 1995–2000, Institute of Forest Investigation and Planning, Hanoi, pp. 145, (in Vietnamese).

Ministry of Natural Resource and Environment (MONRE). (2008). National Target Program for Climate Change Adaptation (NTP), Hanoi, Vietnam.

Ministry of Natural Resource and Environment. (2009). *Climate change and sea level rise scenarios in Vietnam*. Ministry of Natural Resource and Environment: Hanoi, Vietnam (draft version).

N.A.P. (2002). *Vietnam national action program to combat desertification*. UNCCD report. Hanoi, Vietnam, 117 pp.

Nicholls, R. J., Hanson, S., Herweijer, C., Patmore, N., Hallegatte, S., Corfee-Morlot, Chateau, J., & Muir-Wood, R. (2007). *Ranking port cities with high exposure and vulnerability to climate extremes: Exposure estimates*. Environmental Working Papers no. 1. OECD (Organization for Economic Co-operation and Development), Paris.

Ninh, N. H. (2007). Vulnerabilities, adaptation and resilience to climate change in Vietnam: Capacity needs. *The workshop on vulnerability, adaptation and resilience to climate risks: An assessment of education and training capacities and needs in Southeast Asia*, Chiang Mai, Thailand.

Socialist Republic of Vietnam (SRV). (2006). *The five years socio-economic development plan 2006–2010 of Vietnam*. Hanoi, Vietnam: National Publisher.

Socialist Republic of Vietnam (SRV). (2009). Vietnam National Progress Report on the implementation of the Hyogo Framework for Action. Ministry of Agriculture and Rural Development. Hanoi, Vietnam.

Vietnamese Government. (2009). *National plan implementation for the National strategy on National disaster prevention, response and mitigation to 2020*. Hanoi, Vietnam: National Publisher.

Viner, D., & Laurens, M. B. (2006). *Linking climate change adaptation and disaster risk management for sustainable poverty reduction: Vietnam country study*. Switzerland: Vulnerability and Adaptation Resource Group.

Waibel, M. (2008). Implications and challenges of climate change for Vietnam, *Pacific News N. 29*, January/February 2008, pp. 26–27.

World Bank. (2006). Issues and dynamics: Urban systems in developing East Asia, p. 2. Disaster risk management programs for priority countries: East Asia and Pacific, Vietnam.

CHAPTER 19

LOCAL VULNERABILITY AND ADAPTATION TO EXTREME CLIMATE EVENTS ALONG THE CENTRAL COAST OF VIETNAM

Bui Duc Tinh, Tran Huu Tuan, Tran Phong, Bui Dung The and Bui Thi Tam

INTRODUCTION

A review of the existing literature on disasters, impact, local vulnerability, and adaptation indicates to the fact that recent increasing frequency of natural disasters (e.g., floods, storms, and drought) have increasingly caused impacts on a diverse set of physical and biological systems, especially for those living in the developing countries where their livelihood strategies rely on natural resources (McCarthy, Canziani, Leary, Dokken, & White, 2001; Selvaraju, Subbiah, Baas, & Juergens, 2006; Kumar, 2007; Cruz et al., 2007).

Located in Southeast Asia – typhoon center of the South China Sea, which is one of the biggest typhoon centers of the world – Vietnam is recognized as one of the top ten countries most prone to the impacts of climate change (CC) in the world. Increasing climate changes and induced impact on local communities have been occurring widely throughout Vietnam, and particularly in the Central region, which is recognized as the most disaster-prone region of the country. Annually, this region suffers five to seven extreme

Climate Change Adaptation and Disaster Risk Reduction: An Asian Perspective
Community, Environment and Disaster Risk Management, Volume 5, 393–416

ISSN: 2040-7262/doi:10.1108/S2040-7262(2010)0000005025

disasters, particularly of typhoons and floods. The situation is projected to worsen in frequency and severity of extreme disaster events. Climate changes, especially extreme disasters, killed and destroyed the livelihoods of many people, and pushed many local communities into poverty annually.

The year 2006 was recognized as the "year of typhoon," with four typhoons, including Xangsane Typhoon Quang Nam, which is recognized as a hundred-year event (Quang Nam People Committee, 2007). Four typhoons caused serious damages to local communities, with 14 dead, 307 injured, and total damages valued at nearly VND 600 billion. Xangsane hit Quang Nam with a wind intensity that reached the 13th level and, along with heavy rain, thus led to extreme floods (Quang Nam People Committee, 2007). It is reported that this storm also caused damages to Dai Loc district, with 7 dead and 295 injured. The total damages cost an estimated VND 578 billions.

It is recognized that the poor would be the most vulnerable group to climate risks, as poor communities often settle in marginal or disaster-prone areas, such as coastal areas, river basins, mountain slopes, and urban slums with limited capacities and resources that constrain them from response and adaptation to extreme disasters, thus heightening their exposure to the impacts of climate hazards. Climate change impacts and variability, therefore, threaten to exacerbate their existing vulnerabilities and adaptation capacity to climate change (IISD, SEI, IUCN, Intercooperation, 2003).

This chapter is developed based largely upon the concept of local vulnerability and adaptation to typhoon and empirical research in one of the most climate change–related disaster areas and poor communities in the South Central Coast of Vietnam. Eight focus group discussions (FGDs) conducted at study site in which 1 FGD at provincial level, 1 FGD at district level, 2 FGDs at commune level and 4 FGD at village level and 20 key informant interviews were conducted, from the provincial government level to the village level. In doing so, it helped to collect information about socioeconomic context, profiles of disasters and the Xangsane typhoon, local responses and adaptation to typhoon, and information about barriers and difficulties that constrained local organizations in coping with disasters. A questionnaire survey with face-to-face interviews was conducted with 420 local households living in the most Xangsane-affected areas in Quang Nam. The survey covered various themes of local vulnerability and adaptation, such as their risk exposure, typhoon problems, damages, and coping mechanisms.

The main objectives of this chapter are to investigate and analyze the empirical-based evidence of associations between local socioeconomic vulnerability and their adaptive coping measures to typhoon. The first part of the chapter is dedicated to presenting an overview of the theoretical concepts of local vulnerability, local socioeconomic context, and adaptation to climate change–related disasters. Regarding investigation of response and adaptation measures to deal with events, particular attention was placed on the coping mechanisms of both the local government organizations and local communities *before, during,* and *after* of the event. In the third part of the chapter, an analysis will also address the local gap and needs in adaptation to climate change–related events.

THEORETICAL PERSPECTIVES ON ADAPTATION CAPACITY AND ADAPTATION TO CLIMATE CHANGES

Local Adaptive Capacity to Climate Changes

The capability to adapt is a fundamental determinant of how vulnerable a specific system is to external or internal stresses (Keskitalo, 2004). In the climate change context, this attribute is referred to as "adaptive capacity," defined as "the ability of a system to adjust to CC, to moderate potential damages, to take advantage of opportunities, or to cope with the consequences" (McCarthy et al., 2001).

Key components of adaptive capacity enhancement include wealth, technology, education, information, skills, infrastructure, access to resources, and stability and management capabilities (IPCC, 2001). The implementation of adaptation strategies requires resources, including social resources, human resources, institutional resources, natural resources, and economic resources. The types of resources required and their relative importance will depend on the context within which adaptation is pursued, on the nature of the hazards faced, and on the nature of the adaptation strategy. Adaptation strategies will not be successful unless there is a willingness to adapt among those affected, as well as a degree of consensus regarding what types of actions are appropriate. Adaptive capacity, therefore, depends on the ability of a society to act collectively, and to resolve conflicts between its members – factors that are heavily influenced by governance (Lim, Spanger-Siegfried, Burton, Malone, & Huq, 2004).

There are many possible measures to adapt to the climate change. Such possible adaptation measures are based on experience, observation, and speculation about alternatives that might be created; they cover a wide range of types and take numerous forms. Smit, Burton, Klein, and Wandel (2000) presented comprehensive lists of potential adaptation measures in coastal communities, including a wide range of engineering measures, improvements, or changes, including agricultural practices that are more flood-resistant; negotiating regional water-sharing agreements; providing efficient mechanisms for disaster management; developing desalination techniques; planting mangrove belts to provide flood protection; planting salt-tolerant varieties of vegetation; improving drainage facilities; establishing setback policies for new developments; developing food insurance schemes; and devising flood early warning systems.

From review of the literature, it can be seen that a conceptual framework for assessing and enhancing adaptive capacity to climate change should include social, economic, and environmental dimensions of communities. It means that this is very hard to find a "one to fit all" framework for community adaptation to climate change because a framework is largely dependent on local context of certain community.

Local Response and Adaptation to Climate Change

Adapting to climate changes entails taking the right measure to reduce the adverse impacts of climate change by making the appropriate adjustment and changes at every level – from community to national and international (UNFCCC, 2007). A couple of definitions of adaptation can be found in the climate change literature. Brooks (2003) describes adaptation as "adjustments in a system's behavior and characteristics that enhance its ability to cope with external stress." Smit et al. (2000) refer to adaptations as "adjustments in ecological-socio-economic systems in response to actual or expected climatic stimuli, their effects or impacts." Pielke (1998) defines adaptations as the "adjustments in individual groups and institutional behavior in order to reduce society's vulnerability to climate." Smit and Pilifisova (2003) refer to "adjustments in management strategies to actual or expected climatic conditions or their effects, in order to reduce risks or realize opportunities."

Adaptation has the potential to reduce adverse impacts of climate changes and to enhance beneficial effects, but will incur costs and will not prevent all damages. In reality, it is stated that people living in the less developing countries, where economic reliance on natural resources and rain-fed agriculture is high, face additional challenges through climate change due to limited access to management options and have fewer resources to reduce impacts (Selvaraju et al., 2006; Kumar, 2007). Regarding the impact of climate changes on food crop production and local responses, the results of previous studies (Selvaraju et al., 2006; Kumar, 2007) indicate that local people tend to be involved in locally managed practices to mitigate and overcome adverse impacts of climate changes on their livelihood practices. Local people can adopt coping mechanisms to withstand climate variability through activities such as the use of drought-resistant or salt-resistant crop varieties, more efficient use of water resources, sowing earlier, adjusting timing of the crop calendar, and improved pest management (Kumar, 2007). Researchers (Frisvold & Ingram, 1995; McCarthy et al., 2001) have also identified that local responses to natural disasters (e.g., floods, storms) vary widely and depend on a list of elements such as geographic location, livelihood practices, and household characteristics.

CASE STUDY AREA AND VULNERABILITY

Quang Nam is located in the heart of Vietnam, about 860 km to the north of the city of Ho Chi Minh and 865 km south of the country's capital, Ha Noi. The province is located in the intersection between the two geographical regions of North and South Vietnam, which are characterized by a slope topography from west to east, with many mountainous ranges, short rivers, and delta and coastal areas that created a diversified ecosystem that has become one of the highly disaster-prone regions in Vietnam.

Located in the disaster-prone area of Dai Loc with two main rivers (Vu Gia and Thu Bon), and about 2 km to the north of Ai Nghia (the town), Dai Hoa and Dai Cuong are identified as low-lying communes that are vulnerable to climate change–related disasters such as typhoon and floods. Large areas of Dai Hoa and Dai Cuong are fields of paddy and other food crops such as maize, watermelon, banana, and various varieties of vegetables, which are the main sources of income for local people. Consequently, these two communes are the most vulnerable areas to disasters in the Dai Loc district.

The profile of Xangsane and its impacts on local communities indicates that the Dai Loc district is one of the most seriously affected districts of northern Quang Nam. It is reported that Dai Cuong and Dai Hoa are two of seven communes most seriously affected by the landfall of the Xangsane typhoon and flood in 2006. Thus, the Dai Cuong and Dai Hoa communes in the Quang Nam province were purposely chosen. This selection also satisfied the Environmental Economics Program of Southeast Asia's

(EPSEA's) cross-countries guidelines in a sense that the event occurred in coastal lowland.

The figures of 8,779 and 14,148 are the total population of Dai Cuong and Dai Hoa, respectively. With total areas of 963 ha and 1,363 ha, respectively, the population density is relatively high, with 954 people/km^2 in Dai Cuong and 1,038 people/km^2 in Dai Hoa. Local communities are mainly dependent on agricultural production activities in which food crops tend to be subsistence; fortunately, with the fertile cultivated land, local communities have diversified their crop pattern to adapt to climate changes by growing short-term varieties such as vegetables, maize, and watermelon. Nonfarming activities such as fishing net and garment making are also important source of income for local communities. However, due to the low average of cultivated land (less than 400 m^2/person), local communities have to address food security. They all produce food to meet family demand rather than for selling; many even have to buy food for family in several months. The study also found that for households living along the Vu Gia and Thu Bon riverbanks, for example, in Giao Thuy village, lost almost all of their cultivated land due to flood-related landslides.

As far as infrastructure is concerned, as a location near a district town, local commune living in Dai Hoa and Dai Cuong provides good opportunities to access infrastructures and public services such as district markets, banks, high schools, and health care systems. The study also found that there are kindergartens, primary and secondary schools, and medical stations located in the center of each commune that meet the basic needs of local communities. The system of intervillage and intercommune roads are paved, connecting to the center of each commune and to the center of the Dai Loc district (see Fig. 1).

LOCAL ADAPTATION CAPACITY AND RESPONSES TO XANGSANE TYPHOON

Adaptive Capacities/Profile of Interviewed Households

The household survey was conducted in two communes: Dai Hoa and Dai Cuong of the Dai Loc district. The number of respondents was 392, of which 47% were male and 53% were female, and most (82%) of them were the heads of the household. All respondents were between 16 and 85 years old, with the average age being 51 years old. Most of the respondents

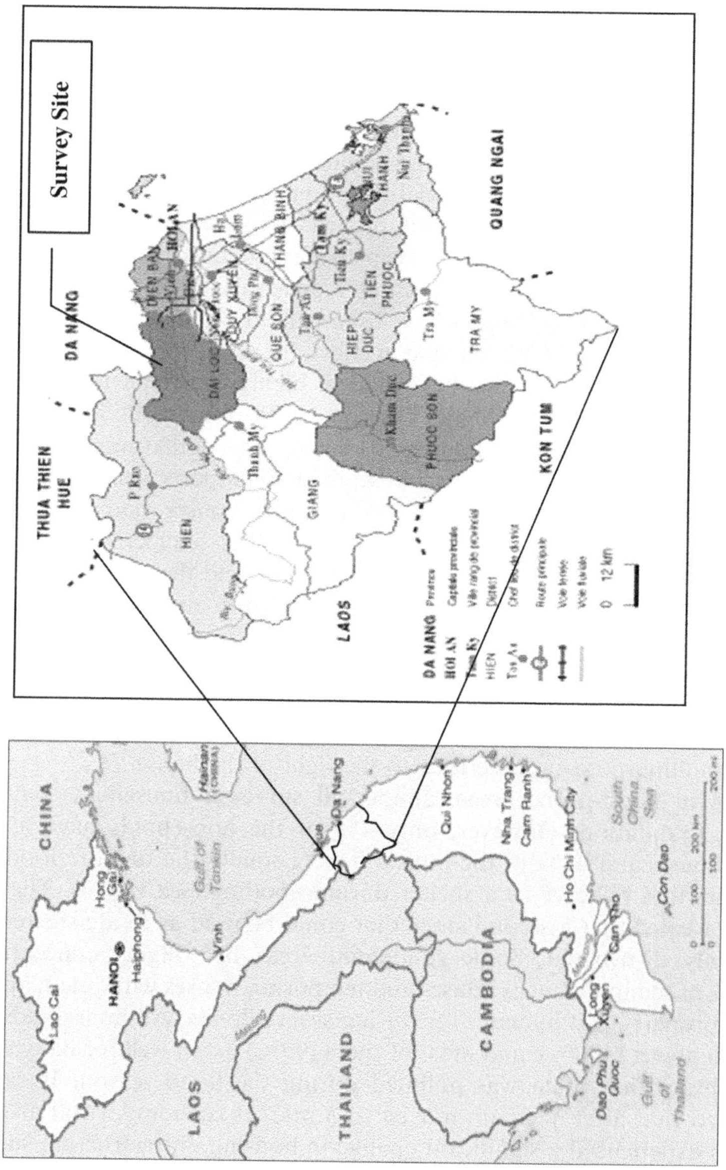

Fig. 1. Maps of Vietnam and Quang Nam Province.

finished their primary and secondary education, and the average number of years of schooling was seven. The average size of the households was 4.6 people, of which 23% and 14% of the household members were children below 15 years old and adults above 60 years old, respectively. Almost all respondents were born and raised in the villages. The average age of respondents in the area was 33 year olds.

One effect of the *doi moi* policy (the economic renovation policy, which has created the improvement of financial situation of households) was that the financial situation for many families improved. A visible effect of this was that families began progressively putting their savings into making stronger houses, rebuilding incrementally with more durable materials, but literally at a price. Worse, few of these newer houses have been built in a manner that will enable them to resist a flood or typhoon, and in effect, the result is that the investment in time and money that families have made is highly vulnerable to damage by floods and typhoons (Tran, Shaw, Chantry, & Norton, 2009). The result of study shown that majority of households having houses that were considered permanent (54%), however, these houses had some parts vulnerable to the storms, particularly the roofs. However, these houses had some parts vulnerable to the storms, particularly the roofs. The semipermanent houses (34%) are easily damaged or unroofed by a storm, and the temporary houses (12%) are extremely vulnerable to flood and storm disasters (see Fig. 2). Unlike in urban areas, 98% of the householders in rural areas own their house. However, the quality of the house varies a lot among households. The average rural house for a family has a main house with a floor area of $35\,m^2$ ($5 \times 7\,m$) divided into three bays, a covered terrace ($2 \times 7\,m$) in front, and a kitchen/multipurpose enclosed area to the right of the house.

Living in flood-prone areas, almost all surveyed households have an elevated foundation. However, only 4% of the households have a two-storied house, and 45% of the houses have a small attic to store food and necessities and to serve as a shelter during flooding (see Fig. 3). The GIS survey revealed that a second story that could be used as a safe shelter was not evenly distributed. Some residential areas had highly concentrated numbers of semipermanent houses and temporary houses without an attic to protect property and human lives during severe floods. All households had access to power (100%), and most of them (99%) had a well for daily water consumption that often was polluted during the flood season. It can be concluded that after Vietnam moved to a market economy, rural areas of Vietnam experienced a significant change in housing, infrastructure such as rural roads, and public services such as electricity. However, people in rural areas are still vulnerable to the impacts of climate-related disasters,

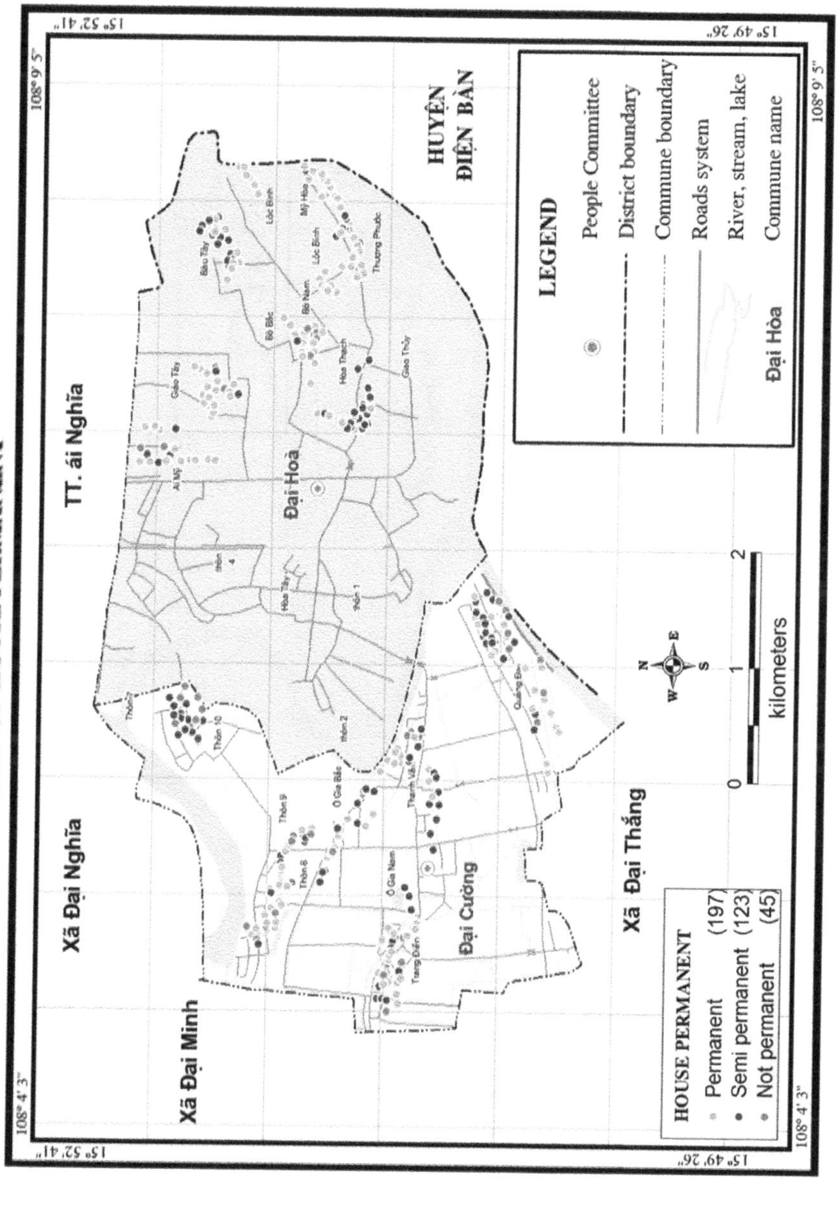

Fig. 2. Location of Housing by Level of Permanence.

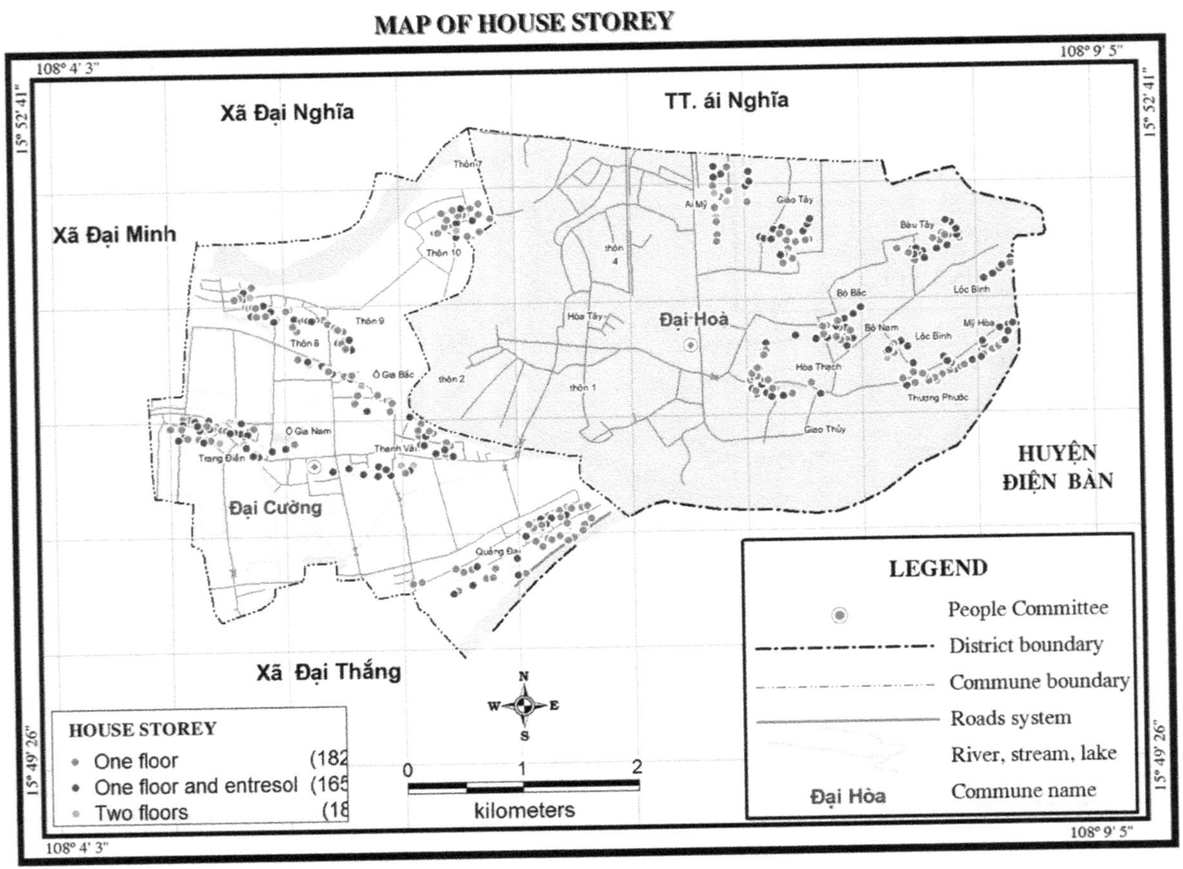

Fig. 3. Location of Housing by Story.

particularly floods and typhoons because a high percentage of housing still does not have a safe shelter to protect their lives and property and the public buildings that can be used as safe shelters do not meet the demand or are inaccessible. The results of GIS survey show that proximity was related to flood risk, because the houses close to rivers are often subject to flash floods, riverbank erosion, and inundation. The survey showed that 7.4% of the houses are located very close to riverbanks (houses located in a 50-m river buffer zone). The families living in these houses particularly need to get as much advance warning about flood events; they also need to have a plan for moving to safer places and, if possible, to protect their belongings. Six percent of households live close to a riverbank (within a 50 to 100 m river buffer zone). This defines areas where preventative measures should be implemented to reduce the impact of flooding inside the houses and to improve access to safe shelter, storage, and escape routes.

Economic Indicators

Within the surveyed households, 27%, 36%, and 37% of households were classified as better off, medium, or poor, respectively, with their average household and per capita income shown in Table 1.

The average of residential land, including garden, of all of those surveyed households was 738 m^2, and 99% of the households have long-term land use rights. In Vietnam, land is the property of the state. Farmland is very limited, with only 1,753 m^2 per household, in which 1,534 m^2 and 219 m^2 of the land with irrigation system and without irrigation system, respectively. This is due to the high population density and the land loss caused by the riverbank erosion during the annual flood disasters. Because the Dai Hoa and Dai Cuong communes are located in the plain areas, only two households have forestland, with the area of 3 ha by 4 ha. There are no

***Table 1.*.** Average Income by Type of Households.

Income Group	Index	Average Income (VND)	Average Income (USD)
Better off	Household income	23,058,131	1,281
	Income per capita	465,362	26
Medium	Household Income	16,208,886	900
	Income per capita	287,865	16
Poor	Household income	12,460,923	692
	Income per capita	2,205,511	12

Source: Authors, surveyed 2009.

aquaculture activities in the surveyed communes. The agriculture in the surveyed areas is very climate sensitive, and therefore, the income sources from agricultural activities are unstable. Moreover, agriculture production is mainly for domestic use. Of respondents, 72% said that they use more than 50% of the agricultural produce that they produced. The main sources of income of all surveyed households are from agriculture (rice production, home garden, banana and peanut crops, etc.) and waged labor in nonfarm activities. There are 75% of households that have their own vehicle (motor bike) to travel/commute to work. However, only 28% of households have their own boat, although the boat is very important for evacuation and crop harvest during the flood season.

Technology Indicators
The experience of severe floods was the most traumatic to the villagers, particularly for those who live in temporary or weak houses. Many of these people have nowhere to refuge because their house is too far from public facilities, often located in commune centers, which can be functioned as safe shelters. For these people, mobility is particularly difficult when the water level rises to the roof of the house and the strong wind caused fierce waves. The study showed that the average distance from the household to the commune center is often more than 2 km.

It is interesting to note that only 54% of buildings were classified as permanent houses that can resist a strong flood and typhoon; however, 70% of respondents believed that their house is strong enough in case of disasters and they stay at home during the disasters. This is a common practice that people often remain in their house during a disaster. There are several reasons for this behavior. First, local communities experience disasters annually and get used to dealing with it. Another reason is that they think their house is safe or that no safer places exist. In fact, among the households that do/will not evacuate to the shelters, only 30% of them stated that they had no safer place to go or could not access the safer places. According to result of the GIS survey, more than 45% of households lived far away from safe shelter more than 300 m, making it very difficult to move to those shelters during a typhoon. This may cause the loss of or damage to human lives if a severe disaster occurs.

Among the households whose do/will evacuate to other places in case of severe disaster, 85% of them move to a neighbor's house (72% of surveyed respondents). The major means of evacuation is by boat (30%) or by walking (70%). It should be noted that mass media such as television and radio are the main sources of early warning information for both local government and communities in the surveyed areas. The possibility of involving mass media for awareness raising or early warning should be considered for future disaster management. It is observed that almost all (95%) households have TV, and this is an important means of receiving disaster-related information during the disaster season. In addition, during a disaster, local government commonly uses loudspeaker systems and megaphones to disseminate early warning information to households. The survey indicated that 74% of households received the disaster information through this system.

Social Capital Indicators

Social capital generally refers to trust, social norms, or networks upon which a community or individual households can count. It is a mutually beneficial collective action, an important factor for sustainability of a community and overall livelihood. In this household survey, some of the questions were asked to understand social capital. In terms of networking, most (64%) respondents thought that, if necessary, they could borrow money to support their needs. Community meetings were used for sharing common issues and building a network. The survey showed that meetings were regularly organized in the communities. Of respondents, 52% said that they attend one meeting a month, and 24% of respondents attended one meeting every three months. In the rural areas, a high percentage of people often interacted and shared the experiences related to disaster and climate change issues. The survey showed that more than 12% of people often shared their experiences, and 58% of people sometimes shared their experiences.

Network capacity is one of the major issues of social capital. The majority (56%) of respondents are members of social organizations such as farmers', women's, elders', or veterans' associations, and more than 70% of them are active members. In Vietnam, formal organizations called mass organizations are set throughout the country and function as a focal point of various issues between people and the local authority or development organizations. In the surveyed areas, two major mass organizations, Farmer's Association and Women's Union, are popular organizations, and many people are active members of them. In addition, it is common for households to receive help when they face problems from the government through those mass organizations.

It is important to note that more than 90% of households have sought help from outside to cope with problems. Most (24%) of the respondents received help from a relative, from friends (6%), from the government (73%), and from others (30%). The help and support are mainly in the form of labor forces, for repairing houses, harvesting, and obtaining construction materials, money for disaster recovery, or food and cloths after a disaster.

Skill/Knowledge Indicator
Most (88%) of the respondents had not been invited to attend any disaster-related trainings organized in the commune in the last five years. Only 12% of respondents attended disaster management trainings. The trainings mainly focus on first aid and rescue, safer housing construction, environment issues related to disasters, and activities needing to be done before, during, and after a disaster. More than 91% of the trainees found that the information they learned was useful during a disaster season in the previous year. Although most of the respondents have not attended training courses on disasters, many of them could learn from other sources such as mass media, relatives, or neighbors.

Local Responses to the Xangsane Typhoon

Adaptation Options Before the Event
Being a disaster-prone area, most people in the surveyed villages were aware of the disasters that come with the seasons. The result of the household survey showed that households engaged in a whole series of activities at least a month before the storm and flood season. The most common adaptation options that people prepared for the Xangsane Typhoon were:

1. Reinforcing and repairing vulnerable parts of their house (for example, a window, wall)
2. Buying and storing foods, drinking water, and other necessities
3. Moving ships, small boats, livestock, and household items to safety
4. Moving family members to a safe place
5. Cutting or trimming trees near the house to prevent damage from falling branches

Commune People Committee (PC) and mass organizations played a major role in encouraging households to make such preparations. As mentioned previously, housing in the surveyed areas was vulnerable to typhoons; hence, 90% of respondents reinforced and repaired their houses before the event. In most of the villages, the households that owned boats brought them to the river or canals nearer to their houses for emergency evacuation. Preparation of drinking water depended on the available sources. In some households, water tanks were used to store rainwater and they were set high enough to protect domestic water supplies from flooding. Because many households relied on water from a dug well, 73% of them prepared and stored water in advance of the flood season.

In terms of collective actions to respond to an event, the result of the FGDs revealed that activating an early warning system, mobilizing organizations involved in climate change management, and informing local communities are three main adaptation strategies that local government units adopted to prevent and mitigate the impacts of events. Additionally, preparing relief and recovery equipment, stocking relief goods, evacuating people, are measures used for prevention and mitigation of the impacts of events.

The result of the FGDs conducted in these two communes also found that households have made great efforts in changing their crop varieties both in their gardens and in paddy fields. They also found that shifting from three paddy crops per year to only two is locally considered to be one of the most effective adaptation measures for mitigating adverse impacts of annual disasters. However, only 2% of surveyed households began growing a new variety of vegetable, such as hot chilly, ground beans, banana, or maize instead of fruit trees as before. This is due to a lack of knowledge and technical availability for changing crop patterns. The agricultural extension services should take this into account when planning adaptation activities in this area.

Household members believed in the effectiveness of many other adaptation measures for mitigating and preventing disaster impacts, but they could not adopt them. Among those options (which should have been commenced before the Xangsane typhoon, but were not), they mainly addressed improving or reinforcing the house (68%). Unfortunately, 64.5% of better-off households, 65.0% of medium-income households, and 72.4% of poor households are not be able to do this. A small percentage of households feel regret that they could not do other options such as protecting family property, preparing evacuation means, evacuating, changing the crop patterns calendar, diversifying the sources of income, etc. There are small percentage of households feel regret that they could not do other options such as protecting family property, preparing evacuation means, evacuation, changing the crop patterns calendar, or diversifying the sources of income etc. in order to mitigate the impact of typhoon due to a lack of resources.

Adaptation Options during the Event

The result of the study showed that 60% of households surveyed adopted adaptation measures during the Xangsane typhoon that mainly involved staying in the shelters and waiting out the disaster. This is due to the extreme and dangerous conditions of the Xangsane typhoon. It was too dangerous to move out of the house during the typhoon. In Central Vietnam, flooding often comes accompanies typhoon. The Xangsane event was not an exception, and most of the houses were flooded right after the typhoon. Thus, the other common options of the households were to clean their house or items when flooding water down (55%), or reinforce the house (34%) to keep the roof from blowing off. It is also important to note that, during the occurrence of the Xangsane typhoon, local communities, government units, and mass organizations could not offer any options for evacuating victims to safer shelters, for providing relief goods to households in isolated areas, etc.

Seven percent of households regretted that they did not prepare or store enough food for their family before the Xangsane event. More than 86% of households believed that they can do nothing during a severe typhoon like Xangsane. Thus, according to them, prevention and disaster risk reduction before the event is more important. During the event, the only option is to protect human lives.

Adaptation Options after an Event

After a typhoon, almost all households (94%) repaired or strengthened their house before asking for any external support. However, for those whose houses were heavily damaged or collapsed, it took long time to recover, and many needed to borrow money to rebuild their houses. As previously mentioned, the typhoon severely impacted the environment and drinking water. The majority of respondents (60%) were aware of this problem and they treated the water source for their family and cleaned the surrounding environment.

It is important to note that local governments have recognized the surveyed villages as "cultural villages." One of the criteria for this recognition is a good relationship among villagers. Community members help one another to overcome day-to-day difficulties, and appear motivated to participate in social and cultural activities to promote good traditional cultural values. Thus, more than 20% of respondents contributed personal days and money to local government to help in disaster damage recovery.

The result of the study found that there is a significant association between risk perception and adaptation measures undertaken. Risk perception plays an important role in shaping coping mechanisms and adaptation to climate change. Risk perception can influence both the design and operational aspects of disaster risk management. If disasters were perceived as hazard-led events, then the coping and management mainly focused on the physical aspects, structure measures, and external relief. In contrast, if the disasters were considered a product of hazard and vulnerability, then the adaptation measures recognized human behavior as an important aspect and coping and management focused on nonstructural measures. Thus, understanding how people perceive disaster risk is important to formulate a disaster risk management policy and plan.

The results of this study showed that the majority (63%) of people perceived the event like the Xangsane typhoon as a hazard-led event that people have little control over. As a result, the common coping strategy was to "live together with the disasters." Houses were often reinforced before the disaster seasons to reduce the impacts of floods and typhoons. Production activities have been adjusted to reduce the adverse impacts as much as possible, and people move livestock or property to higher places before the disaster season. Local adaptation options adopted to prevent and mitigate the impacts of Xangsane are summarized in Table 2.

Barriers Constraining Local Community from Adaptation to Climate Change

High Percentage of Non-Disaster Resistant Houses and Infrastructures

The "*doi moi*" moved Vietnam toward a market economy, and one effect was that the financial situation of many families improved. A visible effect of this was that families progressively began to rebuild their houses with more durable and costly materials. However, changes in housing construction materials and techniques to "modern" styles without appropriate adaptation of the safety measures, have threatened the integrity of the environment. Few of these new houses have been built with safety designs to enable them to resist floods or typhoons. The unfortunate result was an increased vulnerability of the built environment. The survey showed that, to avoid the damage of normal floods to housing, most of the houses needed to be constructed on a high plinth elevating the ground floor an average of 71 cm from the garden level. This was not enough to prevent damages during the Xangsane event, during which floodwater reached an average height of 110 cm. For that disaster, the average loss to housing was about VND 6.95 million, about US$460 per household, whereas the GDP per capita of the province was only $193. However, a high percentage of the population still is not prepared for disasters. In fact, a high percentage of

Table 2. Summary of Adaptation Options Before, During, and After a Disaster.

Adaptation Options BEFORE Disaster	Percentage of Households That Took This Option
1. Reinforcing and repairing vulnerable parts of the house (e.g., window, wall)	90%
2. Buying and storing foods, drinking water, and other necessities	73
3. Moving ships, small boats, livestock, and household items to a safe place	44
4. Moving family members to a safe place	29
5. Cutting or trimming trees near the house to prevent damage from falling branches	28
6. Reinforcing breeding facilities (animal husbandry facilities), pond, and dikes	15
7. Building a mezzanine	9
8. Preparing evacuation means (buying a boat for moving the family during disasters)	7
Adaptation Options DURING Disaster	
1. Staying in shelters, waiting out the disaster	60&
2. Cleaning house or items when flooding water down	55
3. Reinforcing the house	34
4. Moving ships, small boats, livestock, and household items to a safe place	25
5. Moving family members to a safe place	24
6. Keeping track of disaster warning information and instructions for prevention	16
7. Cutting or trimming trees near the house to prevent damage from falling branches	7
8. Helping injured neighbors, helping neighbors to evacuate to safer places	6
Adaptation Options AFTER Disaster	
1. Repairing/strengthening houses and damaged family items	94%
2. Treating water sources for family and cleaning the environment	60
3. Recovering production activities (repairing dams, paddy fields, and ponds)	36
4. Reconstructing houses using more durable materials/more resilient structures	21
5. Contributing personal days and money to local government disaster damage recover	20
6. Consolidating dikes, roads, etc.	11
7. Asking for aid/support from the government	8
8. Asking for support from relatives	7

Table 2. (*Continued*)

Adaptation Options BEFORE Disaster	Percentage of Households That Took This Option
9. Migrating to other areas for jobs; working more to earn extra income for family recovery	5
10. Coping with financial shortage (withdrawing from savings, selling stocks, borrowing money)	4
11. Selling assets: gold, motorbike, land, livestock; selling means of production: seeds, castles, machines	3

Source: Author, surveyed 2009.

households (63%) does not have any plan to cope with a future disaster, due to their limited capacity and perception of the risks.

Insufficient Access to Disaster Risk Management Information and Planning
The study revealed that households had insufficient access to information, especially regarding natural disaster planning and mitigation, or severe weather preparedness and planning, to allow them to properly respond. Beyond the household and village level, at the commune, district, or province level, the disaster management planning process (mainly flood and storm control) takes place every year, but technical experts mainly do the planning, without the community's participation. Once a plan is approved, the commune authorities disseminate the decision to people on loudspeakers or through hamlet leaders during community meetings before the flood and storm season. This annual plan on flood and storm control is not detailed and mainly focuses on organization, structure, roles, and responsibilities of members, rather than a comprehensive preparedness, mitigation, and response plan. This is one of the barriers preventing many household members from properly responding to the event.

Lack of Mutual Help among People
The move of Vietnam from central planning to a market economy had three prominent impacts on disaster management. Although Vietnam's transition from state central planning is often heralded as a macroeconomic success story, the policy transition had negative impacts on social vulnerability. Social capital such as cohesion or bonding, which is very important in disaster management, has been gradually reduced due to the rapid processes of urbanization and privatization. In the surveyed areas, local communities

already had wisdom and social institutions to cope with traditional risks, yet with the current shift of government policy, which speeds up the privatization process and gradually moves from central planning to a market economy, these practices have broken down in many areas (e.g., creating lack of mutual help among people due to out-migration and privatization). For example, whereas each community previously had to donate labor, equipment, and material toward the maintenance and emergency repairs of the flood protection infrastructures, it now receives payment for these efforts from the government.

Unfortunately, with fewer and fewer voluntary contributions and limited capacity of the government to pay for it, less money will be available for upgrading and maintaining infrastructures. As flood protection work is perceived more and more as a government's duty, fewer people will become identified with the program of maintaining the flood protection infrastructure. It is already becoming increasingly difficult to mobilize the population for maintenance and preparedness during the dry season and emergency repairs during the floods, and this indicates that the public support for infrastructure monitoring and repair is gradually reducing.

Traditional Coping Mechanisms under Pressure

This research revealed that in the studied areas besides the traditional risks, such as flood and typhoons, new risks are rapidly increasing, for example, environmental and epidemic problems. However, although floods had been a part of human life for centuries in Vietnam, and local communities have evolved ways of coping with flood's annual cycle, they completely lack experience to deal with the new risks. In addition, in the past, floodwater quickly drained away through the network of rivers and canals built in the flood-prone area. In recent years, the increasing degradation of the natural environment through deforestation and the conversion of agricultural land to urban areas have made the impact of floods more serious and longer-lasting in the lowland areas. For example, because of the expansion of the road network and other barriers to drainage, the flood level has increased beyond people's anticipation. Although most interviewed households (60%) agreed that disasters are becoming worse both in severity and frequency, they were still preparing for disaster at a level commensurate with the worst disaster they experienced, and not for the greater disasters that are likely to strike in the future. In the Xangsane typhoon, for example, individual and collective preparation for the typhoon was all based on previous experiences. Some respondents admitted that they did not begin to make any preparations until the typhoon was virtually upon them. As a result, extreme losses and damage occurred.

Furthermore, the disaster risk management methods in villages still follow "old customs," mostly dependent on large-scale infrastructure once implemented by the government, but too costly to maintain nowadays. However, although these old practices are under pressure, the new methods necessary to deal with the modern risks have yet to be materialized. Ultimately, the worsening environmental problems of Vietnam appear to be the outcome of many different interacting factors including the natural setting, traditional practice, and institutional structure. These factors provide a backdrop for the more recent rapid environmental change brought about by the expansion of the market economy, population pressure, urbanization and industrialization, inappropriate or poorly implemented disaster mitigation policies, and lack of adequate knowledge and understanding of the environment. Hence, as Vietnam is entering this new critical phase, traditional hazards remain high, new risks are rapidly appearing, and traditional disaster coping mechanisms are becoming inadequate. Without scaling up existing programs and initiating new and more innovative ones, the future of many cities and rural regions will be subjected to severe environmental deterioration and increased vulnerability to disasters.

Communities did not have access to proper rescue equipment when the Xangsane occurred. The survey revealed that there were no such activities before and during the event. Respondents said that the lack of necessary rescue equipment to effectively respond to the Xangsane, such as rescue boats, life vests, communications equipment, emergency medical supplies, and facilities, are the major obstacles. Besides the previously mentioned emergency equipment and materials, trained personnel such as doctors, health workers, and engineers might be required to deal with emergencies such as dike breaching, or to repair damaged community water systems. However, such personnel are often unavailable to assist communities.

Limited Access to Financial Resources and Low Income

Private loans (borrowing money from a relative or local money lenders) were a common way for poor people to bridge the food gap during the months before the harvest of the winter–spring crop. These loans were paid back directly after harvest, and normally were not a big problem if they did not accumulate. However, after the Xangsane, the situation became more difficult, as people have taken out more private loans than usual. Income during the years after the event had been low and people have had difficulties repaying the loans.

The local moneylenders are often local traders or more well-off households in the community or neighboring communes. There is a degree of social pressure on these households not to claim too high interest rates. But there is little open discussion about the problem at the commune level. Households who take out loans do not normally talk about it. Loans from traders are more difficult to control, as they are outside community social pressure. In the studied communes, the formal credit fund (from mass organizations, nongovernmental organizations [NGOs], Government Social Policy Bank, etc.) for disaster recovery had not been sufficient for recovery. The credit was largely used to replace lost resources, mainly rice, and did not contribute to the accumulation of new resources. It is therefore difficult for many people to repay. The combination of old debts, which were difficult to pay back after the disasters, and new debts, both formal and informal, put some households in a situation of negative debt spiral, which is difficult to get out of. These households would need special attention to help them out of their difficulties. Normally, the Vietnam Bank for Agricultural and Rural Development (VBARD) does not give new loans if the old loans are not yet repaid. The bank also does not allow loans to be used to repay private debt.

The main economic activities in the surveyed communities are mainly from agriculture. The other services and nonfarm economic activities (for example, small-scale manufacturing) account for a very small part of the local economy. Thus, household income is typically not very high and is very sensitive to the weather and climate. As a result, they have few opportunities to invest in infrastructure development, or to increase their capacity to plan and prepare for natural disasters.

CONCLUSIONS

Climate change, impact, vulnerability, and adaptation have become compelling issues attracting the concerns of social actors from developed countries and developing ones, and for both government and nongovernment organizations. It is recognized that the poor communities are the most vulnerable, with low adaptation capacity for climate changes, as they lack resources and options to adapt to climate changes. The result of the study concluded that local communities had made a great effort and adopted various adaptation measures, including structural adaptation measures, behavior adaptation measures, financial measures, and information ones.

Such adaptation measures are mainly traditional options for traditional risks of typhoons, but not much appropriate to new risks induced by climate changes. Thus, the effectiveness of local adoption of such measures is still limited to reduce the impacts of typhoons. This is consistent with what Kumar (2007) and Selvaraju et al. (2006) stated, that local people tend to become involved in local managed practices to mitigate and overcome adverse impacts of climate changes on their lives. However, the implementation of these adaptation options often require resources, such as social resources, human resources, natural resources, economic resources, and institutional ones (IPCC, 2001). In the context of Quang Nam, local communities are largely dependent on agricultural production and natural resources, thus constraining them from adaptation of these. As a consequence, local communities suffer tremendous damages from typhoons.

It is also important to note that local conventional adaptation measures such as using local material to reinforce houses, building facilities, moving assets and cattle to safer places are less effective measures in the context of extreme events. For example, there is no difference in terms of damage costs between households who did and did not act upon local early warnings. However, there is a significant association between risk perception, adaptation measures, and impact reduction. Risk perception plays an important role in shaping the coping mechanism and adaptation to climate change–induced events. Risk perception can influence both the design and operational aspects of disaster risk management. If disasters are perceived as hazard-led events, then the coping and management mainly focuses on the physical aspects, structure measures, and external relief. In contrast, if disasters are considered as a product of hazard and vulnerability, then the adaptation measures recognize human behavior as important aspects and the coping and management focuses on nonstructural measures. Thus, understanding how people perceive the disaster risk is important for formulating a disaster risk management policy and plan.

There is a significant association between local barriers and their adoption of adaptive measures in response to typhoons. A lack of budget, materials, finance, insufficient access to disaster risk management information, lack of mutual help among people, and low income are constraining local communities from adaptation to climate change. Thus, there is a big gap between the expected adaptation options and the actual implementations to respond to the extreme event. Thus, extreme typhoons like Xangsane cause tremendous impact on local communities, particularly on local housing and livelihoods.

REFERENCES

Brooks, N. (2003). *Vulnerability, risk and adaptation: A conceptual framework.* Working Paper no. 38. Tyndall Centre for Climate Change Research, University of East Anglia, Norwich.

Cruz, R. V., Harasawa, H., Lai, M., Wu, S., Anokhin, Y., Punsaimaa, B., Honda, Y., Li, C., & Huu Ninh, N. (2007). Asia. Climate change 2007: Impacts, adaptation and vulnerability. *Contribution of Working Group II to the Fourth Assessment Report on the Intergovernmental Panel on Climate Change.* Cambridge, UK: Cambridge University Press, Cambridge, UK, (pp. 469–506).

Frisvold, G., & Ingram, K. (1995). Sources of agricultural productivity growth and stagnation in sub-Saharan Africa. *Agricultural Economics, 13*, 51–61.

IISD, SEI, IUCN, Intercooperation. (2003). *Climate change and livelihood.* A conceptual framework paper prepared by the task force on climate change, vulnerable communities and adaptation. International Institute for Sustainable Development. Winnipeg, Manitoba Canada.

IPCC. (2001). *Climate change 2001: The scientific basis.* Cambridge University Press.

Keskitalo, E. C. H. (2004). A framework for multi-level stakeholder studies in response to global change. *Local Environment, 9*(5), 425–435.

Kumar, M. (2007). The impacts of climate change on people and food. Available at www.fijitimes.com.

Lim, B., Spanger-Siegfried, E., Burton, I., Malone, E., & Huq, S. (2004). *Adaptation Policy policy Framework framework (APF) for climate change: Developing strategies, policies and measures.* Cambridge, UK: Cambridge University Press.

McCarthy, J. J., Canziani, O. F., Leary, N. A., Dokken, D. J., & White, K. S. (Eds). (2001). *Climate change 2001: Impacts, adaptation, and vulnerability.* New York: Cambridge University Press/Intergovernmental Panel on Climate Change.

Pielke, R. A. J. (1998). Rethinking the role of adaptation in climate policy. *Global Environmental Change, 8*, 159–170.

Quang Nam People Committee. (2007). Annual report on disasters in 2007 and planning for disaster prevention in 2008. Tam Ky town, Quang Nam, Vietnam.

Selvaraju, R., Subbiah, A. R., Baas, S., & Juergens, J. (2006). *Livelihood adaptation to climate variability in drought-prone areas of Bangladesh.* Rome, Italia: FAO.

Smit, B., Burton, I., Klein, R., & Wandel, J. (2000). An anatomy of adaptation to climate change and variability. *Climatic Change, 45*, 223–251.

Smit, B., & Pilifisova, O. (2003). From adaptation to adaptive capacity and vulnerability reduction. In: S. Huq, J. Smith & R. T. J. Klein (Eds), *Enhancing the capacity of developing countries to adapt to climate change.* London: Imperial College Press.

Tran, P., Shaw, R., Chantry, G., & Norton, J. (2009). GIS and local knowledge in disaster management: A case study of flood risk mapping in Viet Nam. *Disasters, 33*(1), 152–169.

UNFCCC. (2007). Climate changes: Impacts, Vulnerabilities vulnerabilities and Adaptation adaptation in Developing developing Countries countries. Note by Climate Change Secretariat (UNFCCC)., Born, Germany. Available at http://unfccc.int/files/essential_background/background_publications_htmlpdf/application/txt/pub_07_impacts.pdf

CHAPTER 20

IDENTIFYING LINKAGES BETWEEN RATES AND DISTRIBUTIONS OF MALARIA, WATER-BORN DISEASES AND INFLUENZA WITH CLIMATE VARIABILITY AND CLIMATE CHANGE IN VIETNAM

Tran Mai Kien, Tran Thi Tuyet Hanh,
Hoang Duc Cuong and Rajib Shaw

INTRODUCTION

Over the last decades, there has been an increasing interest among scientists on the linkage between population health and climate and environmental factors, as well as health impacts of climate change and climate variability. Numerous studies have been done and substantial results achieved, but mostly in the developed countries, and not much quantitative evidence or assessment of the impacts at national and local levels has been provided for developing countries.

Climate Change Adaptation and Disaster Risk Reduction: An Asian Perspective
Community, Environment and Disaster Risk Management, Volume 5, 417–449

ISSN: 2040-7262/doi:10.1108/S2040-7262(2010)0000005026

The Fourth Assessment Report of the Intergovernmental Panel for Climate Change (IPCC) has highlighted substantial improvements in understanding the health problems from climate change since the previous reports (Confalonieri et al., 2007). In many developed countries, significant results were gain in the national health impact assessments of climate change (Patz et al., 2000; Kovats, Menne, Ahern, & Patz, 2003a; Furgal & Seguin, 2006), but few studies were done in developing countries. One of these works used empiric statistical methods and elaborated on a special index to facilitate the description of association between the weather, climate variability, and health outcomes in Cuba, which can be used to predict outbreaks of some climate-sensitive diseases (Bulto et al., 2006). Another study pointed out the changes in the distribution and burden of climate-sensitive diseases in different locations of small Pacific island states in light of regional climate pattern changes (Ebi, Nancy, & Lewis, 2006). The results have highlighted the importance of such studies for informing decision makers and reducing the health and socioeconomic impacts of climate variability and change.

Scientists' interests in the relationship between infectious diseases and climate change have been raised since the mid-1980s. Infectious disease transmissions are influenced by many extrinsic factors (such as climatic and environmental factors, socioeconomic conditions, health services and interventions, water and sanitation services, population density, personal hygiene behavior, and so on) and the intrinsic factors (such as population immunity) (Haines, Kovats, Campbell-Lendrum, & Corvalan, 2006). Many infectious agents, vector organisms, reservoir host species, and the rates of pathogen replication within the vector organisms are sensitive to climatic conditions (Kovats et al., 2003b). For example, in regions where low temperature, low rainfall, or lack of vector habitat limit vector-borne disease (including malaria) transmissions, climate changes may alter the ecological balance and trigger epidemics (McMichael, Woodruff, & Hales, 2006). Epidemics can also result from climate-related migration of reservoir hosts or human populations (Campbell-Lendrum, Pruss-Ustun, & Corvalan, 2003; McMichael et al., 2004).

A number of studies have found quantitative evidences on the correlations between malaria outbreaks and incidences with main climate factors, and more importantly, with the interannual climate fluctuations such as El Niño Southern Oscillation or ENSO (Kovats et al., 2003b; Khasnis & Nettleman, 2005; Bouma et al., 1997; Hales et al., 1996; Hales, Weinstein, Souares, & Woodward, 1999; Cazelles, Chavez, McMichael, & Hales, 2005). For example, Bouma et al. had analyzed the correlation between El Niño, monsoon rainfall, and the historic malaria epidemics in the Punjab area of the India subcontinent and in Sri Lanka before the introduction of residual insecticides, and came to the conclusion that the malaria epidemics related to excessive rainfall in one place (Punjab) and to failing monsoon and drought in the other (Sri Lanka), but both of these associations were related to El Niño years.

Other important study subjects are sanitation and waterborne diseases (WBDs), not only because they are major contributors to the morbidity (burden of diseases) and mortality (death rate) in the world but because water-related disasters, water scarcity, and quality deterioration are predicted to be some of the most serious public health problems of the coming century (Pruss & Havelaar, 2001; Confalonieri et al., 2007). The main climate change impacts on the epidemiology of WBDs were considered under three major mechanisms: (1) the impact of water-related extreme events and disasters (for example, floods, droughts, storms, inundations, and so on); (2) the impacts of degraded water quality, including brackish and coastal marine water; and (3) the impact of increased temperature. It is well known that salmonella and cholera bacteria proliferate faster at higher temperatures, and heavy rainfall can facilitate the dissemination of the pathogens to either surface or groundwater bodies (Hunter, 2003). Among various WBDs, many studies have focused on the relationships between cholera incidences (historical or recent) and climate change and ENSO, using quantitative analysis and nonlinear time series approaches (Rodó, Pascual, Fuchs, & Faruque, 2002; Koelle & Pascual (2004); Pascual, Bouma, & Dobson, 2002). Other studies have examined the effects of temperature and ENSO to other socially important WBDs such as diarrheal diseases and dysentery (Checkley et al., 2002).

In most of the developing and low-income countries, including Vietnam, climate change and disaster issues have not been properly integrated into health policies, strategies, and plans due to the weak linkages between policy development and scientific evidence and the lack of reliable quantitative and qualitative data. For example, in the latest Vietnam National Strategy on Preventive Medicine and Master Plan on Health Care System Development (Vietnam MOH, 2007), only a few words in one unique paragraph reminded about climate and environmental change and its consequences; no impact and vulnerability assessment plans or adaptation and response programs were considered for the health sector. A recent review by Hanh, Hill, Kay, and Quy (2009) shows that, in Vietnam, although there is a large number of legal documents on disaster preparedness, response, and recovery developed by different ministries, very little attention has been paid on developing policies, guidelines, and plans for health management before, during, and after disasters. The coordination of disaster preparedness, response, and recovery within the health sector as well as between relevant government agencies, UN agencies, NGOs, and donors are still very limited.

With approval by the government and nation-wide implementation of the National Target Program on Climate Change (the current national climate change strategy for Vietnam), the Vietnam Ministry of Health is facing challenges in developing a Climate Change Action Plan for the health sector at the central subnational and provincial levels. Therefore, there is a crucial need for integrating climate change issues into health strategies and policies, making them "climate proof," as well as for research and studies in health implications (impacts and risks) of climate factors and the health–climate linkage.

Despite concerns and needs as well as strong interests of Vietnamese public and policy makers in climate change and public health issues, so far there have been very few studies providing reliable scientific evidences on the impacts of climate change/variability to human health in general and infectious diseases in particular. As a first step to fill the gap, this chapter aims to identify the quantitative evidence on the impacts of climate variability and change to the spatial and temporal changes of several selected infectious diseases in different regions and provinces of Vietnam. Methods and data are described further, as are current situations of each of these infectious diseases in the country, with respective findings and discussions provided.

STUDY LOCATION AND CHARACTERISTICS

Located on the Indo-China peninsula of Southeast Asia or the Greater Mekong Subregion, Vietnam is a densely populated country (with approximately a population of 86 million in an area of about 326,000 km^2). The terrain is diverse with low-lying floodplain and deltas in the south and north; highlands in the center; and hilly mountains in the northwestern region. With a narrow and long S-shape and more than 3,200 km of coastline, Vietnam has been recognized as one of the most vulnerable countries in the world to potential impacts of climate change and sea level rise.

Having tropical climate in the south and monsoon subtropical climate in the north and central part, Vietnam experiences regular tropical typhoons and storms with extensive flooding, especially in the central coast and southern Mekong River Delta. In the south, the influence of southwest monsoons is stronger, leading to two distinct seasons: dry season and the rainy or wet season. The north has warm, wet summers and cool, humid winters, with stronger influences from the northeast monsoons.

Administratively, Vietnam is divided into three regions: North, Central, and South, with relatively differing geography and climatic conditions; and into provinces, of which the number was recently changed from 61 to 64 provinces and cities. Epidemiological and socioeconomical data in the country are usually grouped and assessed by these three regions and provinces.

However, more precisely, from the climatic and agro-ecological point of view, the country can also be divided into eight regions: Northeast, Northwest, Red River Delta, North Central Coast, South Central Coast, Central Highlands, Southeast, and Mekong River Delta (see Fig. 1). We used these three administrative regions, eight geographic regions, and

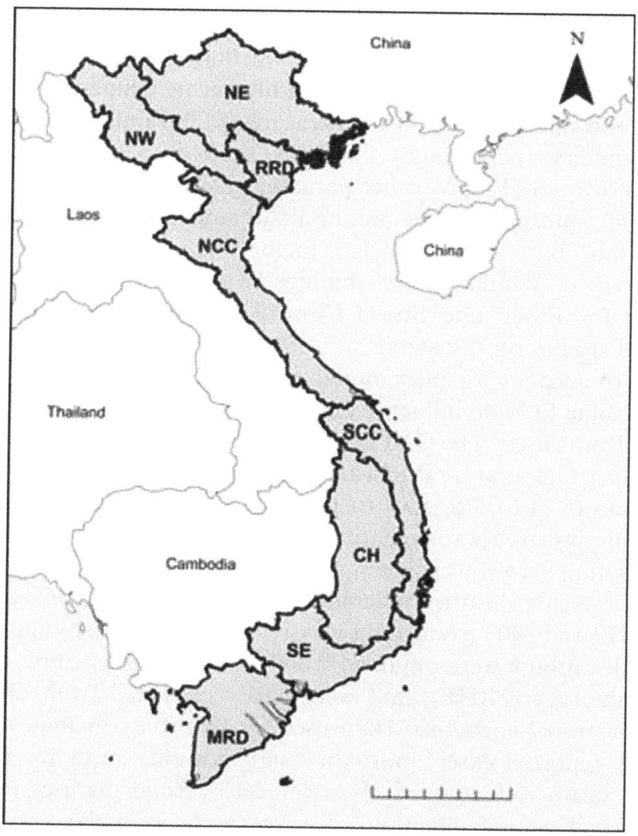

Fig. 1. Vietnam and its Eight Regions: Northeast (NE); Northwest (NW); Red River Delta (RRD); North Central Coast (NCC); South Central Coast (SCC); Southeast (SE); Central Highlands (CH); and Mekong River Delta (MRD).

the former structure of 61 provinces as a basis for our temporal and spatial analyses because, most of the time, series and data were collected during the period before the acknowledgment of 64 provinces.

METHODS AND DATA SOURCES

Data Set and Data Gaps

Climate data from 61 stations representing 61 provinces in Vietnam were collected and used for analyses. Most of the stations had enough time series of monthly climate data, provided by the Vietnam Institute of Meteorology, Hydrology and Environment (IMHEN), that spanned from 1975 to 2004. The data included series of monthly mean temperatures, monthly maximum and minimum mean temperatures (°C), rainfall (in millimeters), relative humidity (percentage) that was used as a primary prioritized climate indicator, and many other parameters such as number of wet days, evaporation, atmospheric pressure, hours of sunlight, and so on, as well as less important but still considered factors. Disaster data for all parts and provinces of Vietnam were obtained from the reports of the Central Committee for Flood and Storm Control (CCFSC) and other available reports and studies on disasters.

Besides, to account for interannual and decadal variability as well as to identify possible El Niño influence on the disease patterns, the Oceanic Niño Index (ONI) was used. The ONI expresses the running three-month mean of the sea surface temperature (SST) anomalies in the El Niño 3.4 region of the Pacific Ocean (5°N to 5°S, 120° to 170°W), and is frequently being used in many climate researches for identifying the El Niño events and monitoring the oscillation (www.cpc.ncep.noaa.gov/products/analysis_monitoring/ensostuff/ensoyears.shtml). Epidemiological data from the available period between 1979 and 2003 (some data was available for diseases until 2005, but not for all locations) were obtained from the National Institute of Hygiene and Epidemiology (NIHE), and were partly extracted from the Vietnam Health Statistical Yearbooks. Data used for this study include the number of monthly reported cases, morbidity and mortality rates (or the rate of illness and death rate respectively, related to certain diseases per 100,000 inhabitants) of major infectious diseases such as malaria (vector-born disease), and some selected waterborne diseases such as cholera, diarrhea/acute diarrheal diseases (ADD), and dysentery, which were expected to be climate-sensitive. Although malaria was of most concern, the monthly malaria data could not be obtained at the time of the study; therefore, the time series analysis for malaria could be done only with the yearly data. Other climate-sensitive vector-born diseases such as dengue fever, dengue hemorrhagic fever, and Japanese viral encephalitis are considered and analyzed in a separate study.

Methodology and Limitation

Two basic methods were used in this study: (1) statistical analysis, including time–series analysis and regression (both linear and nonlinear) applying Statistical Package for the Social Sciences software (SPSS), and (2) spatial analysis using GIS tools (ArcGIS–ESRI software). Time series of climate and disease variables were analyzed for trends and changing patterns of those variables over time. This can be used for highlighting the seasonal and interannual fluctuations (or oscillations) of the climate parameters, SSTs, or disease incidence, and so on.

Regression analysis is a statistical tool for the investigation of relationships between variables. Usually, regression is applied when investigating the causal effects of one variable upon another, like the effect of El Niño on disease incidence, for instance. To explore such issues, one needs to assemble data on the variables of interest and employ regression to estimate the quantitative effect of the causal variables on the consequent variable, as well as define the "statistical significance" of the estimated relationships (the degree of confidence that the true relationship is close to the estimated relationship). Epidemiologists and climate researchers extensively use regression techniques, which have long been central to the field of socioeconomic and political studies.

Further in the analysis process, from the nonlinear formula that characterized certain relationships and the derived value of coefficients, we can estimate the actual quantitative relationship of disease burden with each climate parameter (temperature or rainfall) and with the Oceanic Niño Index (ONI). All the calculations and analyses were automatically conducted by applying the SPSS software, which is a powerful program for many types of statistical analysis.

For spatial analysis, we input data of main climate variables and concerned disease rate (morbidity) of all the provinces in Vietnam for each year from 1997 to 2005, using the ArcGIS software to transfer and highlight these data into the maps. Then we compare the delivered maps of consecutive years to assess the changes of geographical distribution of the disease over time. The monthly reported data of major infectious diseases for the certain period were extracted and analyzed regarding correlation with running monthly ONI, and with monthly temperature and rainfall of the station representing this region.

Only data from the period between 1997 and 2005 were included for spatial analysis due to certain reasons. Firstly, there were changes in the administrative division of Vietnam into provinces over time. Before 1989, Vietnam was divided into 40 provinces and cities. In the period between 1990 and 1991, Vietnam was divided into 45 provinces and cities and then from 1993 to 1996, into 53 provinces. Between 1997and 2003, it had 61 provinces, but since 2004 to now, 64 provinces (including large cities such as Hanoi, Da Nang, and Ho Chi Minh city). Secondly, the full-range provincial epidemiological data for spatial analysis can be included only for the period between 1997 and 2005; therefore, for the years 2004 and 2005, data of some provinces had to be merged.

RESULTS AND ANALYSIS

Climate data analyses have shown that climate change already occurs and has impacted Vietnam. From climate time–series analysis, we saw the obvious increasing trend of the mean temperature in most of the stations representing different regions of Vietnam (see Fig. 2), from Northern Vietnam (Hanoi) to Central (Danang) and Southern (Vung Tau). There is clear evidence that global warming already occurs and has impacts on Vietnam. However, the rainfall pattern and relative humidity's changes varied depending on locations.

Besides, to see the influences of global change to the climate parameters in Vietnam, we have done the regression analysis for the monthly mean temperatures and rainfall in several typical cities as representative for different regions of Vietnam regarding ENSO phenomenon (that is, Hanoi city for the North, Hue city for the Central, and Ho Chi Minh city – Vung Tau station – for the South). From the results we saw that, in many locations, the correlations were obvious and significant but varied, depending on the provinces/locations, and could be both linear or nonlinear (cubic). For example, in Hue and Danang, the two coastal cities of central Vietnam, this correlation between ENSO phenomenon and the climate variables is very significant ($r^2 = 0.18$ and $p < 0.01$). These results were consistent with previous studies showing quite obvious influences of ENSO phenomenon to the climate variables in different parts of Vietnam.

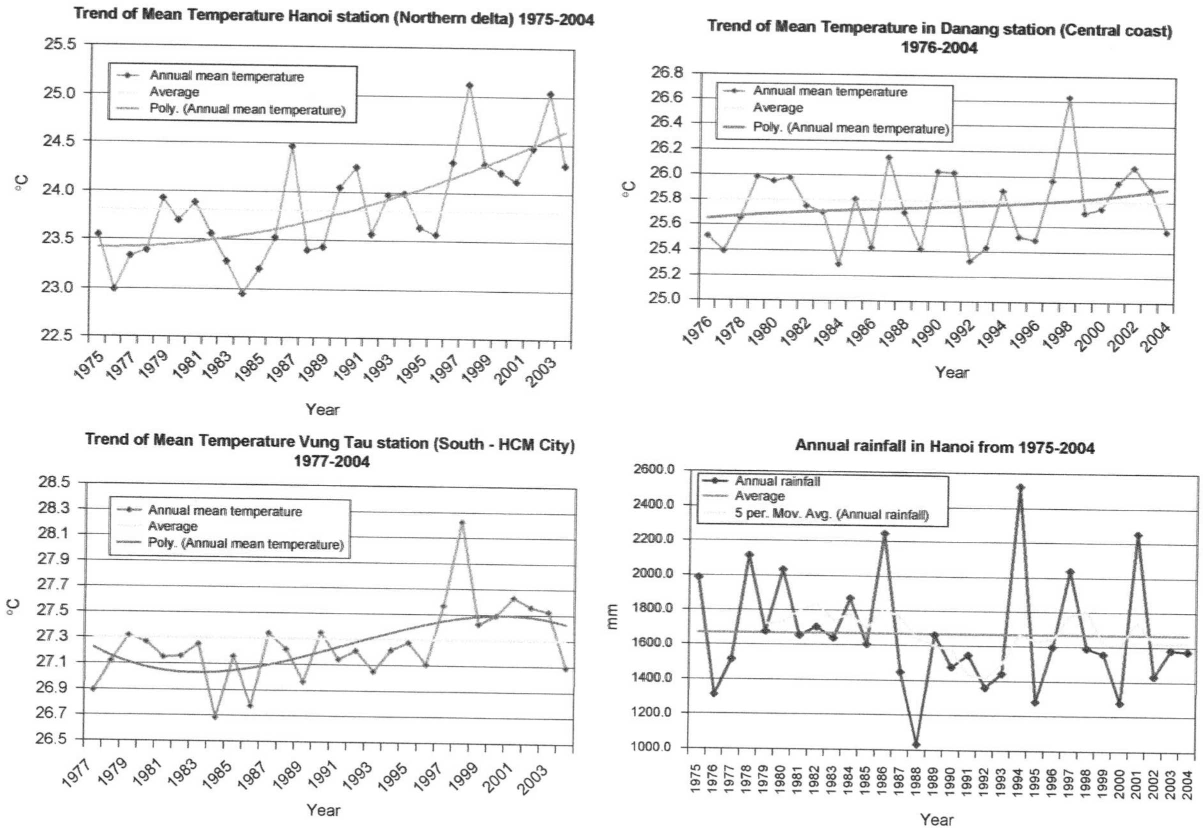

Fig. 2. Changing Patterns of Some Climate Parameters in the Past in Some Representative Locations in Hanoi, Danang, and Vung Tau stations, Vietnam.

Malaria

Malaria is the most important vector-born disease in the world, and remains a serious threat to almost 50 percent of the world's population, with 200 million estimated new cases and 1 to 2 million deaths per year (Le et al., 2002). The morbidity and mortality rates of malaria worldwide were on increasing trends from the 1900s to 1930, then sharply decreased during the period from the 1930s to the 1970s, and have been rapidly increasing since then, mostly in Africa. In many developing countries in Africa and Asia, malaria can exacerbate poverty, disasters, and conflicts, and remains a serious constraint for development. There are a number of conditions in favor of spreading malaria, in which global warming plays a significant role because malaria parasite development expands worldwide in the areas with temperatures above 16°C. From the very beginning of the 1990s, epidemiologists had found malaria to be particularly influenced by climate change. Malaria is endemic in Vietnam, predominantly caused by the parasites *Plasmodium falciparum* and *P. vivax,* and transmitted from infected people to those uninfected by the main vectors—*Anopheles dirus and A. minimus* (mosquitoes). In Vietnam, malaria has been a serious public health problem, with increasing morbidity and mortality rates during the late 1980s and early 1990s. In 1991, a total of 1,642,000 cases of malaria, 32,000 cases of severe malaria, and 4,650 deaths from the disease were reported, and resistances to drugs—chloroquine and sulfadoxine/pyrimethamine—were widespread (Nguyen, 1999).

Spatial Distribution

The spatial analysis results showed that the predominant geographical locations of malaria in Vietnam are the mountainous and tropical forest areas, especially the Central Highlands and the mountainous provinces of the North. Few cases were reported in other regions (Red and Mekong River Deltas, the lowlands of the Central Coast) and very few in the urban centers. In the Mekong River Delta, the southern rural provinces with extensive canal systems and large forest areas have more malaria incidences than the northern and eastern parts of the delta. High rates of malaria have been reported in the Central Highland area below 1,500 m, including the four highland provinces of Dak Lak, Dak Nong, Gia Lai, Kon Tum, and the adjacent Binh Phuoc province, as well as in the western parts of the central coast provinces that usually have mountainous or hilly terrain and forest (see Fig. 3).

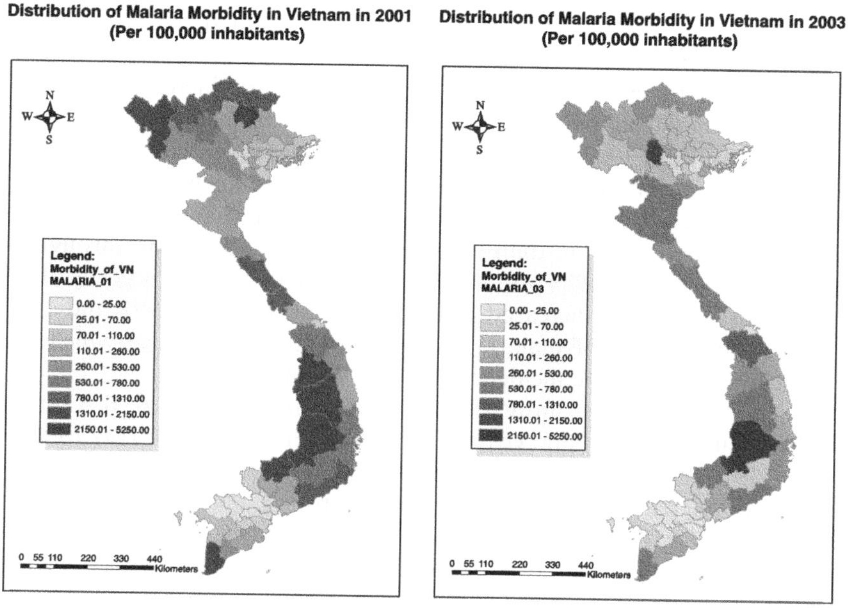

Fig. 3. Change on Spatial Distribution of Malaria through Provinces and Regions of Vietnam Seasonal and Temporal Changes.

During the last decade, there has been a trend of malaria reemerging in Northwestern mountainous provinces, spreading out to transition hilly areas such as the Vinh Phuc province, Red River Delta, and in the North Central Coast (Thanh Hoa, Quang Tri, and Quang Binh provinces), resulting in a minor increase of malaria rates in those provinces. Similar circumstances appear in the South Central Coast provinces such as Khanh Hoa and Ninh Thuan, where malaria spread out from Central Highland. Although the increase in rate is not very significant, there may be serious risk for the population in those marginal areas where people have lower immunity to malaria.

In the Central Highland where malaria incidences are highest, and in southern rural provinces of the Mekong River Delta, the number of reported cases peaked from October through December/January, that is, during and after the rainy season. In the northern mountainous provinces, malaria outbreaks appear to occur periodically throughout the year without specific seasonality. To identify possible impacts of climate factors to malaria in Vietnam, we decided to analyze only the period between 1976 and

1994, to exclude the strong influence of intervention activities and malaria control programs.

During this 19-year period, the incidences of malaria increased to double peaks between the period of 1977 and 1979, then reduced to the lowest point between 1980 and 1982. Since 1983, there was a sudden increase of more than 50 percent, and a steady increase during the following years to reach the highest peaks between 1986 and 1988, when the total number of incidences increased more than double and totaled 1.4 million throughout the country. Between 1989 and 1990, incidence had been reduced by about 40 percent and again increased to another peak between 1991 and 1993, until the first results of malaria control programs took place which had significantly reduced the incidences from 1994 onward.

All the increases and most of the peaks of malaria during this period occurred during the El Niño years: 1977 to 1978, 1982 to 1983, 1987 to 1988, and 1991 to 1992, as described in Figs. 4 and 5. The decrease during between 1989 and 1990 had been coincident with the strong La Niña event of 1988 to 1989. Moreover, the fluctuation of malaria morbidity in Southern Vietnam and Central Highland appears to follow closer to the ENSO phenomenon than in the North and Central parts. Analyzing the acute malaria cases and number of deaths from malaria for the same period, we saw that the increase of deaths and acute cases in Central Highland and the South also coincided with the El Niño years, while there was a lag of about one year in the North and Central Vietnam (see Fig. 5).

From 1976 to 1994, the number of malaria incidences all over the country was high and the magnitude was changing over the years; however, from 1994 onward, the number decreased dramatically. Up to 2004, the malaria rate in the country was reduced almost eight times in comparison with 1994. The main reason of such a significant malaria rate decrease in Vietnam over the last 15 years is the intensive and successful national malaria control program. From the period between 1991 and 2000, the Vietnam government and public health services carried out a large-scale malaria control effort and have gained substantial success, reducing mortality by 97.3 percent and morbidity by 77.0 percent, compared to the 1991 rate (Le, 2002).

It seems that there was no clear explanation for the sudden rise in acute malaria cases in Central Highland to the highest peak (about 40,000 cases) of 1989 (see Fig. 6), or for why the malaria death rate (with more than 3,000 deaths)in Central Highland and the South was that high in 1977. However, one of the suggestions could be the poor socioeconomic conditions and lack of access to health care and treatment after reunification of Southern

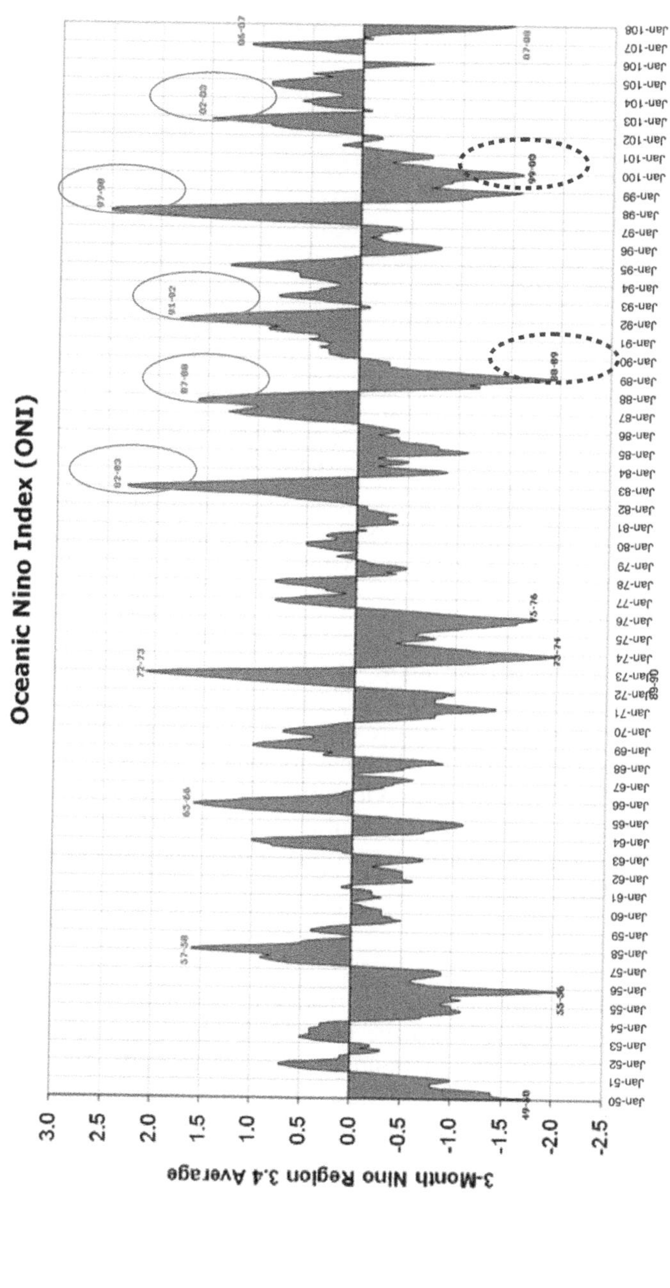

Fig. 4. El Niño (solid circle) and La Niña (dashed circle) Events in the Last Century According to the Oceanic Niño Index.

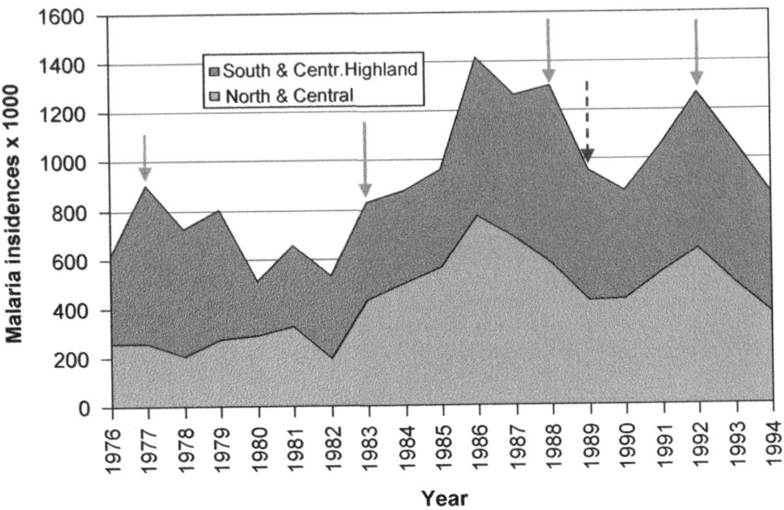

Fig. 5. Coincidence of Malaria and El Niño Events in Vietnam in the Period, 1976–1994 (El-Niño Years Marked as Solid Red Arrow and La Niña as a Blue Dashed Arrow).

Vietnam in 1975. The reason might be the changes in composition of the severe versus normal form of malaria.

All the results showed that, before the massive intervention and national malaria control program, the El Niño phenomena had significant influenced the onset of malaria epidemics in Vietnam, including the number of acute cases and deaths from malaria. This influence seems to be most obvious in the Central Highland and Southern part of Vietnam but also can be detected at the national level.

National Malaria Control Efforts

Malaria control is a public health priority in Vietnam, with a national control program integrated into the large basic health care system. Every year in the last ten years, the Vietnam government spent between US$3 and 6 million on malaria control activities, along with financial and technical support from international organizations. The main measures for malaria control include mass early treatment and behavior-changing health education, vector control by residual insecticide spraying, and using insecticide-treated bed nets (ITBNs).

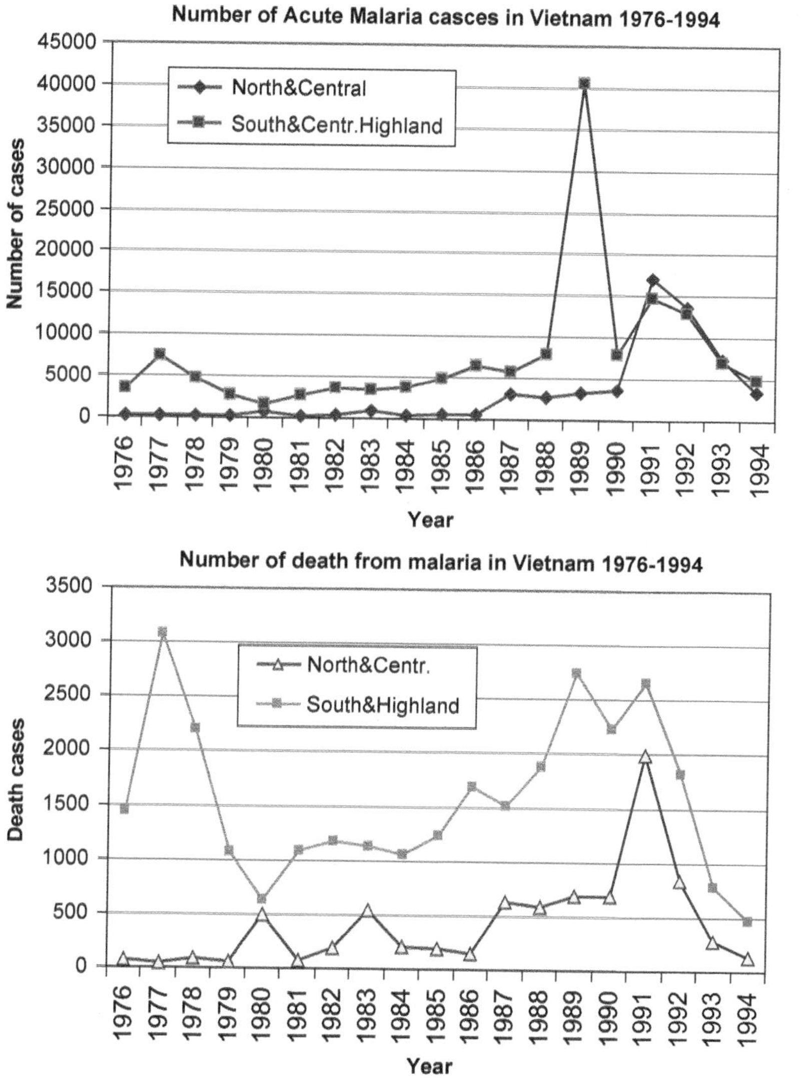

Fig. 6. Number of Acute Cases and Deaths from Malaria in Different Regions in Vietnam, 1976–2004.

In Vietnam, malaria control is considered a responsibility of the community, to be dealt with locally through the primary health care units (horizontal structure), with guidance and assistance from the district provincial and national health authorities (vertical control). The whole system includes the provincial malaria stations, three National Institutes of Malariology, Parasitology, and Entomology (one for each region), and ultimately, the National Malaria Control Program of the Ministry of Health (Le et al., 2002). The nationwide decline of malaria in Vietnam during the last decade can be attributed to the application of the combined approach and effective measures through the whole system. However, with the changing climate and environment and global warming, there is an increasing risk of reemergence and extension of malaria and other infectious diseases outsides the typical areas. This risk has not been considered and addressed in the national health policies and plans so far.

WATERBORNE DISEASES: CHOLERA, DYSENTERY, TYPHOID FEVER AND DIARRHEAL DISEASES

Waterborne or enteric infectious diseases cause considerable morbidity and mortality worldwide, especially among children in developing countries. Cholera, dysentery, and typhoid fever are severe diseases caused by the pathogens *Vibrio cholera, Shigella spp.,* and *Salmonella typhi,* respectively. Diarrheal diseases that are usually reported in the Vietnam Health Statistical Yearbook include a range of diseases with diarrheal symptoms caused by infectious pathogens other than the previously mentioned.

The pathogens or causative microbes of these diseases are environmentally determined, with transmission occurring through fecal contamination of water or food or by person-to-person contact. These enteric diseases are also usually called waterborne diseases (WBDs), although this might not be a correct name. Infection rates are highest in regions with low or inadequate living standards, water supply, and sanitary conditions. The precise burden of these WBDs is difficult to establish as they occur in resource-poor countries where substantial underreporting takes place (Kelly-Hope et al., 2007).

Waterborne diseases are very common in tropical Vietnam and spread throughout the country. Many factors influence the development and spread of these infections, including behaviors and habits, immunity, water, and sanitation, but climate factors are also suggested to play an important role.

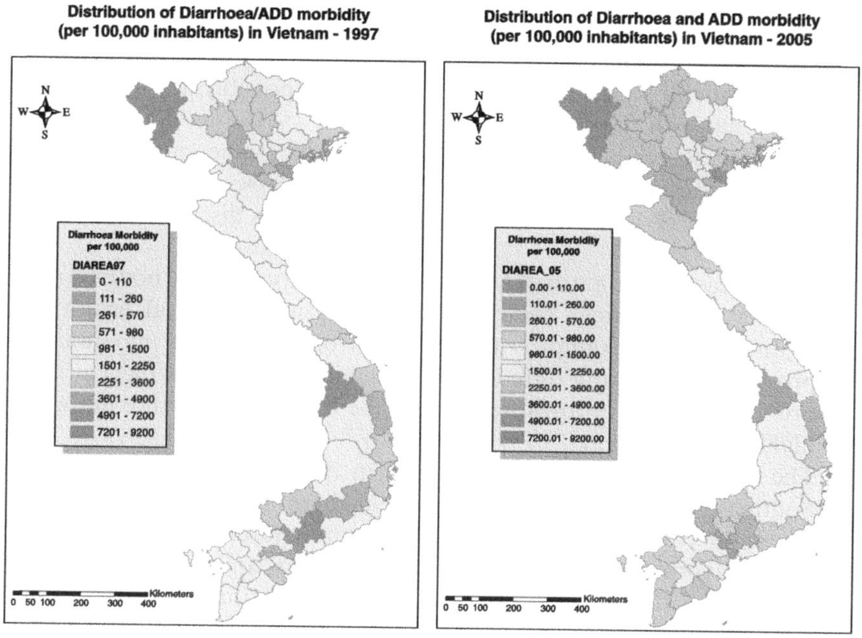

Fig. 7. Change on Spatial Distribution of diarrheal diseases morbidity by provinces and regions of Vietnam, 1997–2005.

In many other tropical countries, diarrheal diseases typically peak during the rainy season and/or rapidly increase during and after disasters (for example, floods, storms, droughts), usually because of a contamination of the water supplies (Kovats et al., 2003b).

Spatial Distribution

Spatial analysis has been conducted only for the two most important and prevalent WBDs—diarrheal diseases and dysentery. The dynamic of changes in spatial distribution of the diarrheal diseases over time, between 1997 and 2005 (see Fig. 7), showed that:

(1) The disease is widespread all over the country but the incidence rate is higher in the mountainous and highland areas (Northwest, some Northern provinces, and Central Highland) than in the coastal and delta

provinces, except for Red River Delta. Some provinces in Red River Delta such as Thai Binh and Ha Nam provinces also have relatively persistent and high rates of diarrheal diseases, possibly due to sanitation and behavioral factors.

(2) The disease tends to occur or persist in the same provinces where existing rates of previous years were high. This observation is consistent with some previous studies on endemic infectious diseases (Kelly-Hope et al., 2007);

(3) The disease rates are higher in the provinces with high poverty rates, low levels of hygiene behavior, and poor sanitation and health care infrastructures. Examples are the cases of remote and mountainous provinces with many different ethnic minorities such as the Lai Chau province in the Northwest region and the Kon Tum province in Central highland.

(4) The disease rates are relatively lower in the Mekong River Delta and Southern Central Coast than in the Northern and mountainous areas, where the tropical climate with higher temperatures would have created more favorable conditions for proliferation and dissemination of diarrheal pathogens. There is no clear explanation about this result, but one of the suggestions is that local people's immunity toward this disease is higher than in the North and mountains. Besides, in provinces with tropical and hot weather, the health authorities are constantly alerted and undertake more active and effective preventive measures and intervention. As well, the division into two clear seasons – a long dry season and periodic rainy season could reduce the natural pathogens' reservoirs and lead to less chances of disease dissemination.

For dysentery, the incidence rates are also typically high in the Northwest and Northern mountainous provinces as well as in the Central Highland. There is a clear trend of extension of prevalent area and increase in dysentery morbidity all over the country. The increasing trend can also obviously be seen in the a time–series analysis later in this chapter (see Fig. 8).

Temporal Changes

Among all enteric infections, dysentery was the most prevalent disease and increased in rates approximately four times during the study period between 1992 and 2003. Due to the absence of dysentery data for Southern Vietnam between 1979 and 1991, we analyzed the temporal changes of this

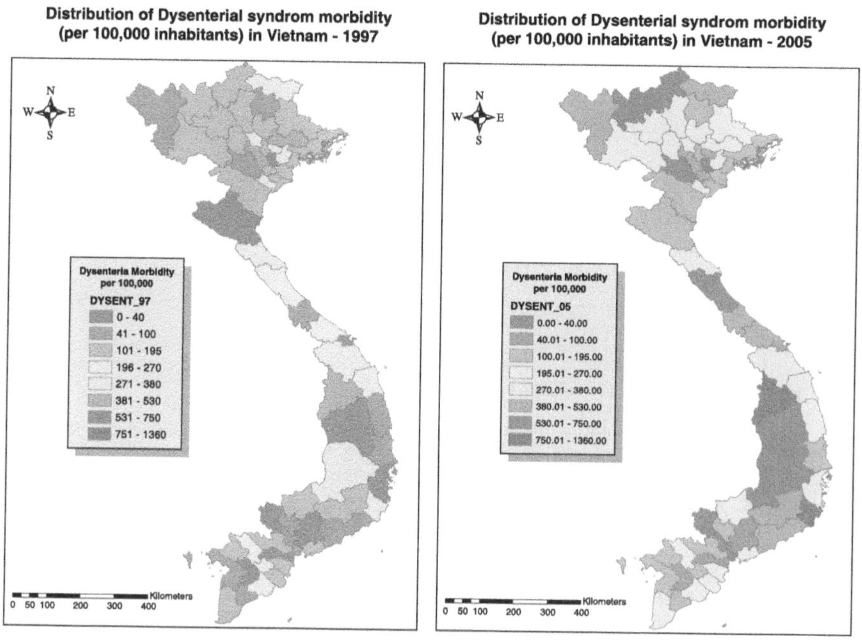

Fig. 8. Change on spatial distribution of Dysentery, 1997–2005.

infectious disease by regions, only for the period between 1992 and 2003, and recognized that the morbidity of the North was almost double that of the South.

The annual number of typhoid fever and paratyphoid incidences in the country was similar at the beginning and at the end of the available period (1979 to 2005), but there was a sudden increase from six times up to nine times between 1993 and 1999, with the peak at 1995 (see Fig. 9). Typhoid fever is typical or endemic for the Mekong River Delta, where the flooding occurs every year. During the period between 1993 and 1999, a dramatic increase in frequency of flooding events occurred nationwide, especially in central and southern Vietnam, perhaps as a result of very strong global activeness of El Niño.

Overall, there were much fewer cholera incidences, which appeared episodically during the whole study period between 1980 and 1996, with four main peaks in 1980, 1983, 1990, and 1995, which were not related with the El Niño events. From 1997 onward, the number of cholera cases reported nationwide decreased dramatically, with only minor peaks in 1999

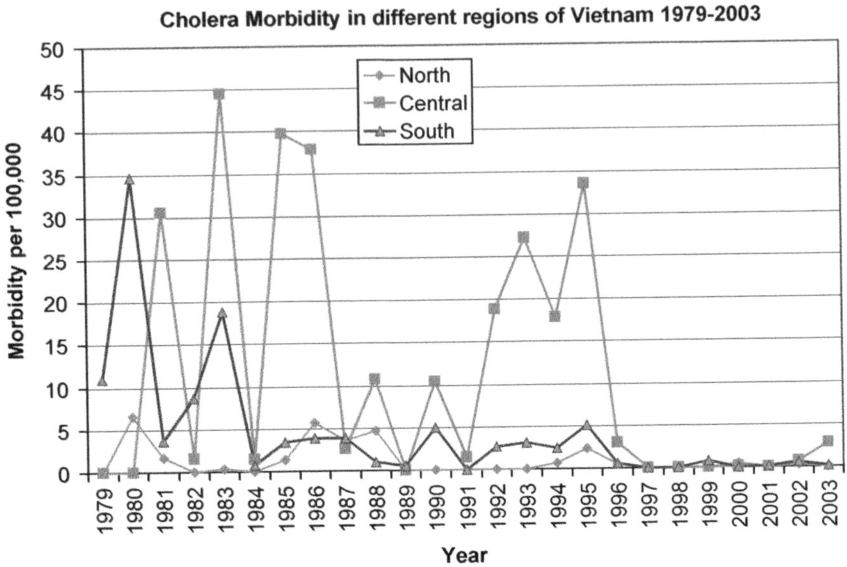

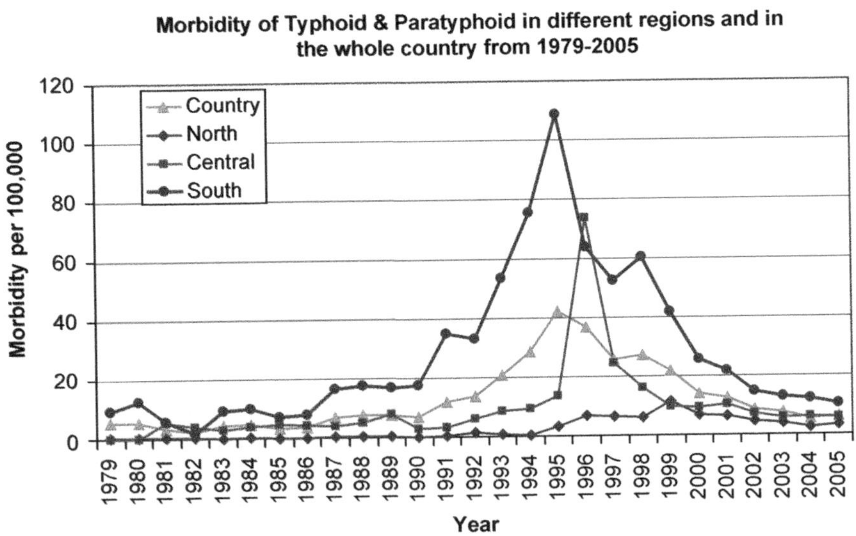

Fig. 9. Morbidity of Cholera from 1979–2003 and Typhoid/Paratyphoid from 1979–2005 in different regions of Vietnam.

and 2003. Major contributions to cholera incidences in the country were from Southern and Central parts, and fewer cases observed in the North (see Fig. 9). In the North, outbreaks occurred in four years: 1980, 1986, 1988, and 1995. In Central Highland during this period, high reported cases of severe outbreaks occurred in only two consequent years, 1994 and 1995. In general, the outbreaks of cholera in Vietnam as whole were neither related with El Niño nor with serious disasters.

The overall rate of diarrheal diseases in the whole country had been increased almost three times from 1979 to 2005 (see Fig. 10). The strongest increase occurred during the period between 1992 and 1996, and then the disease rate remained high until 2005, with equal contribution from all regions. The disease is widespread all over the country, with relatively high morbidity, and the efforts of the Vietnamese government to reduce the rate of this widespread disease did not succeed very much. Another important observation is that many provinces, such as Thua Thien Hue and Danang in the central coast, suffered from severe floods and storms during this period, and in many cases, the diarrheal outbreaks occurred within one to two months of the disasters.

Seasonal Variation

The seasonal variation of diarrheal diseases and dysentery in Vietnam is not equal in different parts of the country (see Fig. 11). At a national level, dysentery had clear seasonal increases in summer and the hot season from May to October; however, in the North, the increase seemed clearer from August to October. In the South and Central part, the onset came earlier in May and reached peaks in the hottest months: July and September. The reason for the decrease of dysentery incidences in all regions in August is unclear.

The diarrheal disease rate in the North clearly increased in the hot summer month (April to September) and decreased in "winter time" or monsoon months (October to February), but in the South and Central part, it increased more in the rainy season (September to December).

Regression Analysis

Regression analysis for cholera in different regions of the country showed that cholera incidences in the North were not statistically correlated with

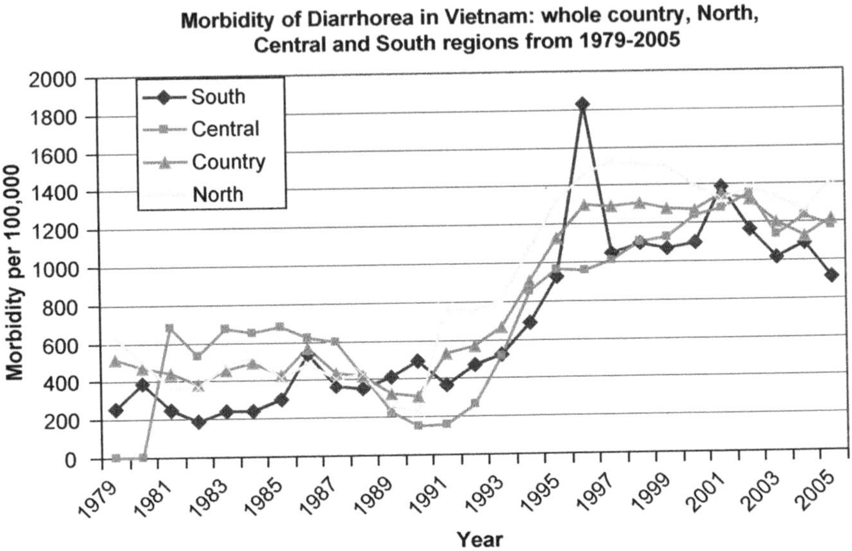

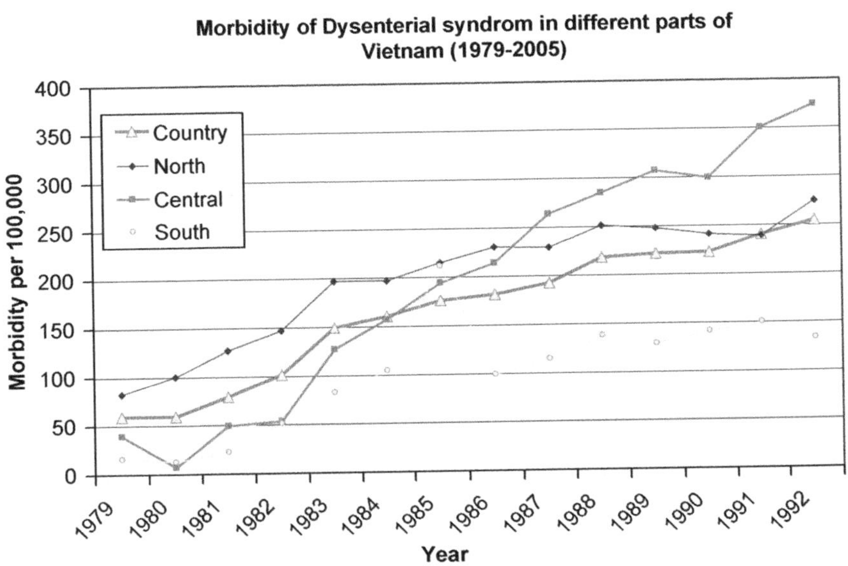

Fig. 10. Changes of Morbidity of Diarrheal Disease and Dysentery in Different Parts of Vietnam, 1979–2005.

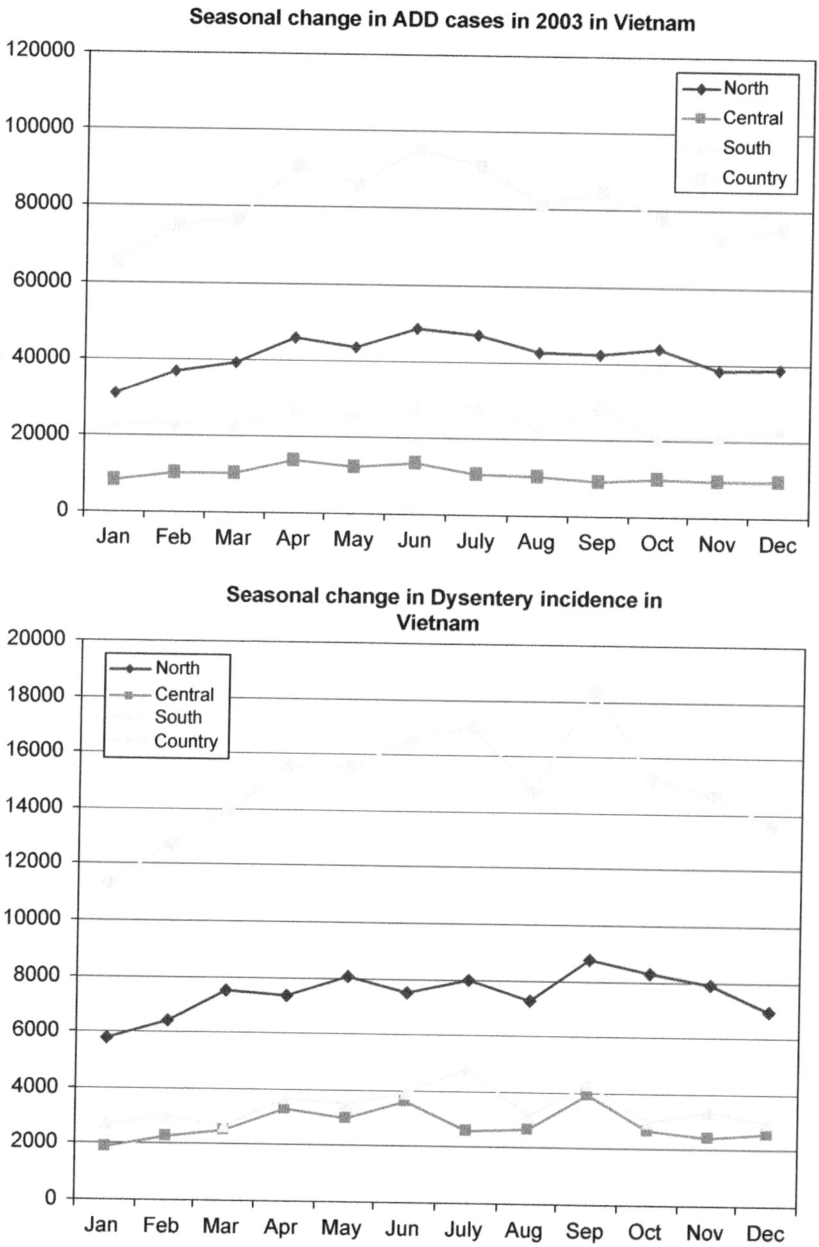

Fig. 11. Seasonal Change of Incidences of Diarrheal Diseases (Average for 1995–2005) and Dysentery (Average for 1980–1990) in Different Parts of Vietnam.

Table 1. Type of Regression Model and Weight/Percentage of Influence and Significance of Each Independent Factor to the Rate of Diarrheal Diseases and Dysentery in Some Provinces of Vietnam.

Diseases and Provinces of Concern	ONI		Temperature		Rainfall	
	Models and % of influence	Significance	Model and % of influence	Significance	Model and % of influence	Significance
Diarrheal diseases in Thua Thien Hue	Cubic 23.9%	High $p<0.001$	Linear 4.9%	High to Medium $P=0.031$	–	No
Diarrheal diseases in Danang	Cubic 9.8%	High to Medium $P=0.04$	–	No	Cubic 68.6%	High $p<0.001$
Dysentery in Thua Thien Hue	–	No	–	No	Cubic 11.3%	High $p<0.013$
Dysentery in Danang	Growth 12.8%	High $p<0.01$	–	No	Cubic 23.9%	High $p<0.001$

both El Niño and climate variables (temperature and rainfall); however, in the Central part, it had a clear correlation with temperature ($r^2=0.411$, $p<0.01$). There was statistically significant correlation between El Niño and cholera in the South (Cubic – $r^2=0.322$, $p<0.001$), moderate to low correlation with temperature (Cubic $r^2=0.119$, $p=0.05$), and no correlation with rainfall. In the country level, cholera is neither associated with El Niño nor with temperature or rainfall.

Regarding cholera and other enteric infections, many debates were raised in the previous studies about the relative roles of climate and environmental factors and the social or human factors like sanitation, water supply, immunity, personal hygiene, and habit (Pascual et al., 2002; Kovats et al., 2003b). For this reason, a decision was made to conduct regression analysis for those diseases in particular provinces, not in the whole region, to avoid incorrect results due to unification of areas/provinces with very different sanitation and socioeconomic conditions. The Thua Thien Hue province and Danang city in Central Vietnam had been selected because they are located in the areas more prone to water-related disaster hazards (typhoons, heavy rains, severe floods). Thua Thien Hue represents the northern provinces of Central Vietnam with very high rainfall and high frequency of typhoon, and Danang represents the southern central coast with less rainfall but still high storm risk. The results are shown in Table 1.

The results from Table 1 and Fig. 12 show that, in both Danang city and Thua Thien Hue province, the monthly rates of diarrheal diseases were significantly correlated with El Niño (cubic model), whereas dysentery rates were significantly associated with El Niño only in Danang, not in Thua Thien Hue. Further, the role of temperature was only significant for diarrheal diseases in Thua Thien Hue, but not in the other cases; at the same time, the roles of rainfall were very high in all cases except for the diarrheal diseases in Thua Thien Hue. One of the possible reasons of having these results was that in the Thua Thien Hue province, frequent and high floods could contaminate the water sources or wash out or dilute the pathogen concentration in the water sources, whereas in the more rapidly urbanized areas of Danang, untreated sewage can be broadly disseminated and water supplies were polluted every time of flooding.

Although there was no statistical significant correlation between diarrheal diseases and dysentery in the Thua Thien Hue province, it was found that there were sudden and strong increases (from two- to fourfold) in monthly incidences of those diseases right after the catastrophic floods in this province, as well as in Danang and some adjacent areas, especially during the historical flood that occurred in November 1999. This association or trend was not statistically significant and could be observed only in some provinces. Although the association was not related to the normal floods, it could highlight some of the consequences of disasters to the population health in affected areas.

Influenza

Spatial Distribution

Spatial analysis of influenza showed that northern mountainous provinces and the Central Highland have highest rates of the disease, whereas the South has insignificant rates. Most of the endemic areas for influenza have lower mean temperature and higher humidity; meanwhile, the disease has clear seasonality with the highest rates occurring during the period of August to October, or during interchange between summer and winter times. There was also an obvious trend of expansion of influenza from the mountainous and highland provinces to the transition areas at Red River Delta and the central coast provinces.

Temporal Change and Seasonality

Although the results from regression analysis showed no correlation between influenza morbidity and major climatic parameters or El Niño, it

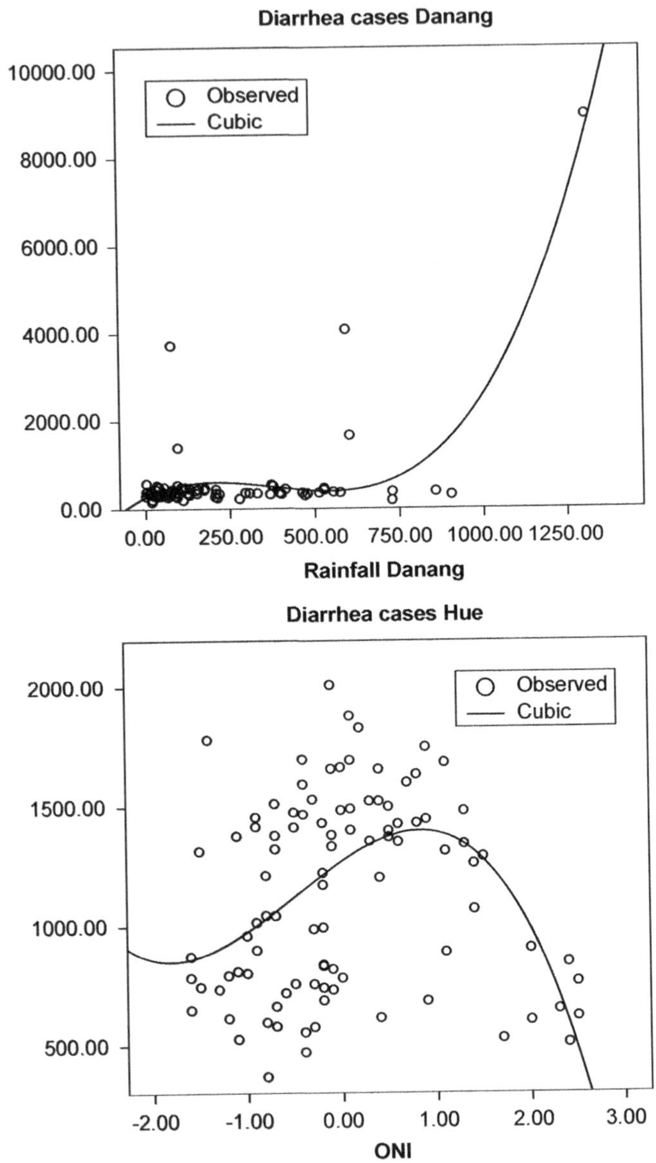

Fig. 12. Graphs of Correlation of Diarrheal Diseases Rate and Rainfall in Danang and the ONI in Thua Thien Hue Provinces.

seems that influenza is relatively climate-sensitive with clear seasonality and persistent location (see Fig. 13). From time–series analysis, the results showed that between 1990 and 2005, the influenza incidences increased more than 16 times all over Vietnam, especially in the Northern part where the disease remained very high up until 2005. These results showed a dangerous trend, especially when the H1N1 global pandemic and H5N1 periodical epidemics posed very serious risks to the population's health in Vietnam. We could not prove the correlation between climate change/climate variability and the influenza dynamic in Vietnam but the increase and extension of the disease definitely have had a certain relation to global change in the last decades.

DISCUSSION

In this study, the surveillance data have been used to highlight the different magnitudes and epidemiologic patterns of each disease in Vietnam during a long period of about 25 to 27 years (the longest time series was between 1979 and 2005), and the authors have tried to link the spatial and temporal changes of the selected infectious diseases with climate variability. Certainly there always are inherent limitations associated with surveillance data, but this long-range data set is probably the most comprehensive available in Vietnam and many other developing countries, providing the basis for more specifics and details in relation to climate and disease.

It was found that the interannual change of several important infectious diseases in Vietnam, including malaria, dysentery, and diarrheal diseases have a significant statistical correlation with changes of climate parameters (monthly rainfall, temperature) and the ENSO phenomenon. In certain areas, some diseases (dysentery and diarrheal diseases) are correlated at a higher level with certain climate parameters, but at a lower level or no correlation with other parameters. Changing patterns of some diseases were supposed to be correlated or coincident with ENSO phenomenon, but the role of socioeconomic factors, especially interventions, preventative and control measures, and so on, is also very high. One of the good examples of successful interventions was the National Malaria Control Program.

The weak association of certain diseases with climate did not mean that those diseases are weakly related with or influenced by climate and other environment factors. The reason for a weak or nonexistent correlation could also be due to the low quality of surveillance data, which in Vietnam

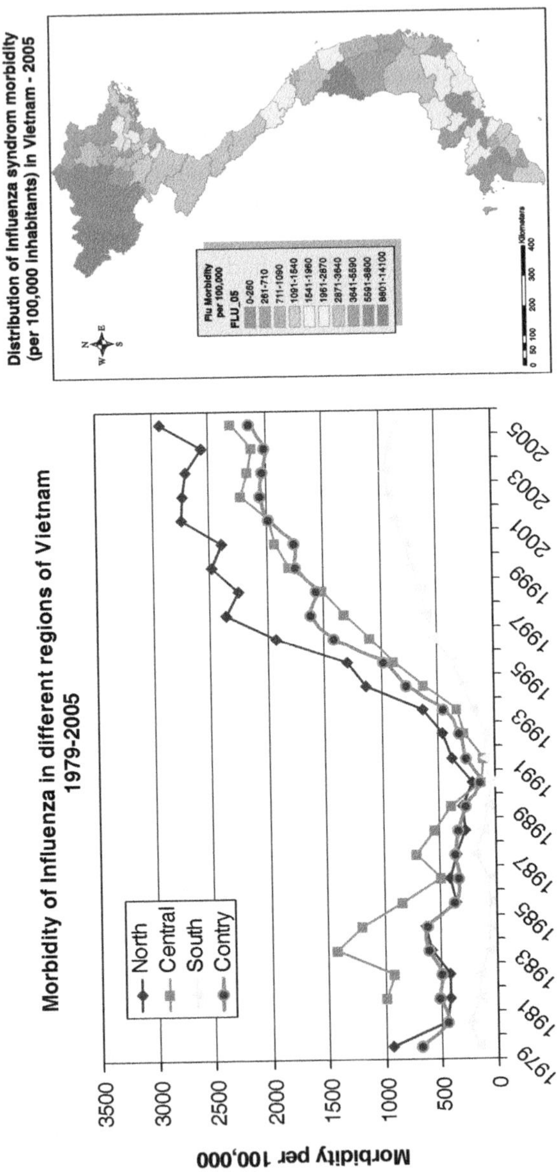

Fig. 13. Change of Morbidity of Influenza from 1979 to 2005 and its Spatial Distribution by Provinces in Vietnam.

are inevitably flawed because of underreporting, misdiagnosis, and misclassification. In Vietnam so far, adequate diagnostic facilities are not universally available, and detection can be difficult and may be biased. Further, other factors such as poor socioeconomic conditions play a role and are also likely to be as important, if not more important, than climate (Kelly-Hope et al., 2008).

Many infectious diseases have obvious seasonal variation but the seasonality differs in different parts of Vietnam and has been changed over time for some diseases, probably under the influence of climate change and global environmental changes. Some infectious diseases such as malaria, dysentery, and influenza have specific spatial prevalence and distribution among provinces in Vietnam, but some others (such as diarrheal diseases) do not have such obvious specific distribution and were widespread throughout the country.

Comparing the spatial distribution over the period between 1997 and 2005, an obvious trend of expansion and increase of morbidity of both diarrheal diseases and dysentery was detected in many mountainous provinces in the North and Northwest, and of dysentery in the Central Highland and Central Coast. Certainly change in such a short period may not be attributed to the climate variability or El Niño influence, but this fact conveys a troubling message that the diseases are obviously expanding beyond the typical geographical border and up to higher-altitude areas, threatening the health of many minority ethnic groups who are already living in poverty.

The dramatic and widespread reduction in the cholera rate after 1997 up to 2005 could be explained by several factors, including interannual variability, changes of population immunity, economic development, and interventions and improvements in hygiene and sanitation as proposed by other similar study in Vietnam (Kelly-Hope et al., 2008). The initial decline might reflect the episodic nature of cholera. Other studies have also shown that interannual variability is common and is affected by climate and events such as the ENSO, as well as by levels of immunity within populations (Pascual et al., 2002; Rodó et al., 2002). However, the fact that cholera numbers remained low from 1997 to 2001 may be related to the introduction of a new locally produced vaccine in 1997 instead of ENSO influence.

Besides, it was found that the onset of the outbreaks of some waterborne diseases like diarrheal diseases and dysentery have occurred after severe disasters like floods and typhoons in some disaster-prone provinces of Central Vietnam.

CONCLUSIONS

There is good epidemiological evidence that climate factors (temperature and rainfall) and climate variability (El Niño) are associated with disease burden of the major infectious diseases in Vietnam. The correlation is higher in the specific geographical areas where climate parameters and anomalies are linked with the ENSO cycle. The associations are obvious with malaria and some waterborne diseases such as diarrheal diseases and dysentery. For influenza, cholera, and typhoid diseases, the correlation is not that obvious, but there is clear linkage between climate factors and the changing pattern of disease over the last 25 years.

More research is needed to determine the nature of the ecological mechanisms of these relationships. Influence of climate variables and El Niño provides an opportunity to illustrate the importance of the ecological basis and climate factors for many diseases, thus the current and potential influences of climate change to the disease burden in Vietnam. These linkages need to be more fully studied, brought to awareness, approbated, and used by health professionals, policy makers, and the general public in disease/epidemic controlling, as well as in health policy and planning activities.

With a projected obvious warming trend and substantial changes in rainfall patterns in Vietnam, there is an increased risk of certain diseases in the future for each region, especially in the marginal areas of current highly endemic zones, in higher elevations in Northern and Northwest mountains, and in the Central Highland and central coast. The spatial extension of diarrheal diseases and dysentery in Northern and Northwest mountains, and in Central Highland in the last decade, point out the higher risk of increase of this disease in those areas. More in-depth studies should be done to confirm the fact that more rapid and stronger warming processes in the mountainous and highland areas of Vietnam could change the seasonal pattern and the local ecosystems and make those areas more vulnerable to the dissemination of infectious-disease outbreaks.

Advances in seasonal climate forecasting as well as El Niño forecasting in Vietnam and in the world provide the opportunity to develop longer-term predictions of epidemic risk in some vulnerable areas. Future research on the incorporation of climate forecasts into epidemic early warning systems in Vietnam may yield benefits in the near future.

Facing the complex threats from disasters, current and future climate change requires interdisciplinary cooperation among health professionals, climatologists, disaster experts, environmentalist, biologists, and social scientists to understand and effectively manage these threats. New collaborative efforts can confront these tough challenges through advances in preventive medicine. Despite already-existing mechanisms and programs, such as flood and storm warning, training, and mobilization for an epidemic early warning system (EWS), there is a clear need to develop and evaluate more effective public health interventions.

For example, various approaches to reduce the potential pressure for and scarcity of water, water quality deterioration as a result of climate change, and increasing pollution due to rapid urbanization and industrialization should include policies for efficient management of available water, improved technology, integrated agriculture policies, water quality, sanitation control, and urban planning and management. Effective purification of water supplies or household storage should be combined with measures aimed at preventing pollution or decontaminating water at its source, especially during a disaster or extreme weather (Khasnis & Nettleman, 2005).

The crucial step to prevent and reduce the adverse health effects of extreme events and to empower the adaptive capacity is raising public awareness and knowledge about the nature of the risk. However, according to practical observation, especially from the low-income developing countries with low education index, the public awareness that always increases after any serious natural disaster, is often short-lived, if the situation returns to the previous state and if no "reminding" or educational measures are maintained (Kien & Shaw, 2009).

As an overall observation, it is clear that urgent actions need to be taken toward further studies on impacts of climate change and climate variability to the whole range of climate-sensitive diseases, including infectious diseases and pathological processes. This will form an important foundation for the comprehensive national public health assessment program toward current and potential climate change impacts, under the National Target Program and Action Plan for Response and Adaptation to Climate Change.

ACKNOWLEDGMENTS

We highly acknowledge the scholarship of Kyoto University Foundation and the Project Fund of the Graduate School of Global Environmental Studies, for the first author, Dr. Tran Mai Kien, in his research in Kyoto University, Japan, and field trip in Vietnam. We highly appreciate the support from the Graduate School of Global Environmental Studies, Kyoto University (Gakudo Project) and MEXT/JSPS Research Fund. We also highly acknowledge the hearty support of Professor Dr. Dam Viet Cuong, Vietnam Health Strategy and Policy Institute, and Professor Dr. Phung Dac Cam, Vietnam National Institute of Hygiene and Epidemiology.

REFERENCES

Bouma, M. J., Proveda, G., Rojas, W., Chavasse, D., Quinones, M., Cox, J., & Patz, J. (1997). Predicting high-risk years for malaria in Columbia using parameters of El Niño southern oscillation. *Tropical Medicine and International Health, 2*, 1122–1127.

Bulto, O. P. L., Rodriguez, A. P., Valencia, A. R., Vega, N. L., Gonzalez, M. D., & Carrera, A. P. (2006). Assessment of human health vulnerability to climate variability and change in Cuba. *Environmental Health Perspectives, 114*, 1942–1949.

Campbell-Lendrum, D., Pruss-Ustun, A., & Corvalan, C. (2003). How much disease could climate change cause? In: A. J. McMichael, D. Campbell-Lendrum, C. Corvalan, K. L. Ebi, A. K. Githeko & J. S. Scheraga, et al. (Eds), *Climate change and health: Risks and responses*. Geneva: World Health Organization.

Cazelles, B., Chavez, M., McMichael, A. J., & Hales, S. (2005). Nonstationary influence of El Niño on the synchronous dengue epidemics in Thailand. *Public Library of Science Medicine, 2*(4), e106.

Checkley, W., Epstein, L. D., Gilman, R. H., Figueroa, D., Cama, R. I., Patz, J. A., & Black, R. E. (2002). Effects of EI Niño and ambient temperature on hospital admissions for diarrhoeal diseases in Peruvian children, *The Lancet, 355*(9202), 442–450.

Confalonieri, U., Menne, B., Akhtar, R., Ebi, K. L., Hauengue, M., Kovats, R. S., Revich, B., & Woodward, A. (2007). Human health. In: M. L. Parry, O. F. Canziani, J. P. Palutikof, P. J. van der Linden & C. E. Hanson (Eds), *Climate change 2007: Impacts, adaptation and vulnerability. Contribution of Working Group II to the Fourth Assessment Report of the Intergovernmental Panel on Climate Change*. (pp. 391–431). Cambridge, UK: Cambridge University Press.

Ebi, K. L., Nancy, D., & Lewis, C. C. (2006). Climate variability and change and their potential health effects in small island states: Information for adaptation planning in the health sector. *Environmental Health Perspectives, 114*, 1957–1963.

Furgal, C., & Seguin, J. (2006). Climate change, health, and vulnerability in Canadian northern aboriginal communities. *Environmental Health Perspectives, 114*, 1964–1970.

Haines, A., Kovats, R. S., Campbell-Lendrum, D., & Corvalan, C. (2006). Climate change and human health: Impacts, vulnerability and public health. *Public Health, 120*, 585–596.

Hales, S., Weinstein, P., Souares, Y., & Woodward, A. (1999). El Niño and the dynamics of vector-borne disease transmission. *Environmental Health Perspectives, 107*, 99–102.

Hanh, T. T. T., Hill, P. S., Kay, B. H., & Quy, T. M. (2009). Development of a framework for evaluating the sustainability of community-based dengue control projects. *American Journal of Tropical Medicine and Hygiene, 80*(2), 312–318.

Hunter, P. R. (2003). Climate change and waterborne and vector-borne disease. *Journal of Applied Microbiology Symposium Supplement, 94*, 37S–46S.

Kelly-Hope, L. A., Alonso, W. J., Thiem, V. D., Anh, D. D., Canh, D. G., Lee, H., Smith, D. L., & Miller, M. A. (2007). Geographical distribution and risk factors associated with enteric diseases in Vietnam. *The American Journal of Tropical Medicine and Hygiene, 76*(4), 706–712.

Kelly-Hope, L. A., Alonso, W. J., Thiem, V. D., Anh, D. D., Canh, D. G., Lee, H., Smith, D. L., & Miller, M. A. (2008). Temporal trends and climatic factors associated with bacterial enteric diseases in Vietnam, 1991–2001. *Environmental Health Perspectives, 116*, 7–12.

Khasnis, A. A., & Nettleman, M. D. (2005). Global warming and infectious disease. *Archives of Medical Research, 36*, 689–696.

Kien, T. M., & Shaw, R. (2009). Climate change adaptation and human health: Linkages of climate, disaster and health issues in Vietnam. In: *Disaster management global challenges and local solutions* (pp. 444–467). Universities Press.

Koelle, K., & Pascual, M. (2004). Disentangling extrinsic from intrinsic factors in disease dynamics: A nonlinear time series approach with an application to cholera. *The American Naturalist, 163*, 901–913.

Kovats, R. S., Bouma, M. J., Hajat, S., Worrall, E., & Haines, A. (2003b). El Niño and health. *Lancet, 362*, 1481–1489.

Kovats, R. S., Menne, B., Ahern, M. J., & Patz, J. A. (2003a). National assessments of health impacts of climate change: A review. In: A. J. McMichael, D. Campbell-Lendrum, C. F. Corvalan, K. L. Ebi, A. Githeko, J. D. Scheraga & A. Woodward (Eds), *Climate change and human health: Risks and responses* (pp. 181–203). Geneva: WHO/World Meteorological Organization/United Nations Environment Programme, 2003.

Le, K. T. (2002). *Report of Vietnam National Malaria Control Project (VNMCP)*. Hanoi, Vietnam.

Le, Q. H., de Vries, P. J., Giao, P. T., Nam, N. V., Binh, T. Q., & Chong, M. T. (2002). Control of malaria: A successful experience from Viet Nam. *Bulletin of the World Health Organization, 80*, 660–666.

McMichael, A. J., et al. (2004). Climate change. In: M. Ezzati, A. D. Lopez, A. Rodgers & C. J. L. Murray (Eds.), *Comparative quantification of health risks: Global and regional burden of disease due to selected major risk factors*. Ch. 20 (pp. 1543–1649). Geneva: World Health Organization.

McMichael, A. J., Woodruff, R. E., & Hales, S. (2006). Climate change and human health: Present and future risks. *The Lancet, 367*, 859–869.

Nguyen, V.K. (1999). Malaria in Vietnam 1957–1997: Strategies and tactics. In: *Malaria course on intensified control measure*. Hanoi, Vietnam.

Pascual, M., Bouma, M. J., & Dobson, A. P. (2002). Cholera and climate: Revisiting the quantitative evidence. *Microbes and Infection, 4*, 237–245.

Patz, J. A., et al. (2000). The potential health impacts of climate variability and change for the United States: Executive summary of the report of the health sector of the U.S. National Assessment. *Environmental Health Perspectives, 108*(4), 367–376.

Pruss, A., & Havelaar, A. (2001). The global burden of disease study and applications in water, sanitation and hygiene. In: L. Fewtrell & J. Bartram (Eds), *Water quality: Guidelines, standards and health* (pp. 43–59). London: IWA Publishing.

Rodó, X., Pascual, M., Fuchs, G., & Faruque, S. G. (2002). ENSO and cholera: A nonstationary link related to climate change? *Proceedings of the National Academy of Sciences of the United States of America, 99*, 12901–12906.

Vietnam Ministry of Health (MOH). (2007). *National strategy on preventive medicine and master plan on health care system development*. MOH.

CHAPTER 21

CLIMATE CHANGE ADAPTATION AND DISASTER RISK REDUCTION: KEY CHALLENGES AND WAYS FORWARD

Juan M. Pulhin, Rajib Shaw and Joy Jacqueline Pereira

INTRODUCTION

Climate change is one of the major challenges confronting human society in the 21st century (Ericksen & O'Brein, 2007; Adger, Lorenzoni, & O'Brien, 2009). Mounting evidence attests that climate change is now happening in many parts of the world as evidenced by increasing mean temperature, changing precipitation patterns, rising sea level, and increasing frequency and growing intensity of extreme weather events (Intergovernmental Panel on Climate Change [IPCC], 2007). In many parts of Asia, these climatic changes have led to massive flooding, landslides, and droughts, resulting in extensive damage to properties, assets, and human life (Cruz et al., 2007). Climate change is also exacerbating water shortages in many areas, constraining agricultural production, and threatening food security and energy supply from hydroelectric source. It is likewise causing forest fires

Climate Change Adaptation and Disaster Risk Reduction: An Asian Perspective
Community, Environment and Disaster Risk Management, Volume 5, 451–476

ISSN: 2040-7262/doi:10.1108/S2040-7262(2010)0000005027

and degradation, damaging coastal and marine resources, and increasing the risk of outbreaks of infectious diseases. Regional climate studies indicate that the worse is yet to come as far as weather-related disaster risks are concerned. If not addressed effectively, climate change could seriously frustrate the region's sustainable development and poverty-reduction efforts (ADB, 2009).

Based on the statistics of the International Strategy for Disaster Reduction (ISDR), Asia registered the highest number of hydro-meteorological disasters at 1,532 occurrences in the period 1991–2005, accounting for about 39% of the total of 3,976 similar type of disasters among all regions in the world. Of these, 672 or 44% are flood-related, while 519 or 34% are caused by windstorm. The other disasters are caused by slides (151 cases or 10%), drought (91 cases or 6%), extreme temperature (66 cases or 4%), and wild fire (33 cases or 2%). At the global level, hydro-meteorological disasters increased by more than 100% from about 100 in 2004 to more than 200 in 2006, while the number of geophysical disasters, such as earthquakes (including tsunamis) and volcanic eruptions, has remained steady throughout the last decade (Birkmann & von Teichman, 2010). These figures provide a powerful evidence of the strong link between the increasing occurrence of hydro-meteorological disasters and the worsening climate.

Recognizing the strong relationship between climate change and disaster risk, international discussions are currently underway to confront both in a more coherent manner. The more current prominent venues for discussions include the Global Platform for Disaster Risk Reduction (DRR) in Geneva 2009 and the Conference of Parties under the United Nations Framework Convention on Climate Change (UNFCCC) in Copenhagen in December 2009 (Birkmann & von Teichman, 2010). Recent reports serve similar purpose, such as the report of the Commission on Climate Change and Development (CCD, 2009) and the Intergovernmental Panel on Climate Change (IPCC) special report on *Managing the risk of extreme events and disasters to advance climate change adaptation*.

The benefits of these international initiatives, however, have yet to trickle down at the country and local areas where the adverse impacts of weather-related disasters are continuously experienced at an alarming degree. At the country level, several states have adopted practical approaches to integrate efforts on climate-change adaptation (CCA) and DRR through the National Adaptation Programs of Action (NAPAs), which provides a process for the least-developed countries (LDCs) to identify areas where urgent activities and projects are needed in order to adapt to climate change.

Even under this initiative, however, the aspect of DRR in the context of climate change and extreme events remains a peripheral topic (Birkmann & von Teichman, 2010). For instance, although disaster risk was identified as an urgent problem by many of the LDCs, only 24 of the 38 NAPAs submitted to the UNFCCC have called for immediate action and projects in the field of disaster management and early warning.

A number of issues and challenges have been identified by various studies that hinder the effective integration of CCA and DRR strategies. Birkmann and von Teichman (2010) summarized and categorized these issues into three general types: (1) differences in functional, spatial, and temporal scales; (2) mismatches regarding norms such as legislative, cultural, and behavioral norms; and (3) knowledge mismatches. Similarly, Romieu, Welle, Schneiderbauer, Pelling, and Vinchon (2010) identified the gaps between vulnerability in the context of climate change and natural hazards in terms of eight types of differences: (1) objective pursued; (2) kind of process; (3) timescale; (4) spatial scale; (5) functional scale; (6) simplified formulation; (7) vulnerability assessment; and (8) level of uncertainty.

Quite a departure from the existing literature, this chapter synthesizes the key issues and challenges facing CCA and DRR based on specific country and local level experiences in Asia as discussed in the earlier chapters of this volume. The focus is not solely on the differences between CCA and DRR per se, although they are part of the problem. Rather, the synthesis centers on the general issues and challenges that constrain the achievement of CCA's and DRR's main objectives in terms of reducing damages and vulnerability and enhancing successful adaptation to weather-related disasters at the country and local levels. Based on the synthesis of issues, a set of strategic actions are discussed to successfully integrate and achieve the objectives of CCA and DRR at the national and local levels.

SYNTHESIS OF KEY ISSUES AND CHALLENGES

The issues and challenges facing the successful integration and outcomes of CCA and DRR at the country- and area-specific levels are varied and multidimensional as reflected in the country-specific cases presented in the preceding chapters. The major objective of this synthesis is to capture the key issues and challenges that limit the achievement of CCA's and DRR's overarching goal to reduce damages and vulnerability and ensure successful adaptation to weather-related disasters at the country and local levels. Given this objective, eight categories of issues and challenges emerged from the synthesis: (1) CCA–DRR integration; (2) institutions and governance; (3) information and communication; (4) assessment methods and tools; (5) natural resources' sustainability and access; (6) building local resilience; (7) education and research; and (8) funding/financial resources. It should be mentioned from the outset that there are major overlaps of the identified issues and challenges, including their categories, due primarily to their interconnected nature. Moreover, it should be noted that it is not the intention of the authors to exhaust all the issues mentioned or implied in all the country-specific cases but to take illustrative examples of the more important ones that represent the key challenges faced by CCA and DRR at the country and local levels.

Table 1 summarizes the key issues and challenges distilled from the 19 cases presented in this volume, which represent 13 countries and regions of Asia. These countries are: Cambodia, India, Malaysia, Nepal, Pakistan, Singapore, Taiwan, and Thailand with one case each; Bangladesh, Japan, the Philippines, and Sri Lanka with two cases each; and Vietnam with three cases. The following discussions of the different issues and challenges drawn from the 19 cases are supplemented with other relevant literature as well as with personal knowledge of the authors on the Asian regional situation.[1]

CCA–DRR Integration

A key barrier to effectively link CCA and DRR in Asia and elsewhere relates to the scale integration. Birkman and von Teichman (2010) identified three types of scale mismatches in relation to this: functional, spatial, and temporal. Functional scale mismatches pertain to the organization and management of disasters and CCA by people with different institutional affiliations, which make it a challenge for them to work together to address both concerns. In many countries, DRR often lies within the responsibility of the ministries of interior, defense, and development, while climate change issues are tackled by the environment ministries and meteorological services (Romieu et al., 2010). In Cambodia, for instance, Nguyen, Shaw, and Prabhakar (2010) noted that the biggest challenge in DRR and CCA is to link relevant government agencies, such as the National Committee for Disaster Management (NCDM) and the Provincial Committee for Disaster Management (PCDM) concerned with DRR and the Ministry of Water Resources and Meteorology and Provincial Department of Water Resources and Meteorology (PDWRAM) working on climate change issues. Such a linkage will enable the establishment of an effective institutional framework

Table 1. Synthesis of the Issues and Challenges on CCA and DRR Based on 19 Country-Specific Cases Representing 13 Asian Countries and Regions.

Issue/Challenge	Illustrative Example and Brief Description from Country Cases
1) CCA–DRR integration • Functional integration	• Cambodia (need to link relevant agencies concerned on CCA and DRR to establish an effective institutional framework to address the disaster risks and climate change impacts in a comprehensive manner; overlap or conflict of mandates among agencies, lack of coordination and information sharing across sectors and agencies) • Pakistan (need for central coordinating authority on disaster-related concerns, need to mainstream DRR into routine planning process) • Thailand (climate change initially classified by government as environmental issue; better coordination among stakeholders needed, especially in government, for effective response to emerge in the water sector)
• Spatial integration	• Philippines (some local government units start to integrate CCA and DRR concerns at municipal and provincial level plans and activities) • Thailand (better integration and implementation of land and water-use planning crucial to reducing vulnerabilities in the water sector)
• Temporal integration	• Japan (consideration of the changing role of forests in the context of climate change, better land use and land management in mountain and coastal areas) • Thailand (increasing uncertainties about future climate in water sector requires flexible and reversible adaptation options) • Taiwan (carry out flood modeling analysis in order to respond to the future conditions of protecting life and property in the coastal regions)
2) Institutions and governance • Enabling policies and mechanisms for implementation	• Cambodia (current policies not responsive to climate change issues; need to mainstream DRR into national and local development planning) • Pakistan (need for effective policy formulation) • Sri Lanka (policy and regulatory enactments need to be reviewed and revised in the light of conservation of coastal resources *vis-a-vis* climate change; although more than 50 laws relate directly or indirectly to coastal resources, implementation has been sluggish) • Vietnam (much of the legislation lacks clear institutional arrangements for enforcement and the current organizational structures, mandates, annual budget earmarks and working agenda focus largely on disaster response rather than prevention; need for a National Platform for DRR, that brings together multiple stakeholders to effectively advocate for DRR at different levels) • The Philippines (presence of Climate Change Act and local ordinances but DRR has not been fully integrated) • Taiwan (establishment of the National Council for Sustainable Development which takes care of UNFCC affairs recently chaired by the Vice Premier; enactment of two superior laws namely, Disaster Prevention and Protection Act and Special Act for Flood Management with corresponding budget allocations and allow participation of different sectors).

Table 1. (*Continued*)

Issue/Challenge	Illustrative Example and Brief Description from Country Cases
• Participation of different stakeholders	• Japan (need to involve forest owners, private industries, government, academe necessary in integrated forest management) • Cambodia (need to ensure community participation and government linkages on disaster-related concerns) • Thailand (meaningful public participation in managing climate risks is crucial to gain acceptance for difficult decisions about zoning or new flood protection infrastructure; stakeholder participation important for observing and monitoring changes in climate, flood regimes, and vulnerabilities; vulnerable groups need to be fully engaged in exploring alternative response options) • Vietnam (need to treat local people directly affected by climate change and disasters as key participants in the planning and implementation of future climate change adaptation measures, particularly where these require relocation or significant dislocation of existing livelihood practices) • Bangladesh (NGOs, donors, and government departments taking specific programs and action plans to address climate change-induced hazards)
• Institutional capacity	• Cambodia (need for policy makers knowledge on climate change issues and capacity to identify eligible program; limited number of professional staff and low commitment of government staff) • Thailand (need for capacity building in government agencies and the wider community to enable appropriate assessment of risks and exploration of adaptation responses) • Vietnam (no professional and specialized cadre of staff who focus on disaster management) • Thailand (projects or policies requires careful monitoring to evaluate whether they meet their climate adaptation objectives and contribute to sustainable development)
• Leadership and political commitment	• Taiwan (National Council for Sustainable Development which takes care of UNFCC affairs recently chaired by the Vice Premier) • The Philippines (Albay Provincial Governor as CCA-DRR champion) • Thailand (key decisions on projects often seem to have their origins in the promises of politicians rather than as outcomes of assessment or longer-term planning strategies)
• Role of the local government	• The Philippines (Some municipal and provincial government units taking the lead in integrating DRR and CCA concerns)
3) Information and communication • Awareness raising/ information access	• The Philippines, Cambodia, Vietnam, Bangladesh (importance of public awareness raising to minimize disasters in various sectors and build local resilience) • Singapore (public awareness as component of national climate strategy) • Vietnam (households had insufficient access to information, especially on natural disaster planning and mitigation to allow them to properly respond) • Cambodia and Pakistan (need for early warning and forecasting system and effective communication) • Japan (communication to specific vulnerable age groups and gender)

Table 1. *(Continued)*

Issue/Challenge	Illustrative Example and Brief Description from Country Cases
• Information sharing and knowledge management	• Thailand (creation of Climate Change Knowledge Management Center in 2009 to collect, synthesize, and disseminate knowledge on climate change to support strategic planning of government agencies, the private sector, and local communities) • Nepal (23-member Climate Change Network (CCN) created to address information, coordination and communication issues related to climate change but suffer from lack of funding)
4) Assessment methods and tools	• The Philippines (potential use of sustainable livelihood framework for assessing local resilience) • Bangladesh (proposed the SIP approach for determining drought resilience in a given area) • Cambodia (importance of Community-Based Disaster Risk Management methodologies in communities and the need continually use it in all future DRR programs) • Singapore, Taiwan, and Thailand (robust scientific assessments and studies including climate change projections central in defining appropriate adaptation strategies)
5) Resource sustainability and access	• Vietnam (increasing degradation of the natural environment through deforestation and the conversion of agricultural land to urban areas results to more destructive floods in the lowland areas) • Thailand (concern on maintaining and restoring the resilience of ecosystems, and securing access to those services crucial to many vulnerable groups in rural areas; need to consider local initiatives to improve soil and water management, including water harvesting and storage for climate-vulnerable communities) • Bangladesh (poor access of common people to natural resources and inequitable sharing of wealth and service-related provisions seriously hamper the development of community resilience) • Sri Lanka (maintaining and supporting healthy functioning of coastal habitats as key strategies to climate change adaptation) • Taiwan (reinforcement of the management and conservation of watersheds, conservation of water sources, and maintenance of the natural ecological function of water systems as part of adaptation strategy) • Sri Lanka (adoption of agroforestry as an important land-use system to resource conservation, help strengthen adaptation, and build the resilience of farming communities and households to climate change)
6) Building local resilience • Integrated vs. piece-meal approach	• Bangladesh (both the National Adaptation Programme of Action (NAPA) and the Bangladesh Climate Change Strategy and Action Plan (BCCSAP) documents failed to foresee the multiscalar nature of climate change problem hence paid lesser attention in building the resilience of local communities. • Japan (working with the local community groups and local governments to enhance resilience)

Table 1. *(Continued)*

Issue/Challenge	Illustrative Example and Brief Description from Country Cases
	• Thailand (integrated approaches to water resources management need to become more learning-oriented) • India (holistic approach, i.e., integrated climate change adaptation (ICCA) should be developed to take into account the multifaceted aspects of rural livelihoods • Sri Lanka (need to adopt an integrated institutional, ecosystem and community approaches of adaptations to minimize risks from climate change and enhance resilience in coastal areas) • Vietnam (need to address climate change in an integrated way with long-term socioeconomic and poverty reduction efforts)
• Sustainable livelihood	• Bangladesh, Cambodia, Philippines, Nepal (need to promote sustainable livelihood as component of CCA and DRR • Vietnam (local communities' limited access to financial resources and low income constrains effective adaptation)
• Local institutions and capacities	• Nepal (importance of existing robust institutions at community level as entry point in tackling climate change issues) • Thailand (local knowledge about managing climate risks is highly relevant to building adaptive capacities and larger-scale adaptation planning in the water sector but often overlooked; watershed organizations provide opportunities for various line agencies to work together across ministerial and jurisdictional lines) • Cambodia and India (training and capacity building is necessary in building local resilience) • Vietnam (need to incorporate direct experience of climate change impacts into future responses and solutions that build upon existing local adaptation practices, where appropriate; in Central coast, local communities have evolved ways of coping with flood's annual cycle but completely lack experience to deal with the new risks)
• Addressing nonclimate-related concerns	• Japan (need to address demographic changes affecting forest management) • Bangladesh (some of the disasters are the results of combined outcomes of many different human interventions other than climate change like unplanned development; marginalization of certain groups of people as a result of exclusion from resources/services cause poverty and many different kinds of problems that limits community resilience) • India (dominance of fish marketing by fish merchants and limited access to fishing grounds undermines fishermen's capacity to adapt to climate change)
7) Education and research • Formal education	• Malaysia (need to mainstream climate change and DRR into the formal education system)
• Nonformal education/public education	• Japan (educating forest owners on climate change and forest management planning, forest-land relation with town watching and mountain watching)

Table 1. (*Continued*)

Issue/Challenge	Illustrative Example and Brief Description from Country Cases
	• Malaysia (non-formal education in climate change and DRR needs to be prioritized for public awareness, enhanced technical knowledge, research, science-governance interphase and strengthened institutional arrangements)
• Inclusion of Indigenous Knowledge System (IKS)	• Bangladesh (little focus on how local knowledge could be documented and disseminated for inclusion in formal education that integrates climate change concerns
• Advancing knowledge through research	• Thailand (up to date scientific knowledge through research needed for informed decision-making in the water sector; strengthening links with technical experts to improve understanding of scientific knowledge and produce new actionable knowledge) • Singapore (comprehensive research by different sectors directed to improve the current state of technology such as in the energy sector, and bring down production costs to a level that would make large-scale adoption commercially viable; research into better vulnerability assessments and adaptation technologies to better understand climate change effects and effectively address vulnerabilities) • Sri Lanka (need to improve the capability of researchers and the availability of research facilities; link with national and international expertise to strengthen and promote research. • Singapore and Taiwan (robust scientific assessments and studies central in defining appropriate adaptation strategies) • Vietnam (climate change requires interdisciplinary cooperation among various fields to understand and effectively manage health risks; research as foundation in the formulation of *National Target Program and Action Plan for Response and Adaptation to Climate Change*)
8) Funding/financial resources	• Cambodia (need to access more resources and longer funding cycles to enable DRR initiatives to be fully integrated into development plans) • Nepal (no clear cut mechanism of funding for joint initiatives or actions of the Climate Change Network) • Vietnam (need to balance limited government budgets to cover both socio-economic development needs and DRR investments in the context of increasing disaster impacts and related damages and losses) • Singapore (allocates considerable funding for R&D to enhance the country's climate change mitigation and adaptation) • Taiwan (allocated NT$116 billion budget to be dispensed over a period of eight years and for the development of a comprehensive Flood-Prone Area Management Plan)

Sources: Various authors of the 18 case studies in this volume.

to address the disaster risks and climate change impacts in a comprehensive manner through both structural and nonstructural measures. Similar challenge holds for Pakistan based on the analysis of Khan (2010) who highlighted the need for central coordinating authority on disaster-related concerns and the mainstreaming of DRR into the routine planning process. Likewise, in Thailand, Lebel, Lebel, and Daniel (2010) noted that climate change was initially classified by the government as an environmental issue and hence, better coordination among stakeholders, especially in government, is needed to develop effective adaptation responses.

On the spatial scale, mismatches are rooted from the fact that up to the present time, climate change issues have been analyzed primarily on a global scale, particularly due to the lack of downscaled data in many Asian countries. On the other hand, disasters have been studied in the countries and localities where they occur. The localization of the prediction of potential climate change impacts of extreme events, such as heat waves, heavy precipitation, storms, floods, landslides, etc., through local, downscaled data for climate change, which could facilitate the preparation of specific CCA and DRR strategies, is therefore one of the major challenges (Birkman & von Teichman, 2010). Limited spatial integration of CCA and DRR is happening on the ground in some places such as in few municipal and provincial government units in the Philippines through the preparation of local development plans and activities that include adaptation and risk management activities (Lasco, Delfino, Pulhin, & Rangasa, 2008; Pulhin et al., 2009). Such an integration, however, is mostly based on the assessment of past and present climate variability and extremes and has yet to benefit from long-term climate projections that will allow a more comprehensive adaptation planning. Along the same vein, in Thailand, better integration and implementation of land- and water-use planning using long-term climate change projections have been assessed to be crucial to reducing vulnerabilities in the water sector (Lebel, Lebel, & Daniel, 2010).

Related to the spatial scale, mismatches in terms of temporal scale limit the successful integration of CCA and DRR concerns at the national and local levels. DRR responses spearheaded by humanitarian assistance, such as the Red Cross, the United Nations Office for the Coordination Affairs, and various other international and national aid agencies and donor organizations, are often event-driven and therefore short-term in their orientation and processes (Birkman & von Teichman, 2010). Consequently, humanitarian agencies primarily provide short-term funding and focus on a specific activity, sector, or outcome, like relief operation or reconstruction of physical infrastructure. In contrast, sustainable CCA strategies require long-term perspectives involving comprehensive vulnerability and adaptation assessments and planning and process-oriented funding beyond the usual political election period.

The implications of this divergent temporal orientation in various sectors (e.g., forestry and water resources sectors) in the Asian context are manifold. Japan's forest management for instance has become more complicated at present compared to the past several decades when climate change was not a major issue (Takeuchi & Shaw, 2010; Fujita & Shaw, 2010). Forest management is now concerned not only with wood production but also with DRR, CCA, and mitigation. In the context of DRR and CCA, the climate change and disaster risk communities are now paying increasing attention to differential vulnerability among demographic groups, particularly children and the elderly living in forest areas. This requires temporal perspective that goes beyond the traditional manner of forest management focused on wood production to incorporate vulnerability reduction and mitigation in the context of changing climate. Similarly, in the Thai water sector, the increasing uncertainties about future climate require flexible and reversible adaptation options (Lebel, Lebel, & Daniel, 2010). This entails long-term orientation in adaptation planning of which a robust assessment of future climate change requiring downscaled data for Thailand is a prerequisite. Thus, in the case of Taiwan (Wen, Huang, Hsu, & Chang, 2010) flood modeling analysis is being carried out in order to respond to the future conditions of protecting life and property in the coastal regions as part of the CCA strategy.

Institutions and Governance

A major challenge that deters the successful integration and outcomes of CCA and DRR in the majority of the 13 Asian countries and regions covered in this volume relates to institutional and governance constraints. These include the following areas of concern: enabling policies and mechanisms for implementation, participation of different stakeholders, institutional capacity, and quality of leadership and political commitment. Enabling policies and mechanisms for effective implementation to successfully address CCA and DRR concerns is usually inadequate if not totally absent, such as in the cases of Cambodia, Pakistan, Sri Lanka, and Vietnam. Appropriate platforms, organizational structure, institutional arrangements, and support systems including budget are lacking in mainstream CCA and/or DRR at the national and local levels of development planning.

Even in the case of the Philippines where a Climate Change Act has been successfully passed into law (Pulhin, Tapia, & Perez, 2010), DRR concerns have not been fully integrated, although referred to in the document. Consequently, a separate DRR law is currently in the process of being drafted. Though this initiative may be laudable, it will likely exacerbate the divide between the CCA and DRR efforts, if approved. Among the Asian countries, Taiwan is relatively in a better-off position with adequate policies and sufficient funding in place for both CCA and DRR.

Promoting meaningful participation of stakeholders at various levels of governance is also a major challenge in many countries. In Japan, there is a need to engage forest owners, private industries, government, and academe in integrated forest management to address the concerns of DRR and CCA. In Cambodia, Thailand, and Vietnam, meaningful community participation is indispensable in various levels of decision-making from planning to implementation as well as in monitoring changes in climate, flood regimes, and vulnerabilities. An important issue is to ensure the full engagement of the more vulnerable groups, whose lives and livelihoods will be affected by certain adaptation options such as relocation, by involving them in exploring alternative response options (Lebel, Lebel, and Daniel, 2010; Nguyen & Shaw, 2010). An example of meaningful stakeholder participation has been noted in the case of Bangladesh where nongovernment organizations, donors, and government departments take part in specific programs and prepare action plans to address climate change-induced hazards.

Limited institutional capacity in terms of human and infrastructural support is also a common challenge in a number of countries. In Cambodia, policy makers lack the knowledge on climate change issues and the expertise to identify eligible programs that will advance CCA and DRR objectives. Cambodia and Vietnam are also constrained by the limited number of professional staff dedicated to do DRR work. In Thailand, the need for capacity building in government agencies and the wider community has been noted to enable appropriate assessment of risks and exploration of adaptation responses. The ability to perform continuous monitoring of CCA and DRR initiatives to evaluate their overall contribution to sustainable development also forms part of the institutional capacity building, a need pointed out also in the case of Thailand.

Quality leadership and political commitment is also crucial in advancing CCA and DRR initiatives and hence an important component of the governance challenge. In Taiwan, the willingness of the vice premier to take over the responsibility of chairing the National Council for Sustainable Development, which takes care of UNFCCC affairs, represents a strong signal of the political commitment of the country to address climate change. Similarly, in the Philippines, strong political will and innovative leadership at the local level as demonstrated by the Governor of the Albay province provides great opportunity for championing CCA and DRR at the local government level (Lasco et al., 2008). In a number of places, however, such as Thailand, key decisions on projects including those on CCA and DRR often seem to have their origins in the promises of politicians rather than as outcomes of assessment or longer-term planning strategies (Lebel, Lebel, & Daniel, 2010).

Information and Communication

Information and communication represents a third set of challenge toward the achievement of successful adaptation and risk reduction at the country and local levels. Relevant aspects under this issue are awareness raising including early-warning system and information sharing and knowledge management. The value of public awareness to minimize disasters in various sectors and build local resilience has been highlighted in the case of the Philippines, Cambodia, Japan, Vietnam, and Bangladesh. Recognizing the importance of well-informed public, Singapore has included public awareness as an essential component of its national climate strategy. In a number of countries, however, such as in Cambodia and Pakistan, the need for an early-warning and forecasting system as well as for an effective communication to reach the public in a timely manner prior to disasters remains an urgent concern. Similarly, in Vietnam, many households have insufficient access to information, especially on natural disaster planning and mitigation, to allow them to properly respond and escape disasters. The Japanese example shows a unique case of the importance of risk communication to vulnerable groups, like aged people and women, and has focused on the gender perspective of the adaptation measures (Takeuchi & Shaw, 2010).

Platforms for information sharing and knowledge management have been created in Thailand and Nepal. In Thailand, the Climate Change Knowledge Management Center was established in 2009 to collect, synthesize, and disseminate knowledge on climate change to support strategic planning of government agencies, the private sector, and local communities. In Nepal, a 23-member Climate Change Network (CCN) has also been created to address information, coordination, and communication issues related to climate change (Subedi, 2010). While the idea of establishing a platform for information sharing and knowledge management may indeed be sound and worth exploring in other countries, the viability and sustainability of their operation should be a major consideration. In Nepal, for instance, the CCN faces an uncertain future due to lack of funding support.

Assessment Methods and Tools

The different methods and tools for assessing vulnerability and resilience provide an entry point for integrating climate change and disaster risk concerns in the local area. For instance, Uy, Takeuchi, and Shaw (2010) demonstrate the potential use of a sustainable livelihood framework for assessing local resilience of coastal communities to disasters based on their study in Albay, the Philippines. Similarly, Habiba, Takeuchi, and Shaw (2010) advocate for the use of an integrated framework termed "SIP approach" (socioeconomic, institutional, and physical) for determining resilience among drought-prone communities in Bangladesh. For Cambodia, Nguyen et al. (2010) highlight an important lesson from the implementation of a DRR program in the country: that community-based disaster risk management methodologies work and should be continually used in all future DRR programs. A major strength of all these methods is their ability to integrate socioeconomic and physical parameters including climatic factors in the process of assessment. However, as illustrated in the three country cases, all of them still face the challenge of incorporating future climate change scenarios using downscaled data at the local level. Indeed, authors in the case of Singapore (Li & Rajola, 2010), Taiwan (Wen et al., 2010), and Thailand (Lebel, Lebel, & Daniel, 2010) are in agreement that robust scientific assessments and studies including climate change projections are central in defining appropriate CCA strategies.

Resource Sustainability and Access

Sustainability and access to natural resources and the environmental services they provide, especially by the poorer sector of the society, are an important factor that is likely to influence the success of CCA and DRR initiatives at the local level. Considering the dismal condition of natural resources (such as forests, water, coastal resources, biodiversity, etc.) in many Asian countries and their potential link to local communities' vulnerability, sustaining the productive condition and environmental functions of these resources (such as water, forests, and coastal resources, etc.) is a major challenge to many nations. In Vietnam, for instance, an increasing degradation of the natural environment through deforestation and the conversion of agricultural land to urban areas are believed to contribute to more destructive floods in lowland areas. Maintaining and restoring the resilience of the forest ecosystem is therefore among its environmental priorities, similar to many of its Asian neighbors. Indeed, in countries like Sri Lanka and Taiwan, maintaining and supporting healthy functioning of coastal habitats and watershed management, respectively, are considered as key strategies to CCA. Adoption of appropriate land-use system, especially agroforestry, is also being promoted in Sri Lanka (as well as in many other Asian countries) to promote resource conservation, help strengthen adaptation, and build the resilience of farming communities and households to climate change.

In addition to resource degradation, lack of access of common people to natural resources and inequitable sharing of wealth and service-related provisions also seriously contributes to their further marginalization and hamper the development of community resilience. In the case of Bangladesh, for instance, Islam, Rahman, and Mallick (2010) noted that forests are destroyed by grabbing forest lands by the dominant groups or by the leasing out by the state authorities to the powerful sector. Forest-dependent communities are excluded in the process and deprived of forest benefits that make them more vulnerable to climate-related disasters. This of course is a common problem in many Asian countries such as the Philippines and Indonesia (Broad, 1995). The poor people's exclusion from resources contributes to poverty and many different kinds of problems, exacerbating their already vulnerable condition. A key challenge in advancing CCA and DRR at the local area is therefore linked to addressing structural barriers such as elite control over natural resources and the political marginalization of the poor sector.

Building Local Resilience

Building climate change resilient communities to reduce damages is a major challenge and ensures successful adaptation at the local level. Authors from at least six country cases, such as from Bangladesh (Islam et al., 2010), Japan (Takeuchi & Shaw, 2010), Thailand (Lebel, Lebel, & Daniel, 2010), India (Iwasaki & Shaw, 2010), Sri Lanka (Wickramasinghe, 2010), and Vietnam (Nguyen & Shaw, 2010), are in agreement for the need to treat climate change issues as multidimensional, which requires a "holistic" or "integrated approach" instead of the traditional piece-meal approach to managing disasters. A common challenge therefore in these countries (as well as in others) is to implement and institutionalize such an approach at the local level, regardless of the sectors involved. For instance, CCA strategies will have to be integrated with the long-term socioeconomic and poverty-reduction efforts as being proposed in Vietnam, or with the integrated water resource management in Thailand, or with the "integrated institutional, ecosystem, and community approaches" to enhance resilience in Sri Lanka's coastal areas. Relevant to the implementation of an integrated approach, the need to promote sustainable livelihood as a component of CCA and DRR is being advocated in the case of Bangladesh, Cambodia, the Philippines, and Nepal. In Vietnam, local communities' limited access to financial resources and low income, and hence by implication, the absence of sustainable livelihood, constrain effective adaptation.

Local institutions and their capacities are also major determinants for building local resilience to climate change. While robust community level institutions (Nepal case) and local knowledge about managing risks (Thailand case) may be present and may serve as an entry point for tackling climate change issues, their potential for large-scale adaptation planning is often overlooked (Lebel, Lebel, & Daniel, 2010). The Cambodia and Indian cases highlighted the need for training and capacity building at the local level as a requisite for building local resilience. One of the Vietnam cases underscored the importance of incorporating direct experience of climate change impacts into future responses and solutions that build upon existing local adaptation practices, where appropriate (Nguyen & Shaw, 2010). However, the other case also points to some limitations of local experience as the basis for future adaptation. Based on the experience from the central coast of Vietnam, Tinh, Tuan, Tran, and The (2010) argued that while local communities have evolved ways of coping with the annual flood cycle, they completely lack experience to deal with the new risks.

Another major challenge that relates to building community resilience pertains to the need to address nonclimate-related concerns, which oftentimes are broader in scope and associated with the structural forces in the society. In Japan, climate change strategies in the forestry sector need to simultaneously address the demographic changes in rural communities affecting forest management. In Bangladesh, unplanned development and exclusion from natural resources result in poverty and limit the development of community resilience. Similar situation holds true for India where the dominance of fish marketing by fish merchants and limited access to fishing grounds undermine fishermen's capacity to adapt to climate change.

Education and Research

Education and research are important pillars in ensuring the attainment of the overall goal of CCA and DRR in terms of reducing disasters and vulnerability and ensuring successful and sustainable adaptation initiatives. In Malaysia and many other Asian countries, mainstreaming climate change in the formal educational system at different levels remains a major challenge. The long and tedious curriculum development process in many countries serves as a major hindrance for offering new curricular programs that focus on climate change and DRR. The traditional approach of "classroom" teaching does not also support the more engaging requirement of climate change and DRR-related curricular programs where students need to be immersed in field realities for meaningful learning to take place. In terms of content, the importance of indigenous local knowledge for inclusion in formal education that integrates climate change concerns was highlighted in the Bangladesh case (Islam et al., 2010), although the authors also noted the little effort spent on the documentation and dissemination of this type of knowledge.

To promote wider public awareness, CCA and DRR also need to be mainstreamed in nonformal education programs in different sectors. In Japan, for instance, Fujita and Shaw (2010) have noted the difficulty faced by forest owners in preparing management plans considering the multiple objectives of timber production, prevention of natural disasters, watershed protection, biodiversity conservation, carbon sequestration, etc., hence the need for training in forest management planning. Similarly in Malaysia, Pereira, Tiong, and Komono (2010) highlighted the value of prioritizing education in climate change and DRR for public awareness, enhanced technical knowledge, research, science governance interphase, and strengthened institutional arrangements.

Advancing scientific knowledge through enhanced research is a priority of many countries with the end view of identifying better adaptation strategies. Yet, there is the necessity in many countries such as Sri Lanka to improve the capability of researchers and the availability of research facilities and to collaborate with national and international expertise to boost research capacity. The Vietnam case also noted the inevitability of interdisciplinary type of research considering the complexity of climate change issues.

Funding/Financial Resources

Most countries, especially the LDCs and developing countries, faced financial constraints and need external assistance to integrate CCA and DRR concerns as part of their priority programs. This is especially considering that national governments need to balance their limited national budgets to cover both socioeconomic development needs and DRR investments in the context of increasing disaster impacts and related damages and losses brought about by climate change. Of the different country cases, only Singapore mentioned that it has allocated substantial funding for research and development to mitigate CO_2 emissions in the country and adapt to climate change in an environmentally sustainable and cost-effective manner. On the other hand, Taiwan allocated NT$116 billion budget to be dispensed over a period of eight years for the development of a comprehensive Flood-Prone Area Management Plan. Other than these two countries, probably only Japan among the 13 countries and regions covered in this volume has the capacity to fully finance its own CCA and DRR programs without relying on external support.

WAYS FORWARD

The key issues and challenges confronting the successful integration and outcomes of CCA and DRR are complex and multidimensional as discussed in the preceding section. As such, dealing with these issues and challenges requires holistic, comprehensive, and sustainable strategies that go beyond the piece-meal or ad hoc approach currently employed in many countries in dealing with CCA and DRR concerns. The following sections briefly discuss eight major strategies to address the different issues and challenges with the end view of reducing disaster risks and promoting successful CCA in the Asian context.

Facilitate CCA and DRR Integration at Various Levels

Integration of CCA and DRR concerns at various levels from national to local should be a major priority that needs immediate implementation in many countries. An appropriate platform should be established to facilitate the functional integration among the different agencies and sectors concerned with climate change and DRR. Such a platform may require the creation of a national-level central coordinating body with authority to harmonize varying institutional priorities and activities and promote synergy among concerned agencies. At the minimum, a common institutional framework to guide the integration of CCA and DRR in the activities and programs of different agencies should be developed to facilitate the integration process.

Climate change and DRR should be mainstreamed in development planning at all levels to address the challenge of spatial and temporal mismatches and produce better outcomes on the ground. In support of this, localized prediction of potential climate change impacts of extreme events (i.e., heat waves, heavy precipitation, storms, floods, landslides, etc.) from local, downscaled data for climate change should be supported to facilitate spatial integration through the preparation of CCA and DRR strategies in specific places. Where limited spatial integration is already happening on the ground, such as in the case of the Philippines, they should be strengthened through the use of long-term climate projections that will allow a more comprehensive adaptation planning. DRR responses spearheaded by humanitarian assistance such as the Red Cross, which has short-term orientation, should be integrated with the long-term perspective of CCA through the adoption of a more comprehensive development framework that takes into account the potential risks from climate change and other forms of disasters and factors these in the development planning.

Strengthen Governance System

At the core of achieving the objective of integrating CCA and DRR concerns is a strong governance system at different levels. Governance is not the sole purview of the state through the government, but rather emerges from interactions of many actors including the private sector and the civil society (Lebel, Sinh, & Elena, 2010). It includes policies, laws, regulations, discursive debates, negotiation, mediation, conflict resolution, elections, public consultations, partnerships between various stakeholders, protests, and other decision-making processes.

To strengthen governance and improve the outcomes of an integrated CCA and DRR, enabling policies should be formulated and effective implementation mechanisms developed from the national down to the local level in many Asian countries. Such policies and implementation mechanisms should be directed toward the mainstreaming of CCA and DRR in all levels of development planning as already pointed out. Major institutional obstacles particularly the "mind your own business" type of organizational culture, which prevails in many Asian bureaucracies, should likewise be replaced with the idea of "adaptive governance" (Folke, Hahn, Olsson, & Norberg, 2005; Pahl-Wostl, 2009; Lebel, Sinh, & Elena, 2010) that puts emphasis on collaborative learning and sharing of resources and responsibilities.

Strengthened governance also means the creation of more political space that allows the participation of different stakeholders in decision-making processes, including the more vulnerable groups. It will also involve a lot of capacity building among various stakeholders at different levels. Such capacity building should go beyond the normal trainings conducted by the government and civil society organizations. It also requires among others beefing-up of organizational infrastructures in terms of modern equipment and facilities, and building up of reliable historical data among meteorological organizations and research institutes to enable the conduct of more robust climate change assessments using sound scientific methodologies. Strengthened governance also calls for quality leadership from various sectors that will champion CCA and DRR integration.

Finally, the great potential of local government units in terms of providing a meso-level organizational platform for promoting DRR and CCA integration at the local level should be recognized (Lasco et al., 2008; Shaw, Pulhin, & Pereira, 2010). An appropriate support and incentive system from the central government and donor organizations as well as from their local political constituents should be provided to sustain and scale up successful initiatives (see Lasco et al., 2008 for example of successful initiative in Albay, the Philippines). Indeed, in many Asian countries where adaptation to weather-related disasters is not only a development but also a survival issue, local politicians can effectively harness CCA and DRA as a political platform for demonstrating responsive governance and hence a potential venue for serving their local constituents while furthering political ambition.

Improve Information System and Knowledge Management

The value of well-informed public to minimize, if not avoid, risk-associated disasters cannot be overemphasized. More reliable and timely information is needed by the public to forewarn themselves of pending disasters so that they can act more responsively. In many countries, equipment, facilities, and trained staff are required to develop a reliable early-warning and forecasting system and effectively communicate information to the public in a timely manner to avoid disasters. Investments should likewise be made through research in generating quality information on DRR and CCA and in making them available to the general public to support decision-making and development planning at all levels.

Where appropriate, formal mechanism for information sharing and knowledge management should be created similar to the Climate Change Knowledge Management Center of Thailand or Nepal's CCN. Permanent budget should be allocated by the government to sustain its operation. The creation of a formal structure, however, should not preclude informal arrangements among institutions to share and manage knowledge about CCA and DRR.

Apply Integrated Assessment Methods and Tools

The application of existing methods and tools that use integrated framework such as the sustainable livelihood framework, "SIP approach," and community-based disaster risk management methodologies should be encouraged for more robust assessments of climate change impacts, vulnerability, and adaptation, including DRR concerns. These tools, however, should be further developed to incorporate future climate change scenarios using downscaled data at the local level. Moreover, participatory approaches that engaged the more vulnerable groups of the community should be employed in using these assessment methods to facilitate local learning but also to ensure appropriateness of the resulting adaptation plans at the end.

Ensure Natural Resources Sustainability and Resource Rights

Developing countries, particularly those in rural areas, are typically dependent on climate-sensitive natural resources for their livelihoods, economic activities, and national income (Leary et al., 2008). As mentioned earlier, most of these resources are in a degraded situation that makes these resources and the people dependent on them, highly vulnerable to damages from climate change. Serious efforts and investment should therefore be made by governments to ensure the sustainability of these resources to promote both their productive functions and environmental services. Moreover, necessary reform should be made to secure the "bundle of rights" of local communities, especially the poor people, over these resources to enable them to access, use, manage, and exclude others from these resources. The ongoing tenure reform (Larson, Barry, Dahal, & Colfer 2010) happening in the forestry sector through community-based forest management is the right direction to achieve this and hence should be promoted by governments in the Asian region.

Build Local Resilience and Reduce Vulnerability

The success of integrating CCA and DRR can be effectively measured only in terms of its outcomes at the local level where the real impacts of changing climate are experienced. Collective efforts by the governments and other stakeholders should therefore be focused on building local resilience of both the people and nature to climatic and environmental threats, reducing social vulnerability of communities, particularly the poor sector. This requires strategic interventions on the following:

- adoption of integrated or holistic approach to development that addresses both the social–political and biophysical (including climate change) related drivers of sustainability with specific focus on sustainable livelihood as a means to build the capital assets (social, natural, physical, financial, and human assets) of the local communities;
- capacity building of local institutions on climate- and disaster-related concerns and incorporation of their knowledge and experience, where appropriate, in larger-scale adaptation planning; and
- addressing the broader structural forces, such as demographic changes, unplanned development, poverty, and inequitable social structure, that contribute to vulnerability and impinge on the development of local resilience.

Advance Innovative Education and Research and Development Initiatives

Innovative education and research and development initiatives that integrate CCA and DRR should be supported and promoted. This should advance both the formal and nonformal educational systems and should go beyond the more conventional classroom-type (or lecture-type) learning to allow the engagement of the learners in actual field conditions for meaningful learning to take place.

To augment meager resources, collaborative education and research programs should be promoted among different educational and research institutions. An example of this approach, which has started to produce more tangible results, is the case of the Asian University Network of Environment and Disaster Management (AUEDM). The Network is composed of 18 universities in Asia working on various aspects of CCA and DRR and is bounded by the following objectives (AEDUM, 2010):

- To share and work together (bilaterally or multilaterally) in promoting environment and disaster management in higher education (focusing on, but not restricted to, postgraduate education)
- To seek possibilities of mutual collaboration on field-based action research (focusing on, but not restricted to, CCA)
- To broaden the scope of education and learning in the environment and disaster management field through collaboration with other stakeholders like NGOs and local governments.

Secure Adequate Funding Support

As already mentioned, lack of financial resources in many Asian countries is a major obstacle to successful integration and results of DDA and DRR initiatives. Governments, universities and research institutions, and other stakeholder groups should continue to seek adequate financial support from various sources, such as international donors, private sectors, etc. A more sustainable funding mechanism would have to be worked out in all countries to sustain various initiatives and produce long-lasting and greater impacts from these initiatives.

NOTE

1. Where reference is not provided as a specific country is mentioned in the discussion, the attribution is credited to the author/authors of the case study of the pertinent chapter in this volume.

REFERENCES

ADB. (2009). The economics of climate change in Southeast Asia: A regional review. Asian Development Bank, Jakarta, 223pp.

Adger, W. N., Lorenzoni, I., & O'Brien, K. (2009). Adaptation now. In: W. N. Adger, I. Lorenzoni & K. O'Brien (Eds), *Adapting to climate change: Thresholds, values, governance* (pp. 1–22). Cambridge: Cambridge University Press.

AEUDM (2010). Asian University Network of Environment and Disaster Risk Management (AUEDM). http://www.auedm.net/ (downloaded July 12, 2010).

Birkmann, J., & von Teichman, K. (2010). Integrating disaster risk reduction and climate change adaptation: Key challenges – scales, knowledge, and norms. *Sustainability Science*, *5*, 171–184.

Broad, R. (1995). The political economy of natural resources: Case studies of the Indonesian and Philippine forest sectors. *Journal of Developing Areas*, *29*(3), 317–339.

Commission on Climate Change and Development (CCD). (2009). Closing the gaps: Disaster risk reduction and climate change adaptation in developing countries. Report of the Commission on Climate Change and Development, Ministry of Foreign Affairs, Stockholm, Sweden, 80pp.

Cruz, R. V., Harasawa, H., Lal, M., Wu, S., Anokhin, Y., Punsalmaa, B., Honda, Y., Jafari, M., Li, C., & Huu Ninh, N. (2007). Asia. Climate change 2007. In: M. L. Parry, O. F. Canziani, J. P. Palutikof, P. J. van der Linden & C. E. Hanson (Eds), *Impacts, adaptation and vulnerability. Contribution of Working Group II to the fourth assessment report of the intergovernmental panel on climate change* (pp. 469–506). UK: Cambridge University Press.

Ericksen, S. H., & O'Brein, K. (2007). Vulnerability, poverty, and the need for sustainable adaptation measures. *Climate Policy*, *7*, 337–352.

Folke, C., Hahn, T., Olsson, P., & Norberg, J. (2005). Adaptive governance of social-ecological systems. *Annual Review of Environment and Resources*, *30*, 441–473.

Fujita, K., & Shaw, R. (2010). Forest management as an adaptation option in mountain areas of Japan. In: R. Shaw, J. Pulhin & J. Pereira (Eds), *Climate change adaptation and disaster risk reduction: An Asian perspective*. UK: Emerald.

Habiba, U., Takeuchi, Y., & Shaw, R. (2010). Drought risk reduction approaches in Bangladesh: An overview. In: R. Shaw, J. Pulhin & J. Pereira (Eds), *Climate change adaptation and disaster risk reduction: An Asian perspective*. UK: Emerald.

Intergovernmental Panel on Climate Change (IPCC) (2007). *Climate change 2007: Impacts, adaptation and vulnerability. Summary for Policy Makers*. Geneva: World Meteorological Organisation.

Islam, T., Rahman, M. A., & Mallick, F. (2010). Impacts of climate change induced hazards and adaptation processes in Bangladesh: An overview. In: R. Shaw, J. Pulhin & J. Pereira (Eds), *Climate change adaptation and disaster risk reduction: An Asian perspective*. UK: Emerald.

Iwasaki, S., & Shaw, R. (2010). Integrated climate change adaptation in Chilika Lagoon fisheries, India. In: R. Shaw, J. Pulhin & J. Pereira (Eds), *Climate change adaptation and disaster risk reduction: An Asian perspective*. UK: Emerald.

Khan, A. (2010). Climate change adaptation and disaster risk reduction in Pakistan. In: R. Shaw, J. Pulhin & J. Pereira (Eds), *Climate change adaptation and disaster risk reduction: An Asian perspective*. UK: Emerald.

Larson, A., Barry, D., Dahal, R. G., & Colfer, C. (Eds). (2010). *Forests for people: Community rights and forest tenure reform*. London, Sterling, VA: Earthscan.

Lasco, R. D., Delfino, R. J., Pulhin, F. B., & Rangasa, M. (2008). The role of local government units in mainstreaming climate change adaptation in the Philippines. AdaptNet Policy Forum 08-09-P-Ad.

Leary, N., Adjuwon, J., Barros, V., Batima, P., Biagini, B., Burton, I., Chinvanno, S., Cruz, R., Dabi, D., Comarmond, A., Dougherty, B., Dube, P., Githeko, A., Hadid, A., Hellmuth, M., Kangawawe, R., Kulkarni, J., Kumar, M., Lasco, R., Mataki, M., MEdany, M., Mohsen, M., Nagy, A., Osama-Elasha, B., Sanjak, E., Seiler, R., Taylor, M., Travasso, M., Maltitz, G., Wandiga, S., & Webbe, M. (2008). A stitch in time: General lessons from specific cases. In: N. Leary, J. Adejuwwon, V. Barroos, I. Burton, J. Kulkarni & R. Lasco (Eds), *Climate change and adaptation* (pp. 1–27). UK: Earthscan.

Lebel, L., Lebel, P., & Daniel, R. (2010a). Water insecurities and climate change adaptation in Thailand. In: R. Shaw, J. Pulhin & J. Pereira (Eds), *Climate change adaptation and disaster risk reduction: An Asian perspective*. UK: Emerald.

Lebel, L., Sinh, B. T., & Elena, N. (2010). Adaptive governance of risks: Climate, water and disasters. In: R. Shaw, J. Pulhin & J. Pereira (Eds), *Climate change adaptation and disaster risk reduction: Issues and challenges*. UK: Emerald.

Li, B., & Rajola, V. (2010). Climate change and disaster risk: The Singapore response. In: R. Shaw, J. Pulhin & J. Pereira (Eds), *Climate change adaptation and disaster risk reduction: An Asian perspective*. UK: Emerald.

Nguyen, H., & Shaw, R. (2010). Climate change adaptation and disaster risk reduction in Vietnam. In: R. Shaw, J. Pulhin & J. Pereira (Eds), *Climate change adaptation and disaster risk reduction: An Asian perspective*. UK: Emerald.

Nguyen, H., Shaw, R., & Prabhakar, S. V. R. K. (2010). Climate change adaptation and disaster risk reduction in Cambodia. In: R. Shaw, J. Pulhin & J. Pereira (Eds), *Climate change adaptation and disaster risk reduction: An Asian perspective*. UK: Emerald.

Pahl-Wostl, C. (2009). A conceptual framework for analyzing adaptive capacity and multi-level learning processes in resource governance regimes. *Global Environmental Change, 19*, 345–365.

Pereira, J., Tiong, T. C., & Komono, I. (2010). Mainstreaming climate change adaptation and disaster risk reduction: A Malaysian approach. In: R. Shaw, J. Pulhin & J. Pereira (Eds), *Climate change adaptation and disaster risk reduction: An Asian perspective*. UK: Emerald.

Pulhin, J., Tapia, M., & Perez, R. (2010). Integrating disaster risk reduction and climate change adaptation: Initiatives and challenges in the Philippines. In: R. Shaw, J. Pulhin & J. Pereira (Eds), *Climate change adaptation and disaster risk reduction: An Asian perspective*. UK: Emerald.

Pulhin, J. M., Lasco, R.D., Delfino, R.J., Rangasa, M., Pulhin, F.B., & Cosio, R. (2009). Climate change adaptation and local governance in the Philippines. In: *Disaster risk reduction: Investment for a safer future. Proceedings of the Fourth International Conference ACT AHEAD, 23-25 February 2009*, Jaypee Vasant Continental, New Delhi, India (organized by the Kyoto University and SEEDS, 47 pages).

Romieu, E., Welle, T., Schneiderbauer, S., Pelling, M., & Vinchon, C. (2010). Vulnerability assessment within climate change and natural hazard contexts: Revealing gaps and synergies through coastal applications. *Sustainability Science, 5*, 159–170.

Shaw, R., Pulhin, J., & Pereira, J. (2010). Climate change adaptation and disaster risk reduction: Overview of issues and challenges. In: R. Shaw, J. Pulhin & J. Pereira (Eds), *Climate change adaptation and disaster risk reduction: Issues and challenges* (pp. 1–19). UK: Emerald.

Subedi, J. (2010). Climate change adaptation in Nepal: Issues and strategies. In: R. Shaw, J. Pulhin & J. Pereira (Eds), *Climate change adaptation and disaster risk reduction: An Asian perspective*. UK: Emerald.

Takeuchi, Y., & Shaw, R. (2010). Climate change adaptation and disaster risk reduction experiences in Japan. In: R. Shaw, J. Pulhin & J. Pereira (Eds), *Climate change adaptation and disaster risk reduction: An Asian perspective*. UK: Emerald.

Tinh, B. D., Tuan, T. H., Tran, P., & The, B. D. (2010). Local vulnerability and adaptation to extreme climate events in central coast of Vietnam. In: R. Shaw, J. Pulhin & J. Pereira (Eds), *Climate change adaptation and disaster risk reduction: An Asian perspective*. UK: Emerald.

Uy, N., Takeuchi, Y., & Shaw, R. (2010). Local adaptation to livelihood resilience in Albay, Philippines. In: R. Shaw, J. Pulhin & J. Pereira (Eds), *Climate change adaptation and disaster risk reduction: An Asian perspective*. UK: Emerald.

Wen, J. C., Huang, S. Y., Hsu, C. C., & Chang, K. C. (2010). The experiences of DRR through climate change adaptation in Taiwan. In: R. Shaw, J. Pulhin & J. Pereira (Eds), *Climate change adaptation and disaster risk reduction: An Asian perspective*. UK: Emerald.

Wickramasinghe, D. (2010). Coastal ecosystems and climate vulnerability in Sri Lanka. In: R. Shaw, J. Pulhin & J. Pereira (Eds), *Climate change adaptation and disaster risk reduction: An Asian perspective*. UK: Emerald.

Printed and bound by CPI Group (UK) Ltd, Croydon, CR0 4YY
16/09/2026
14963402-0004